R for Programmers:
Advanced Techniques

R for Programmers: Advanced Techniques

Dan Zhang

CRC Press
Taylor & Francis Group
Boca Raton London New York

CRC Press is an imprint of the
Taylor & Francis Group, an **informa** business

CRC Press
Taylor & Francis Group
6000 Broken Sound Parkway NW, Suite 300
Boca Raton, FL 33487-2742

© 2017 by Taylor & Francis Group, LLC, under exclusive license granted by China Machine Press for English language and throughout the world.
CRC Press is an imprint of Taylor & Francis Group, an Informa business

Library of Congress Cataloging-in-Publication Data

Names: Zhang, Dan, 1983-
Title: R for programmers : advanced techniques / Dan Zhang.
Description: Boca Raton : CRC Press, [2017] | Includes bibliographical references and index.
Identifiers: LCCN 2016046363| ISBN 9781498736879 (pbk. : acid-free paper) | ISBN 9781138627185 (hardback : on demand) | ISBN 9781498736886 (ebook)
Subjects: LCSH: R (Computer program language)
Classification: LCC QA76.73.R3 Z43 2017 | DDC 005.13/3--dc23
LC record available at https://lccn.loc.gov/2016046363

Visit the Taylor & Francis Web site at
http://www.taylorandfrancis.com

and the CRC Press Web site at
http://www.crcpress.com

Printed and bound in the United States of America by
Edwards Brothers Malloy on sustainably sourced paper

This book is dedicated to my dearest family and the fans of R.

Contents

SECTION II R PROGRAMMING IN DEPTH

SECTION III   DEVELOPING R PACKAGE

Preface

Why This Book

This book is the second one of the series *R for Programmers*. It focuses on R's core techniques, the advanced developing applications, and the cross-discipline integration of knowledge of R and other fields.

Early in my first book *R for Programmers: Mastering the Tools*, I have introduced how to work with 30+ tooling packages of R and how to efficiently use third-part R packages for applying existing IT knowledge to the learning process of R. However, due to the limited pages, the book discussed only the usage without the underlying principles.

This book will make up for the regret. I will mainly discuss the core techniques of R itself, including the environments, object-oriented, file management, mathematical calculations, R package development, and so on. I hope that by walking through this book my readers can deeply learn about the R language, master these core techniques of R, understand the characteristics of third-part packages, and even have capability to develop excellent packages with their own styles. Perhaps in the near future, I will be amazed at a lot of efforts because of the packages developed by you.

Another highlight of this book is the cross-discipline integration of knowledge of R and other fields. In the book I will show the readers without reservation how I incorporate R and other knowledge to maximize the power of R in different fields. I believe that this part of content will enlighten many readers who would be surprised at the ways how I use R. I also hope that this part of content will inspire my readers and that R can be learned and used by people from various industries and knowledge domains. Nowadays, R is no longer the laboratory language that was used by only scientists. Instead, it already has the capabilities of actual development and application. It is intelligent and creative in terms of mining data value, discovering data rule, and creating data wealth.

If we compare R to Kong Fu, then the book *R for Programmers: Mastering the Tools* is like guidelines for tool usage. It helps you promote productivity easily and effectively in short time. An obvious improvement of your R skills can be achieved quickly. But in a long run, you will encounter your bottleneck for various reasons and it is difficult for you to break through.

Well, this book is like the inner strength of Kong Fu. It brings you the core techniques of R itself that make you get the elementary logic of R. The book pays more stress on how to integrate R and other disciplines and fields in practice. With what you learn from this book, you will get a clear picture of R. Then it is possible for you to win tricks without tricks, or even to create your own Kong Fu style to become a great master in the future! (Um, I am too far from the topic …)

Here, I must stress that this book is not an introductory of R. Anyone with blank background should learn some basic knowledge of R before reading this book. This book contains advanced contents for R development, which require you to have experience of working with R and basic computer knowledge. Else you cannot understand the output of my experience in this book.

The content of this book is the summary of my usage of R in practice. It is a record of my working experience with R. This book puts stress on R's advanced development, involving the knowledge from fields of computer, statistics, mathematics, and finance.

The core content of this book consists of two topics: the advanced programming techniques of R and the cross-discipline application. For the first topic, this book discusses in detail the definition and usage of R's environments, the file management, and the new features of R version 3.1.1, which get you an experience to the low-level design of R. This book comprehensively introduces the design and usage of the four object-oriented architectures of R. The object-oriented programming architectures make R be able to develop complicated applications that follow rules in real world. Besides, the book introduces a complete process for developing R packages with cases of Daily China Weather app and game developments, which enlighten readers to develop own packages and open the door to productization with R.

As for the topic of cross-discipline application, R can easily handle the bothering mathematical calculations involving elementary or advanced mathematics, probability theory or statistics. Mathematics is no longer various models. The algorithms in the book include the collaborative filtering model, the PageRank model based on matrix computation, the trading strategy model used in finance, and the usage of genetic algorithms. With just several code lines and a few minutes, R turns our ideas into executable algorithm prototype.

Another thing I want to say is that although R is not adapted to develop games, no language can be competitive against R with just 200 lines of code to complete the 2048 game. Someone may ask me "why do you want to use R to develop games," "why not with Java," "is it the same way I use Java instead of R to development?" In fact, I just want to take game development as an instance to show the style of simplicity, the idea of freedom, and the creative of full of imaginations that belong to R. I hope my attitude of playing as an "R geek" can inspire your unlimited thoughts about R! Finally, we can turn our model into product and publish our own R package that can be used by people all over the world. How exciting it is!

When communicating with R users with different backgrounds, I find that on one hand, users with programming background are able to write clean and efficient code but due to lack of statistics, they have no idea on how to optimize the model. On the other, those with statistics background are able to design and optimize a model, but they don't know how to implement the model into product.

This book introduces several cases where I not only design the model from the perspective of academy but also implement the model into product. By studying practical cases, user with different discipline backgrounds can think from each other's perspective to find new ways for solving problem. This is another highlight!

For most programmers, it is easy to learn R but difficult to use. R has no complex programming syntax like C/C++, no need for consideration of global architecture like Java, and no flexible usage like Javascript. However, the data-oriented programming of R is totally different from other languages, which makes many programmers confusing how to use R although they have mastered its syntax.

In my opinion, learning the R language is to customize yourself, to find your real position and to make cross-discipline innovation by integrating your knowledge. It is not to copy other's idea. To use R across disciplines, you are required to combine knowledge of basic disciplines (elementary/advanced mathematics, linear algebra, probability theory, statistics) and IT technologies (R syntax, R packages, database, algorithms). So you cannot master R until you promote your comprehensive knowledge level. In other words, once you have mastered the R language, you are outstanding.

Again, I have to emphasize that this book is not an introductory of R. It is an advanced development book. Neither the introductory syntax nor the usage of third-part packages is introduced in this book. Instead, if you have had certain basics on R and want to productize your R model, then I will tell you how to enhance the reliability and extensibility of your R program and how to publish your own packages.

This book is the second one of the series *R for Programmers*, while the third book *Quantitative Investments* will introduce the applications of R in finance. That book uses the R language to create trading models and to implement the process of automatic trading, which makes technical engineers turn their knowledge into real values.

The development environments of this book include Linux Ubuntu and Windows 7, which are declared in each section. All the programs have passed the test against R version 3.1.1.

The R language is being improved and updated. It will lead a revolution of data. Cross-discipline integration is the trend of the times and it is also the opportunity for us to grasp!

Potential Readers of This Book

This book will be helpful to the following people working with R:

- Software engineers with a computer background
- Advanced users of R
- Data scientists with a data analysis background
- Scientific researchers with a statistical background
- Students in universities and colleges

How to Read This Book

The content of this book is divided into three sections:

Section I discusses R's application to calculations and algorithms (Chapters 1 and 2), which introduces R's knowledge system and R' support for basic disciplines. By implementing various algorithms of basic disciplines with R, this part helps readers easily learn the approaches of mathematical calculations and the development of customized modeling algorithms.

Section II discusses the in-depth development of R (Chapters 3 and 4), which introduces programming related to R's kernel-related programming skills including the definition and usage of environments, and the design and application of object-oriented programming. The aim of this section is to help readers to have learn in depth R's low-level knowledge and to design complicated application structure using object-oriented programming skills.

Section III is about developing own R packages (Chapters 5 and 6). A complete development process of R package is introduced in this section. The section provides cases including Daily China Weather app and games development, which show readers how to establish their own R packages to open the door to productization using R.

There many cases that incorporate different knowledge, so it is best for readers to follow the chapters in sequence.

Correction and Support

Because the time spent writing this book and the author's knowledge of R are both limited, there will inevitably be some errors or incorrect viewpoints in this book. I sincerely hope that readers will point out and comment on any errors. For this purpose, I have created an online communication website for readers of this book (http://fens.me) to use to communicate. If you encounter any problems reading this book, please heave notes on this website, and I will try my best to provide a satisfactory solution. All of the source code of this book can be downloaded from the official website of CRC Press (https://www.crcpress.com) or from the online communication website, where I will update the codes in time. This book is printed in black and white, so all the colored pictures can only be achieved by running the codes of this book. I sincerely hope that you can send your valuable feedback and advice on this book to bsspirit@gmail.com.

Acknowledgments

I wish to thank He Ruijun, the acquiring editor at CRC Press, who helped promote the publication of this book. Thanks are also extended to the translator Wang Tao for his efforts on this book. I give special thanks to my parents and my wife for their support and care.

About the Author

 **Dan Zhang** is the founder and former CTO of Qutke.com and he now works at China Minsheng Bank Corp. Ltd. as a data scientist in the company's Big Data Center. Dan has 10 years of programming experience and obtained a number of technical certificates from Sun and IBM. With rich knowledge of developing Internet application architectures, he is proficient with languages including R, Java, Nodejs. Dan masters the techniques of big data and data mining and he has accumulated lots of knowledge on statistics and finance. He is the author of *R for Programmers: Mastering the Tools and R for Programmers: Advanced Techniques*. His blog is available at http://fens. me, with global Alexa ranking of 70 k.

About the Translator

 **Wang Tao** is a senior data analyst and software engineer. He is also an R user. He is currently working at an auto e-commercial company in China. He specializes in the research of user portrait and auto price system. He has 10+ years of experience in information technology. Tao is proficient at C# and SQL and familiar with R and Python. He is very interested in translation. He always spends his spare time in translating articles of computer technology such as software testing, database, virtual machines, programming languages. His blog is available at http://cauwt.me.

Translator's Words

I started the translation work in May 2016. And now in November 2016, the work has been completed and delivered. During the translation, one word left me the most profound impression: geek. So at this moment, I would like to take some space to talk about this book from the perspective of the geek spirit.

What is the geek spirit? In my opinion, it stands for the curiosity to explore the world and the actions to change the world. "Stay hungry, stay foolish." Jobs said. Geeks are full of curiosity and keep exploring and discovering new objects. The attraction of the geek spirit lies that geeks keep breaking existing rules and creating changes. Geeks hope the world is like what they think and promote the change of the world with what they think.

This book can be considered as a representation of the author's geek spirit. Let me take several examples with leading questions.

Can we perform object-oriented programming with R? In version 3.1.2, there are four types of object-oriented programming architecture. An ordinary author maybe stresses only on 1 or 2 architectures of them. But as a geek, the author discusses all in detail, including the principles, the implementations, and the advantages vs. weaknesses. Besides, there are different implementing cases for each type of architecture. This allows readers to understand the object-oriented programming of R from different aspects.

Is R applicable for developing apps? The author performed experiments and trials on this topic. In Chapter 5 for R package development, the author provides a complete solution to develop a weather app. The solution including how to retrieve and process weather data, how to do data visualization, and how to capsulate the R code and build R package. Readers can see the way the author uses R to solve practical problem and eventually to change the world.

Is R capable to develop computer games? Many people would say no. However, the author breaks the existing point of view and gives a different answer. In Chapter 6, you will learn that how the author uses R to implement a general game framework and two classical grid games, despite the limitations of R. I believe that the author's geek spirit of exploring can inspire you to discover the unknown world of the R language.

Another name of this book is *A Geek Ideal of* R. In fact, the author's geek spirit exists everywhere in the book, not only the sections mentioned above. Walking through the book, you will believe that the author is exploring and trying various advanced programming techniques of R with the geek spirit. By reading this book, you will get a new understanding for R, and if you can apply the knowledge learned to practice, your R level will be promoted dramatically, and even you can become a geek of R.

Hereby, I extend my sincere thanks to Dan Zhang and CRC Press for the opportunity to translate this book. And I hope I can contribute my efforts to the spread of China's R techniques to the world.

APPLYING R IN MATHEMATIC CALCULATIONS AND ALGORITHMS

1

Chapter 1

The Knowledge System and Mathematical Functions of R

This opening chapter of the book mainly introduces the knowledge system and learning materials of R. The mathematical and statistical calculations and visualization of continuous distribution functions were implemented using R in this chapter, which help readers fully understand the R language and quickly handle calculating problems in basic disciplines.

1.1 Overview of the Knowledge System of R

Question

How do we learn the R language efficiently?

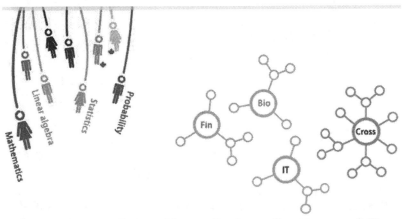

Overview of the Knowledge System of R
http://blog.fens.me/r-overview/

Introduction

Recently, I met a lot of programmers who want to move to data analytics and begin to learn the R language. Having mastered programming other languages, they tend to consider learning R a simple thing. They always pursue learning speed but never understand it deeply. Some even say it only takes two weeks to master R. However, what they mastered is just the syntax of R—a beginning actually.

The knowledge system of R is not as simple as its syntax. If one does not understand the whole picture of R, how could one master the language? This section will show you the knowledge system of R and tell readers how to master it efficiently.

1.1.1 The Knowledge System of R

R is a statistical language, mainly used in aspects including mathematical modeling, statistical computing, data processing, and data visualization. The R language is inherently different from other programming languages. It encapsulates the calculating functions of various basic disciplines. By calling these functions during programming, we can build complex mathematical models orienting to various domains and business. Mastering the syntax of R is just the beginning of learning R. To master it completely, you need to combine the capabilities of basic disciplines (elementary mathematics, advanced mathematics, linear algebra, discrete mathematics, probability theory, and statistics), the knowledge of specific industrials (finance, biology, and Internet), and IT technique (R syntax, R packages, database, and algorithms). In a long-term perspective way, you are capable of mastering the R language completely only if you enhance your level of comprehensive knowledge. In other words, once you have mastered the R language, you are irreplaceable.

1.1.1.1 Overview of the Knowledge System of R

The knowledge system of R is complex. You must integrate the knowledge from multiple disciplines in order to master it completely. The most difficult thing is not in the language itself, but the user's knowledge basics and the capability of using knowledge comprehensively.

First of all, let us take an overall look at the knowledge system as shown in Figure 1.1. Then, I will demonstrate the details of every part.

In Figure 1.1, I divided R's knowledge system into three parts: IT technique, business knowledge, and capabilities in basic disciplines. This is just my own understanding of the R language, which may be one-sided due to my limited experience.

IT technique is one of the essential skills in the computing era and the R language is such a skill which we must master.

Business knowledge is the marketing experience and rules. Any company has products, sales, and markets. You need to know what your company's products are, who your clients are, and how to sell the products to your clients.

The basic disciplines are the theoretical knowledge we studied in school. We did not know why we studied this when we were at school. However, if you are still able to master some of the basic disciplines and can use them in practical work, then you possess the most valuable competitive advantage.

Every individual part has its limitations, but the combinations of the knowledge of any two parts constitute various technical innovations in current society.

IT technique + business knowledge creates Alibaba—the e-commerce empire and Tencent—the full-eco-chain social network.

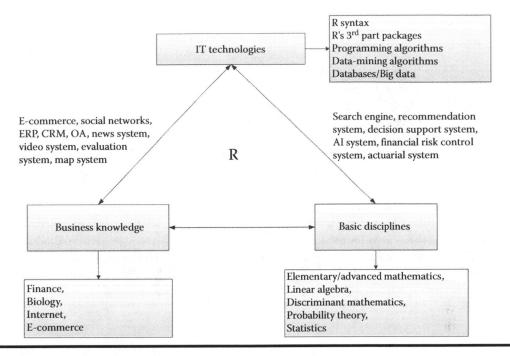

Figure 1.1   The knowledge system of R.

IT technique + basic disciplines creates Google—the searching myth, and Wall Street—the undefeated financial empire.

Of course, as a computer language technology, R cannot be responsible for rewriting history by itself. But it indeed brings us a huge space for imagination, which allows us to move forward to understand the world's laws, find the endless intersections, and create new empires.

If you can learn and use the R language from the above perspective just as I do, surely we can become fellows who go forward side-by-side. Welcome to my team where we are trying to change the future.

1.1.1.2 The Basic Knowledge of the R Language

The blueprint is great and beautiful; however, the specific implementation is difficult. Next, I will review the ideas and map all the knowledge points to operational documents. I hope it will be helpful for you to grasp the outlines of the R language!

The basic knowledge of the R language includes the syntax, the usage of core packages, the kernel programming, the development of packages, and the virtual machines.

1. The Syntax of R

    Understanding the syntax is the first step toward the R language. As all people did, I was adaptive to R's syntax rules, data structures, primary types, commonly used functions, and so on in a short time. But actually the traps in R's syntax are much more than what you have known.

    Let me take a question for instance. Who clearly knows what the differences are between the operators =, <-, and <<- and when to use them? Don't whisper that the question is too partial and we don't need to know this in actual work. The three signals are used anywhere in my code and you just didn't know them yet. When you learn R, don't directly apply experiences from C, Java, and Python to the syntax of R, otherwise you will fall into the trap. Learn from scratch, with no shortcut.

R is a functional language, which is free of syntax, free of naming, and simple to use. But this is just for average users. How can you stay in syntax as a geek with great resolutions? R is fully object oriented. Do you know what object-oriented programming is? R's object-oriented programming breaks the original freedom of syntax, with which the design needs to be compatible. Can you feel such a tangled design? Just knowing about the syntax doesn't mean mastering it. There are various traps. You would really grow up only if you stepped in and climbed out by yourself.

2. Using the Core Packages of R

As other programming languages, R provides us seven core packages when launched, including many basic functions such as mathematical calculation functions, statistical calculation functions, date functions, package loading functions, data-processing functions, function operating functions, graphics devices functions, and so on. By calling the function search(), we can view the seven core packages loaded by default when launching R.

```
# View the loaded packages in current environment.
> search()
[1] ".GlobalEnv"          "package:stats"      "package:graphics"
[4] "package:grDevices"   "package:utils"      "package:datasets"
[7] "package:methods"     "Autoloads"          "package:base"
```

The seven core packages are the basics for us to build complex models. However, they are also the obstacles to learning R. That is because they are at a low-level and many functions in them are encapsulated in C and there is no source code in R. Besides, there is no document detailing them, except for the official documents. Never simply think that it's OK to master some of the functions. There is a deeper meaning behind them.

Another question: since all operations in R are functional, then what function is the statement "a<-1:10" resolved to?

```
# Assignment.
> a<-1:10;a
 [1]  1  2  3  4  5  6  7  8  9 10
```

The answer is that "1:10" is corresponding to seq() and "<-" to assign().

```
# Assignment through functions.
> assign('b',seq(1:10));b
 [1]  1  2  3  4  5  6  7  8  9 10
```

The meaning of the correspondence is that because R is an interpreting language, we can make function B dynamically call function A by passing the handler of function A, which is the usage of the feature of closure of dynamic languages. This idea has been widely used in JavaScript, while only a few of the functions in core packages employ the syntax. There are many similar scenarios that require computer background knowledge in R, especially when considering how to enhance the performance of R code So, do not easily say you have mastered the R language. Keep thinking how to bring the basics of the other languages into R's world.

3. The Kernel Programming of R

The kernel programming of R is another complex problem of computer science. What contents does it contain? Besides the syntax and the core packages mentioned earlier, it also contains object-oriented programming, vectorized computing, special data types, environments, and so on. This book will focus on these aspects.

Object-oriented programming (OOP) is kind of a methodology of understanding and abstracting the real world, mainly used for solving the design and implementation of complex problems. There was no OOP implementation in R until R 2.12 released in 2011, which finally provided RC types with OOP implementation. The ability for OOP implementation marked R has had the capability of building complex and large-scaled applications. But it seems that statistic scientists are not good at OOP. Few people are able to write OOP code like Hadley Wickham in the circumstance of R.

Vectorized computing is a kind of parallel computing way that characterizes the R language. In R, vector is a primary data type. When you operate a vector, each element in the vector is calculated respectively and the result is returned in the form of a vector. Here is an example where two vectors with equal length are added.

```
# Two vectors are added.
> 1:10+10:1
 [1] 11 11 11 11 11 11 11 11 11 11
```

Vectorized computing has a wide range of application scenarios in R. It can replace the cycled computing in almost all aspects and finish computing tasks efficiently. Let us define two vectors, add them, and get the sum. Function run1() implements the algorithm using vectorized computing and run2() with cycled computing.

```
> a<-1:100000
> b<-100000:1

# Vectorized computing.
 > run1<-function(){
+    sum(as.numeric(a+b))
+ }

# Cycled computing.
> run2<-function(){
+    c2<-0
+    for(i in 1:length(a)){
+       c2<-a[i]+b[i]+c2
+    }
+    c2
+ }

# Measure the execution time of function.
# run1().
> system.time(run1())
User System Elapse
    0    0    0
# Measure the execution time of function.
# run2().
> system.time(run2())
User System Elapse
0.14 0.00 0.14
```

By running the programs, we can clearly get that vectorized computing is faster than loops. The time cost difference becomes more obvious with the growth of the complexity of the algorithm and the amount of data. One of experimental rules in R programming is to use vectorized computing rather than the cycled.

Special data types. Except for the basic data types, there are also some advanced data types in R. They are rarely seen in your code not because they are unusual. It is just because you don't know them. S3, S4, and RC types are corresponding to three kinds of object-oriented programming data structures in R, respectively. Environment type is a data structure defined in kernel, comprised of a series of frames with hierarchical relationships. Each environment corresponds to a frame, used to distinguish different runtime scopes.

The environment is an essential knowledge point when developing R packages. Every environment is an instance of the environment type. Every R package is loaded into an environment, forming a hierarchical and callable scope structure.

The functions and variables we defined are stored in R's environments. Calling ls() can view these variables in the current environment, such as the variables and functions defined in the vectorized computing program.

```
# View the variables in current environment.
> ls()
[1] "a"      "b"      "run1" "run2"
```

In addition to the customized variables and functions, many others exist in environments, such as sum(), length(), system.time(), and so on. These functions can be called directly but are not located in the current environment, so we cannot see them by simply calling ls(). We can find the definition of sum() after we have switched to the environment of base.

```
# View the variables in base environment.
> ls(pattern="^sum$",envir=baseenv())
[1] "sum"
```

Just as other languages, R's kernel programming has many knowledge details which contain not only the points I mentioned above. But due to lack of documentation and spread of R core technologies, these details are not well known by people and few people can use them. I keep exploring these everyday and expects to find more secrets.

4. The development of R packages

The development of R packages is a difficult but inevitable problem in R language programming. It not only needs to integrate the technologies mentioned above, but also comply with the programming standards of developing R packages and is documented in LaTex and then submitted to CRAN for publishing. Although the technical problems are difficult, we are able to take time to solve them. However, publishing your packages on CRAN is even harder than climbing up to heaven. In the past 20+ years with R's growth, only 8000+ packages were published on CRAN. The review is much more strict than usual! The game package gridgame and the weather package chinaWeather have been revised many times but still cannot pass the review. I almost want to give up.

On the other hand, only strict review can make sure that there is no error for users installing the third part R package. Because CRAN's review is too strict, Hadley Wickham cannot stand it. He developed devtools package, which not only provides the utilities that simplify R package development, but also supports GitHub community publishing. As a result, we can get rid of the bondage of CRAN and publish R packages with various ideas (even packages that "don't do honest jobs") in the name of individuals.

5. The Virtual Machines of R

Finally, it's time to talk about a topic that I am not familiar with. I didn't encounter the virtual machines of R language during my 5+ years' experience of using R. But on the Internet, I saw a lot of outline masters are able to recompile R software in the production environment. For instance, they use OpenBLAS to speed up R's matrix computing, implement the parallel computing of matrix in the layer of the virtual machine, or use a Graphic Processing Unit (GPU). Some masters use C++ to re-implement the algorithms that are already implemented in R, encapsulate them through Rcpp, and then directly connect to the virtual machines of R to call.

1.1.1.3 The Third Part Packages of R

The third part packages of R language mainly include the 8000+ packages on CRAN and packages from other R communities, which play important roles in various fields. In *R for Programmers: Mastering the Tool*, I introduced more than 30 packages, containing basic utility packages (fortunes, formatR, rjson, RJSONIO, Cairo, and CaTools), time series packages (zoo, xts, and xtsExtra), cross-platform communication packages (Rserve, Rsession, and rJava), R server packages (Rserve, RSclient, FastRWeb, and Websocket), data-accessing packages (RMySQL, rmongodb, rredis, RCassandra, and RHive), Hadoop-operating packages (rhdfs, rmr2, and rhbase), and so on.

Other frequently used packages include data-processing packages (lubridate, plyr, reshape2, stringr, formatR, mcmc, and data.table), machine-learning packages (nnet, rpart, tree, party, lars, boost, e1071, BayesTree, gafit, and arules), visualization packages (ggplot2, lattice, and googleVis), map packages (ggmap, RgoogleMaps, and rworldmap), and so on.

R language supports the finance business as well with packages such as time series packages (zoo, xts, chron, its, and timeDate), financial analysis packages (quantmod, RQuantLib, portfolio, quantstrat, blotter, PerformanceAnalytics, TTR, sde, and YieldCurve), risk management packages (parma, evd, evdbayes, evir, extRemes, and ismev), and so on. Meanwhile, the writer is involved in the venture of quantitative investment, with the R language in the core position as the algorithm engine of the system, in the most valuable business. I will have a complete introduction to R language in the application of quantitative investment system in the following book *R for Programmers: Quantitative Investment Applications*.

1.1.1.4 The Basic Knowledge of Mathematics

The basic knowledge of mathematics contains elementary mathematics, advanced mathematics, linear algebra, probability theory, statistics, and so on. What radically determines your level of mastering the R language is the basic knowledge of the kind of mathematics that we have studied in college, which we don't know why we learn and we just study for examinations.

After R spreads into a popular programming language, getting started becomes more and more easy and calling third part packages becomes more and more simple. Finally, people struggle for their foundation of basic disciplines, of which mathematics is the most difficult for all.

Elementary mathematics. The Chinese have been stressing that mathematics is their advantage, much more so than people in other countries. In fact, the advantage is just limited to elementary mathematics. The Chinese, using the addition and multiplication tables, are able to perform the four arithmetic operations in mind with numbers less than 100.

Advanced mathematics. This may be the discipline that fails most students in colleges. Teachers just clone what is printed in textbooks to blackboards and students don't understand at all what their teachers say. I didn't know why least squared methodology could perform optimized computing until I met R. It is the essential to learn R to pick up advanced mathematics again.

Linear algebra. Many years later after I finished reading Google's PageRank paper, I understood clearly the reasons why matrix can perform the calculation for a huge amount of data, and implement the consistency of distributed algorithms, and do this on a single machine.

Probability theory. By performing the random tests that follow various distributions through R language, and applying the probability density curve functions to practical business, we understand that only probability is the indicator that measures the happening of objective things.

Statistics. Through R we can easily build various statistical models, identify junk mail using the Bayes discriminant device, or predict house price tendencies using a regression model.

It is the R language that makes me truly feel the application of the basic knowledge of mathematics in our daily life. Again, it is the R language that reduces the distance between academia and industry. If we could relate all the knowledge we learned since childhood together, I believe that everyone would have a knowledge structure different from each other and create great innovations in all kinds of industries.

1.1.1.5 Business Knowledge

Business knowledge involves a wide range of industries. Everyone should have the knowledge in their industry, and combine this with the fields at which R language is good to discover new opportunities. R is good at such fields as statistical analysis, financial analysis, data mining, Internet, bioinformatics, bio-pharmaceuticals, global geographic science, data visualization, and so on.

I worked in the software and Internet industry for 10 years, and witnessed rapid development and change in the two industries. New technologies emerge wave by wave. New subjects arise year by year. However, there are fewer and fewer people following. Although fresh blood continuously joins in, their capability and experience are far from meeting the requirements.

Looking at China's capital market, the key to making money is grasping business knowledge. When a business has grown up and all know the rules of the game, the competition becomes fierce. Examples include e-commerce, Groupon, tourism, hoteling, and the game industry. What is worth young people struggling for are the new business areas. The vigorously growing fields, such as O2O, Internet finance, the Internet of Things, and robots, have became the breaking points in 2016. If you have mastered both technology and business, and study hard, then you will be the creator of a new empire.

1.1.1.6 Comprehensive Skills across Disciplines

Again, as long as you are able to integrate the knowledge from various disciplines, you will become a new master of the R language, and furthermore, you will surely realize your own value.

When integrating IT technologies and business intelligence, you will find opportunities in emerging markets. Once the markets mature, business competition turns to capital competition and no opportunity opens again.

When integrating IT technologies and basic disciplines, you can establish the technical barriers through innovations and keep technical advantage to become an industrial leader.

When integrating all three parts together, you will be irreplaceable. As long as you find the team that you should belong to, get your products developed and promote them to your customers, you will have succeeded!

The R language can help you succeed from the perspective of IT. In the meantime, your success is also the success of R!

1.1.2 Learning the R Language

I have eventually explained the knowledge system of R language with my understanding, which however takes up much space. Then, how shall we learn the R language efficiently? Learning is tough and there is no shortcut. If you want to succeed, you must face the most difficult things. The correct methodologies of learning can save us twists and turns. Learning from others can speed up our growth.

Through the above description of interdisciplinary knowledge systems, I believe that you have understood that the most difficulty to master R is not in the language itself, but in the knowledge basics and the integrative capability of users. Of course, the integrative capability is built on good basic knowledge. Despite the business knowledge and basic disciplines, what knowledge should we have only regarding IT technologies?

1.1.2.1 The Basic Knowledge of IT

For the R language itself, we need to take the foundation of R, including the syntax, the usage of its core packages, the kernel programming, the development of R packages, and the usage of third part packages that relate to specific business areas.

If you have got a lot of experience of other programming languages like Java and Python before you learn R, you will be familiar with R in a short time. What you need is to add some knowledge of data analysis and data-mining algorithms, and you can use R in your practical work soon.

If you were a data scientist in SAS or Matlab before, then as long as you get familiar with the programming syntax of R and third part packages, you are able to finish all tasks of SAS and Matlab.

If you are BI programmer with daily works for data processing and visualization, you can learn R with supplementing some statistical knowledge to find the value of data from dull ETL (extraction, transformation, and loading) processes.

If you are a college student of statistics, the R language helps you put the boring knowledge in textbooks into programs. You will find the laws of society from learning R.

If you have been using Excel and complain the lack of its functionalities, you can try using R and your idea will be quickly realized.

If you are a Quant but have not heard about R, you will soon be eliminated by the market.

If you are a Hadoop algorithm engineer, with thousands of lines of code when implementing an MR algorithm in Java, you can try using RHadoop, which takes one-tenth of the lines to finish the same task.

More examples can be taken to explain. The R language can be integrated with a wide range of technologies and ideas. Make collision with R and the knowledge you have mastered, and you will become different from others.

1.1.2.2 The Chinese Books about R

Yishuo Deng wrote a blog naming *the books on learning routines of R language*, which is a good reference. The article introduces books that fit different levels and business fields including beginners, advanced learners, graphics and visualization, econometrics, time series analysis, finance, and so on. More than 30 books and booklets were mentioned, most of which are written in English.

As time passed, a lot of new books about R were published in recent years and Chinese books are getting more and more. I redefined the learning routines for R's Chinese books, shown as in Figure 1.2.

There are market segmentations for users at different levels. The beginners can start from *The Art of R Programming*. Users with certain basics can read *R in Action*. Those who want to expand the range of knowledge can read *R for Programmers: Mastering the Tool*. After mastering the beginning technologies, the advanced R developers can read *R for Programmers: Advanced Technologies* (this book). Those involving visualization using R can read *ggplot2: Elegant Graphics for Data Analysis*. Students in statistics can read *Statistical Modeling and R*. Those who want to step into finance can read *Time Series Analysis with Applications in R* (Second Edition) and *An Introduction to Analysis of Financial Data with R*.

All of the recommended books above were read by me, so the quality is guaranteed. I will keep updating the book list in my blog. I would like to share the excellent books with everyone.

1.1.2.3 The Chinese Communities about R

In addition to books, the Chinese R communities and personal blogs grow vigorously. Capital of Statistics is the most authoritative R organization in mainland China. It not only accumulated a large number of high-quality R language articles, but also hosted seven Chinese R language conferences. The team members also participated in the translation of many books such as *The Art of R Programming, R in Action: Data Analysis and Graphics with R, ggplot2: Elegant Graphics for Data Analysis, R in a Nutshell* (Second Edition), *R Graphics Cookbook*, and *Introductory Statistics with R* (Second Edition).

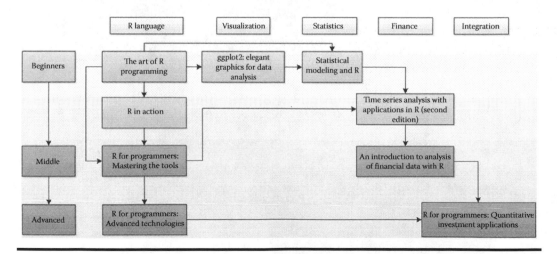

Figure 1.2   The learning routines for R's Chinese books.

Dataguru, focusing on data analysis, has a board for R language, and provides online training for R beginners. The explanation of algorithms by Huang Zhihong (with the nickname Tigerfish) is super excellent.

The economics forum of Renming University focuses on the education of business and economics, with an R language board opened. Its main business is offline training.

1.1.2.4 The Chinese Blogs about R

The writer's personal blog, Fens Log (http://fens.me), originates lots of articles on the practical technologies of R language, including the serial articles of *R for Programmers*, the series of *RHadoop in Action*, the series of *R Swords for NoSQL*, and the series of books *R for Programmers*.

Yihui Xie's personal blog (http://yihui.name) has various interesting technical and teasing articles on site. Xie is the founder of Capital of Statistics and currently works at RStudio as a programmer.

SiZhe Liu's personal blog, the planet of Vegeta (http://www.bjt.name), mainly involves articles about enterprise applications in R. Liu worked at Jingdong as a recommendation algorithm manager.

Jian Li's personal blog (http://jliblog.com) involves articles about modeling using R. Li is the data director of the China Region at Mango Solutions.

Lastly, I wish that all of you master and use the R language well, find breaking points in your own fields to realize your values. I also hope you can give your feedback to R language communities to accelerate the growth and development of R.

1.2 The Mathematical Calculations of R

Question
How do we perform mathematical calculations using R?

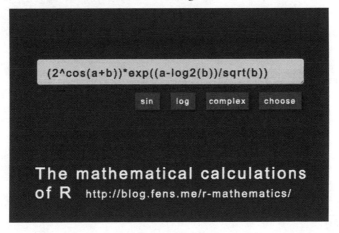

Introduction
R is a statistical language with a natural support for mathematics. It is very convenient to solve mathematical calculation problems using R. A calculator with R's calculating functions embedded will surely be a hi-tech product. I really take R as my calculator!

1.2.1 The Basic Calculations

R language has good support for mathematical calculations. This section will completely introduce various calculation operations in elementary mathematics.

The system configuration in this section

■ Win7 64bit
■ R: 3.1.1 x86_64-w64-mingw32/x64 (64-bit)

Implement the arithmetic operations, including addition, subtraction, multiplication, division, remainder, exact division, absolute value, and signal judgment.

```
# Define two variables.
> a<-10;b<-5

# The operations of addition, subtraction, multiplication, and division.
> a+b;a-b;a*b;a/b
[1] 15
[1] 5
[1] 50
[1] 2

# Remainder, exact division.
> a%%b;a%/%b
[1] 0
[1] 2

# Absolute value.
> abs(-a)
[1] 10

# Signal judgment.
> sign(-2:3)
[1] -1 -1  0  1  1  1
```

Implement the mathematic calculation operations, including power, the power of the natural constant e, square root, and logarithm.

```
# Define 3 variables.
> a<-10;b<-5;c<-4

# Power operations.
> c^b;c^-b;c^(b/10)
[1] 1024
[1] 0.0009765625
[1] 2

# Get the natural constant e.
> exp(1)
[1] 2.718282

# The power of the natural constant e.
> exp(3)
[1] 20.08554
```

```
# Square root.
> sqrt(c)
[1] 2

# Binary logarithm.
> log2(c)
[1] 2

# Denary logarithm.
> log10(b)
[1] 0.69897

# Logarithm with customized base.
> log(c,base = 2)
[1] 2

# Logarithm with base of the natural constant e.
> log(a,base=exp(1))
[1] 2.302585

# Logarithm of power.
> log(a^b,base=a)
[1] 5

> log(exp(3))
[1] 3
```

Implement the comparison operations using R, including ==, >, <,!=, <=, >=, isTRUE, and identical.

```
# Define two variables.
> a<-10;b<-5

# Comparisons.
> a==a;a!=b;a>b;a<b;a<=b;a>=c
[1] TRUE
[1] TRUE
[1] TRUE
[1] FALSE
[1] FALSE
[1] TRUE

# Check whether a is TRUE.
> isTRUE(a)
[1] FALSE
> isTRUE(!a)
[1] FALSE

# Exactly compare two objects.
> identical(1, as.integer(1))
[1] FALSE
> identical(NaN, -NaN)
[1] TRUE
```

```
> f <- function(x) x
> g <- compiler::cmpfun(f)
> identical(f, g)
[1] TRUE
```

Implement the logic operations using R, including &, |, &&, ||, and xor.

```
# Define two vectors.
> x<-c(0,1,0,1)
> y<-c(0,0,1,1)

# &&, || compare only the first elements of the vectors.
> x && y;x || y
[1] FALSE
[1] FALSE

# &, |, the logical operations for S4 objects, comparing all elements.
> x & y;x | y
[1] FALSE FALSE FALSE  TRUE
[1] FALSE  TRUE  TRUE  TRUE

# The operation of exclusive or.
> xor(x,y)
[1] FALSE  TRUE  TRUE FALSE
> xor(x,!y)
[1]  TRUE FALSE FALSE  TRUE
```

Implement the roundabout operations, including ceiling, floor, trunc, round, and signif.

```
# Round up.
> ceiling(5.4)
[1] 6

# Round down.
> floor(5.8)
[1] 5

# Get the integer part.
> trunc(3.9)
[1] 3

# Round about.
> round(5.8)
[1] 6

# Rsound with 2 decimal places reserved.
> round(5.8833, 2)
[1] 5.88

# Round with the leading 2 decimals reserved.
> signif(5990000,2)
[1] 6e+06
```

Implement array operations, including solving maximum, minimum, range, sum, mean, weighted mean, multiple multiplications, differentiate, rank, median, quantile, any, and all.

```
# Define a vector.
> d<-seq(1,10,2);d
[1] 1 3 5 7 9

# Maximum, minimum, and range.
> max(d);min(d);range(d)
[1] 9
[1] 1
[1] 1 9

# Sum, mean.
> sum(d);mean(d)
[1] 25
[1] 5

# Weighted mean.
> weighted.mean(d,rep(1,5))
[1] 5
> weighted.mean(d,c(1,1,2,2,2))
[1] 5.75

# Multiple multiplications.
> prod(1:5)
[1] 120

# Differentiation.
> diff(d)
[1] 2 2 2 2

# Rank.
> rank(d)
[1] 1 2 3 4 5

# Median.
> median(d)
[1] 5

# Quantile.
> quantile(d)
0%  25%  50%  75% 100%
1    3    5    7    9

# Any condition, all conditions.
> any(d<5);all(d<5)
[1] TRUE
[1] FALSE
```

Implement permutation and combination operations, including factorial, combination, and permutation.

```
# 5! Factorials.
> factorial(5)
[1] 120
```

```
# Combination. Choose 2 from 5 objects.
> choose(5, 2)
[1] 10

# List all the combinations for choosing 2 from 5.
> combn(5,2)
     [,1] [,2] [,3] [,4] [,5] [,6] [,7] [,8] [,9] [,10]
[1,]    1    1    1    1    2    2    2    3    3     4
[2,]    2    3    4    5    3    4    5    4    5     5

# Calculate the combinations for 0:10.
> for (n in 0:10) print(choose(n, k = 0:n))
[1] 1
[1] 1 1
[1] 1 2 1
[1] 1 3 3 1
[1] 1 4 6 4 1
[1]  1  5 10 10  5  1
[1]  1  6 15 20 15  6  1
[1]  1  7 21 35 35 21  7  1
[1]  1  8 28 56 70 56 28  8  1
[1]   1   9  36  84 126 126  84  36   9   1
[1]   1  10  45 120 210 252 210 120  45  10   1

# Permutation. Choose 2 from 5 objects.
> choose(5, 2)*factorial(2)
[1] 20
```

Implement the cumulative operations, including cumulative addition, cumulative multiplication, cumulative minimum, and cumulative maximum.

```
# Cumulative addition.
> cumsum(1:5)
[1]  1  3  6 10 15

# Cumulative multiplication.
 > cumprod(1:5)
[1]   1   2   6  24 120

# Define a vector.
> e<-seq(-3,3);e
[1] -3 -2 -1  0  1  2  3

# Cumulative minimum.
> cummin(e)
[1] -3 -3 -3 -3 -3 -3 -3

# Cumulative maximum.
> cummax(e)
[1] -3 -2 -1  0  1  2  3
```

Implement set operations between two arrays, including union, intersection, differentiation, judging where two arrays are equal, making array elements unique, finding the indices of matching elements, and finding the indices of duplicated elements.

```
# Define two arrays of vectors.
> x <- c(9:20, 1:5, 3:7, 0:8);x
 [1]   9 10 11 12 13 14 15 16 17 18 19 20  1  2  3  4  5
[18]  3  4  5  6  7  0  1  2  3  4  5  6  7  8
> y<- 1:10;y
 [1]  1  2  3  4  5  6  7  8  9 10

# Intersection.
> intersect(x,y)
 [1]  9 10  1  2  3  4  5  6  7  8

# Union.
> union(x,y)
 [1]   9 10 11 12 13 14 15 16 17 18 19 20  1  2  3  4  5
[18]  6  7  0  8

# Differentiation, exclude y from x.
> setdiff(x,y)
 [1]  11 12 13 14 15 16 17 18 19 20   0

# Check whether the 2 arrays are equal.
> setequal(x, y)
[1] FALSE

# Make array elements unique.
> unique(c(x,y))
 [1]   9 10 11 12 13 14 15 16 17 18 19 20  1  2  3  4  5
[18]  6  7  0  8

# Find the indices of x elements that exist in y.
> which(x %in% y)
 [1]   1   2 13 14 15 16 17 18 19 20 21 22 24 25 26 27 28
[18]  29 30 31

# Same as %in%.
> which(is.element(x,y))
 [1]   1   2 13 14 15 16 17 18 19 20 21 22 24 25 26 27 28
[18]  29 30 31

# Find the indices of duplicated elements.
> which(duplicated(x))
 [1]  18 19 20 24 25 26 27 28 29 30
```

1.2.2 Calculations of the Trigonometric Functions[*]

1.2.2.1 The Trigonometric Functions

Only the trigonometric functions for acute angles (with degrees >0 and <90) are defined in a right triangle. Given an acute θ, one can draw a right triangle with θ as one of the interior angles. Suppose that the lengths of the opposite side, the adjacent side, and the hypotenuse side of θ are a, b, and h, respectively, as shown as in Figure 1.3.

---

[*] https://en.wikipedia.org/wiki/Trigonometric_functions

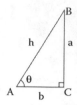

Figure 1.3   A right triangle.

Six relationships in trigonometric functions: sine, cosine, tangent, cotangent, secant, and cosecant.

- The sine of θ is the length ratio of its opposite side to hypotenuse.
- The cosine of θ is the length ratio of its adjacent side to hypotenuse.
- The tangent of θ is the length ratio of its opposite side to adjacent side.
- The cotangent of θ is the length ratio of its adjacent side to opposite side.
- The secant of θ is the length ratio of its hypotenuse to adjacent side.
- The cosecant of θ is the length ratio of its hypotenuse to opposite side.

Some special values of the trigonometric functions.

Function	0	$\pi/12$	$\pi/6$	$\pi/4$	$5\pi/12$	$\pi/2$
sin	0	$\left(\sqrt{6}-\sqrt{2}\right)/4$	$1/2$	$\sqrt{2}/2$	$\left(\sqrt{6}+\sqrt{2}\right)/4$	1
cos	1	$\left(\sqrt{6}+\sqrt{2}\right)/4$	$\sqrt{3}/2$	$\sqrt{2}/2$	$\left(\sqrt{6}-\sqrt{2}\right)/4$	0
tan	0	$2-\sqrt{3}$	$\sqrt{3}/3$	1	$2+\sqrt{3}$	/
cot	/	$2+\sqrt{3}$	$\sqrt{3}$	1	$2-\sqrt{3}$	0
sec	1	$\sqrt{6}-\sqrt{2}$	$2\sqrt{3}/3$	$\sqrt{2}$	$\sqrt{6}+\sqrt{2}$	/
csc	/	$\sqrt{6}+\sqrt{2}$	2	$\sqrt{2}$	$\sqrt{6}-\sqrt{2}$	1

Implement the calculation of basic trigonometric functions, including sine, cosine, and tangent.

```
# Sine.
> sin(0);sin(1);sin(pi/2)
[1] 0
[1] 0.841471
[1] 1

# Cosine.
> cos(0);cos(1);cos(pi)
[1] 1
[1] 0.5403023
[1] -1
```

```
# Tangent.
> tan(0);tan(1);tan(pi)
[1] 0
[1] 1.557408
[1] -1.224647e-16
```

Next, let us draw the graphics of trigonometric functions.

```
# Load the library of ggplot2.
> library(ggplot2)
> library(scales)
```

Draw the graphics of trigonometric functions. The following codes generate the curves of trigonometric functions, shown as in Figure 1.4:

```
# x coordinates.
> x<-seq(-2*pi,2*pi,by=0.01)

# y coordinates of sine.
> s1<-data.frame(x,y=sin(x),type=rep('sin',length(x)))
# y coordinates of cosine.
> s2<-data.frame(x,y=cos(x),type=rep('cos',length(x)))
# y coordinates of tangent.
> s3<-data.frame(x,y=tan(x),type=rep('tan',length(x)))
# y coordinates of cotangent.
> s4<-data.frame(x,y=1/tan(x),type=rep('cot',length(x)))
# y coordinates of secent.
> s5<-data.frame(x,y=1/cos(x),type=rep('sec',length(x)))
# y coordinates of cosecant.
> s6<-data.frame(x,y=1/sin(x),type=rep('csc',length(x)))
> df<-rbind(s1,s2,s3,s4,s5,s6)

# Draw graphics using ggplot2.
> g<-ggplot(df,aes(x,y))
> g<-g+geom_line(aes(colour=type,stat='identity'))
```

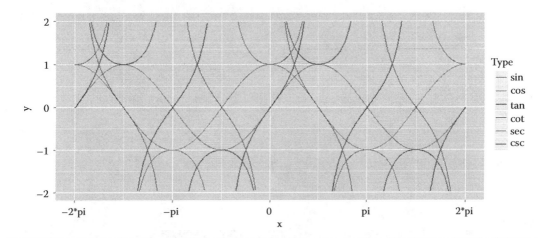

Figure 1.4   Curves of trigonometric functions.

```
> g<-g+scale_y_continuous(limits=c(-2, 2))
> g<-g+scale_x_continuous(breaks=seq(-2*pi,2*pi,by=pi),
  labels=c("-2*pi","-pi","0","pi","2*pi"))
> g
```

1.2.2.2 The Inverse Trigonometric Functions

Definitions of the inverse trigonometric functions:

Function	Definition	Value Field
arcsin(*x*) = *y*	sin(*y*) = *x*	$-\pi/2 \le y \le \pi/2$
arccos(*x*) = *y*	cos(*y*) = *x*	$0 \le y \le \pi$
arctan(*x*) = *y*	tan(*y*) = *x*	$-\pi/2 < y < \pi/2$
arccot(*x*) = *y*	cot(*y*) = *x*	$0 < y < \pi$
arcsec(*x*) = *y*	sec(*y*) = *x*	$0 \le y \le \pi, y \ne \pi/2$
arccsc(*x*) = *y*	csc(*y*) = *x*	$-\pi/2 \le y \le \pi/2, y \ne 0$

Implement calculation of the inverse trigonometric functions, including arcsin, arcos, and arctan.

```
# Arc-sine.
# pi/2=1.570796.
> asin(0);asin(1)
[1] 0
[1] 1.570796

# Arc-cosine.
# pi/2=1.570796.
> acos(0);acos(1)
[1] 1.570796
[1] 0

# Arc-tangent.
# pi/4=0.7853982.
> atan(0);atan(1)
[1] 0
[1] 0.7853982
```

Draw the graphics of inverse trigonometric functions. The following codes generate the curves of inverse trigonometric functions, as shown as in Figure 1.5:

```
# x coordinates.
> x<-seq(-1,1,by=0.005)
# y coordinates of arcsine.
> s1<-data.frame(x,y=asin(x),type=rep('arcsin',length(x)))
# y coordinates of arccosine.
```

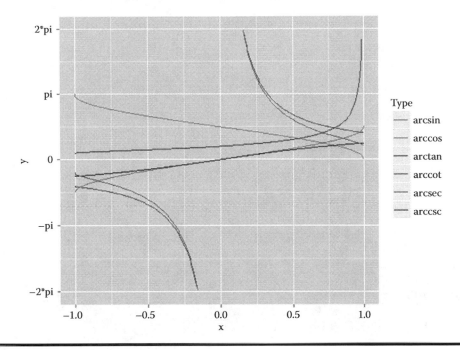

Figure 1.5   Curves of inverse trigonometric functions.

```
> s2<-data.frame(x,y=acos(x),type=rep('arccos',length(x)))
# y coordinates of arctangent.
> s3<-data.frame(x,y=atan(x),type=rep('arctan',length(x)))
# y coordinates of arccotangent.
> s4<-data.frame(x,y=1/atan(x),type=rep('arccot',length(x)))
# y coordinates of arcsecent.
> s5<-data.frame(x,y=1/asin(x),type=rep('arcsec',length(x)))
# y coordinates of arccosecent.
> s6<-data.frame(x,y=1/acos(x),type=rep('arccsc',length(x)))
> df<-rbind(s1,s2,s3,s4,s5,s6)

# Draw graphics using ggplot2.
> g<-ggplot(df,aes(x,y))
> g<-g+geom_line(aes(colour=type,stat='identity'))
> g<-g+scale_y_continuous(limits=c(-2*pi,2*pi),breaks=seq(-2*pi,2*pi,
by=pi),labels=c("-2*pi","-pi","0","pi","2*pi"))
> g
```

1.2.2.3 The Trigonometric Formulae

Next, let us describe the trigonometric formulae by the method of unit test. The left side is equivalent to the right side. We perform unit tests using package testthat. Refer Section 5.2 for the installation and usage about package testthat.

Pass the expressions at both sides of the formula to function expect_that(right,left) as parameters, and then run this function. No output indicates that the two parameters are equal; otherwise, we need to investigate the reasons according to the output.

```
# Load the testthat package.
> library(testthat)
# Define The variables.
> a<-5;b<-10
```

Pythagorean formula

■ $\sin^2 x + \cos^2 x = 1$

```
> sin(a)^2+cos(a)^2
[1] 1
# Check whether the two sides are equal using unit test method.
> expect_that(1, equals(sin(a)^2+cos(a)^2))

# If not equal, error is shown.
> expect_that(2, equals(sin(a)^2+cos(a)^2))
Error: 2 not equal to sin(a)^2 + cos(a)^2
Mean relative difference: 1
```

Sum-and-difference formulae:

■ $\sin(\alpha + \beta) = \sin \alpha \cos \beta + \sin \beta \cos \alpha$

■ $\sin(\alpha - \beta) = \sin \alpha \cos \beta - \sin \beta \cos \alpha$

■ $\cos(\alpha + \beta) = \cos \alpha \cos \beta - \sin \beta \sin \alpha$

■ $\cos(\alpha - \beta) = \cos \alpha \cos \beta + \sin \beta \sin \alpha$

■ $\tan(\alpha + \beta) = (\tan \alpha + \tan \beta)/(1 - \tan \alpha \tan \beta)$

■ $\tan(\alpha - \beta) = (\tan \alpha - \tan \beta)/(1 + \tan \alpha \tan \beta)$

The following shows the unit tests against the sum and difference formulae:

```
> expect_that(sin(a)*cos(b)+sin(b)*cos(a),equals(sin(a+b)))
> expect_that(sin(a)*cos(b)-sin(b)*cos(a),equals(sin(a-b)))
> expect_that(cos(a)*cos(b)-sin(b)*sin(a),equals(cos(a+b)))
> expect_that(cos(a)*cos(b)+sin(b)*sin(a),equals(cos(a-b)))
> expect_that((tan(a)+tan(b))/(1-tan(a)*tan(b)),equals(tan(a+b)))
> expect_that((tan(a)-tan(b))/(1+tan(a)*tan(b)),equals(tan(a-b)))
```

Double-angle formulae:

■ $\sin 2\alpha = 2 \sin \alpha \cos \alpha$

■ $\cos 2\alpha = \cos^2 \alpha - \sin^2 \alpha = 2 \cos^2 \alpha - 1 = 1 - 2 \sin^2 \alpha$

The following shows the unit tests against the double-angle formulae:

```
> expect_that(cos(a)^2-sin(a)^2,equals(cos(2*a)))
> expect_that(2*cos(a)^2-1,equals(cos(2*a)))
> expect_that(1-2*sin(a)^2,equals(cos(2*a)))
```

Triple-angle formulae:

■ $\cos 2\alpha = 4\cos^3 \alpha - 3\cos \alpha$

■ $\sin 3\alpha = -4\sin^3 \alpha + 3\sin \alpha$

The following shows the unit tests against the triple-angle formulae:

```
> expect_that(4*cos(a)^3-3*cos(a),equals(cos(3*a)))
> expect_that(-4*sin(a)^3+3*sin(a),equals(sin(3*a)))
```

Half-angle formulae:

■ $\sin\dfrac{\alpha}{2} = \pm\sqrt{\dfrac{1-\cos\alpha}{2}}$

■ $\cos\dfrac{\alpha}{2} = \pm\sqrt{\dfrac{1+\cos\alpha}{2}}$

■ $\tan\dfrac{\alpha}{2} = \dfrac{1-\cos\alpha}{\sin\alpha} = \dfrac{\sin\alpha}{1+\cos\alpha}$

The following shows the unit tests against the half-angle formulae:

```
> expect_that(sqrt((1-cos(a))/2),equals(abs(sin(a/2))))
> expect_that(sqrt((1+cos(a))/2),equals(abs(cos(a/2))))
> expect_that(sqrt((1-cos(a))/(1+cos(a))),equals(abs(tan(a/2))))
> expect_that(abs(sin(a)/(1+cos(a))),equals(abs(tan(a/2))))
> expect_that(abs((1-cos(a))/sin(a)),equals(abs(tan(a/2))))
```

Product-to-sum formulae:

■ $\cos\alpha\sin\beta = \dfrac{1}{2}(\sin(\alpha+\beta) - \sin(\alpha-\beta))$

■ $\sin\alpha\cos\beta = \dfrac{1}{2}(\sin(\alpha+\beta) + \sin(\alpha-\beta))$

■ $\sin\alpha\cos\beta = \dfrac{1}{2}(\sin(\alpha+\beta) + \sin(\alpha-\beta))$

■ $\sin\alpha\cos\beta = \dfrac{1}{2}(\sin(\alpha+\beta) + \sin(\alpha-\beta))$

The following shows the unit tests against the product-to-sum formulae:

```
> expect_that((sin(a+b)+sin(a-b))/2,equals(sin(a)*cos(b)))
> expect_that((sin(a+b)-sin(a-b))/2,equals(cos(a)*sin(b)))
> expect_that((cos(a+b)+cos(a-b))/2,equals(cos(a)*cos(b)))
> expect_that((cos(a-b)-cos(a+b))/2,equals(sin(a)*sin(b)))
```

Sum-to-product formulae:

■ $\sin\alpha + \sin\beta = 2\sin\dfrac{(\alpha+\beta)}{2}\cos\dfrac{(\alpha+\beta)}{2}$

■ $\sin\alpha - \sin\beta = 2\cos\dfrac{(\alpha+\beta)}{2}\sin\dfrac{(\alpha-\beta)}{2}$

■ $\cos\alpha + \cos\beta = 2\cos\dfrac{(\alpha+\beta)}{2}\cos\dfrac{(\alpha+\beta)}{2}$

■ $\cos\alpha - \cos\beta = -2\sin\dfrac{(\alpha+\beta)}{2}\sin\dfrac{(\alpha-\beta)}{2}$

The following shows the unit tests against the sum-to-product formulae:

```
> expect_that(sin(a)+sin(b),equals(2*sin((a+b)/2)*cos((a-b)/2)))
> expect_that(sin(a)-sin(b),equals(2*cos((a+b)/2)*sin((a-b)/2)))
> expect_that(2*cos((a+b)/2)*cos((a-b)/2),equals(cos(a)+cos(b)))
> expect_that(-2*sin((a+b)/2)*sin((a-b)/2),equals(cos(a)-cos(b)))
```

The universal formulae:

■ $\sin 2\alpha = \dfrac{2\tan\alpha}{1+\tan^2\alpha}$

■ $\cos 2\alpha = \dfrac{1-\tan^2\alpha}{1+\tan^2\alpha}$

■ $\sin 2\alpha = \dfrac{2\tan\alpha}{1+\tan^2\alpha}$

The following shows the unit tests against the universal formulae:

```
> expect_that(sin(2*a),equals(2*tan(a)/(1+tan(a)^2)))
> expect_that((1-tan(a)^2)/(1+tan(a)^2),equals(cos(2*a)))
> expect_that(2*tan(a)/(1-tan(a)^2),equals(tan(2*a)))
```

The trigonometric square difference formulae:

■ $\sin(\alpha+\beta)\sin(\alpha-\beta) = \sin^2\alpha - \sin^2\beta$

■ $\cos(\alpha+\beta)\cos(\alpha-\beta) = \cos^2\alpha - \sin^2\beta$

The following shows the unit tests against the trigonometric square difference formulae:

```
> expect_that(sin(a)^2-sin(b)^2,equals(sin(a+b)*sin(a-b)))
> expect_that(cos(a)^2-sin(b)^2,equals(cos(a+b)*cos(a-b)))
```

Power-reduction formulae:

- $\cos^2\alpha = \dfrac{1 + \cos 2\alpha}{2}$

- $\sin^2\alpha = \dfrac{1 - \cos 2\alpha}{2}$

The following shows the unit tests against the trigonometric power-reduction formulae:

```
> expect_that((1+cos(2*a))/2,equals(cos(a)^2))
> expect_that((1-cos(2*a))/2,equals(sin(a)^2))
```

The auxiliary-angle formula:

- $a\sin\alpha + b\cos\alpha = \sqrt{a^2 + b^2}\,\sin\left(\alpha + \arctan\dfrac{b}{a}\right)$

The following shows the unit tests against the auxiliary-angle formula:

```
> expect_that(sqrt(a^2+b^2)*sin(a+atan(b/a)),equals(a*sin(a)+b*cos(a)))
```

1.2.3 Calculations of Complex Numbers

Complex numbers are the extension of real numbers, which make any polynomial gets root(s). The imaginary unit of a complex is i, the square root of -1, that is, $i^2 = -1$. Any complex can be expressed as $x + yi$, where x and y are real numbers, called the "real part" and "imaginary part," of the complex respectively.

1.2.3.1 Newing a Complex

```
# New a complex directly.
> ai<-5+2i;ai
[1] 5+2i
# View the type of the complex.
> class(ai)
[1] "complex"

# New a complex using the function complex().
> bi<-complex(real=5,imaginary=2);bi
[1] 5+2i
> is.complex(bi)
[1] TRUE

# The real part.
> Re(ai)
[1] 5

# The imaginary part.
> Im(ai)
[1] 2
```

```
# Get the modulus.
# Sqrt(5^2+2^2) = 5.385165.
> Mod(ai)
[1] 5.385165

# Get the argument.
> Arg(ai)
[1] 0.3805064

# Get the conjugate.
> Conj(ai)
[1] 5-2i
```

1.2.3.2 The Arithmetic Operations of Complex Numbers

- Addition: $(a + bi) + (c + di) = (a + c) + (b + d)i$
- Subtraction: $(a + bi) - (c + di) = (a - c) + (b - d)i$
- Multiplication: $(a + bi)(c + di) = (ac - bd) + (ac + bc)i$
- Division: $(a + bi)/(c + di) = [(ac + bd)+(bc - ad)i]/(c^2 + d^2)$

```
> a<-5;b<-2;c<-3;d<-4
> ai<-complex(real=a,imaginary=b)
> bi<-complex(real=c,imaginary=d)
```

Following shows the unit test against the complex arithmetic operations:

```
> expect_that(complex(real=(a+c),imaginary=(b+d)),equals(ai+bi))
> expect_that(complex(real=(a-c),imaginary=(b-d)),equals(ai-bi))
> expect_that(complex(real=(a*c-b*d),imaginary=(a*d+b*c)),equals(ai*bi))
> expect_that(complex(real=(a*c+b*d),imaginary=(b*c-a*d))/
  (c^2+d^2),equals(ai/bi))
```

1.2.3.3 The Square Roots of Complex Numbers

```
# Try to evaluate the square roots of -9 in real domain.
> sqrt(-9)
[1] NaN
Warning message:
In sqrt(-9) : NaNs produced

# Try to evaluate the square roots of -9 in complex domain.
> sqrt(complex(real=-9))
 [1] 0+3i
```

1.2.4 The Equation Calculations

Equation calculation is a basic form of mathematic calculation. R language can help us solve equations easily. In the following, I will introduce the solution to polynomial equations with one unknown and linear equations with two unknowns.

We can use the function uniroot() to solve the polynomial equations with one unknown.

1.2.4.1 Linear Equations with One Unknown

Given a linear equation with one unknown: $ax + b = 0$, let $a = 5$ and $b = 10$, solve x.

```
# Define the function for the equation.
> f1 <- function (x, a, b) a*x+b
# Assign the constants a and b.
> a<-5;b<-10
# Evaluate the root in the interval of (-10,10), with tolerance = 0.0001.
> result <- uniroot(f1,c(-10,10),a=a,b=b,tol=0.0001)
# Print the equation's root x.
> result$root
[1] -2
```

It is very easy solving the linear equation with one known. The root of the equation is −2. Show the function $y = 5x + 10$ in graphics, as in Figure 1.6.

```
# Create data points.
> x<-seq(-5,5,by=0.01)
> y<-f1(x,a,b)
> df<-data.frame(x,y)

# Draw the graphics using ggplot2.
> g<-ggplot(df,aes(x,y))
# The red line.
> g<-g+geom_line(col='red')
# The points.
> g<-g+geom_point(aes(result$root,0),col="red",size=3)
# The axes.
> g<-g+geom_hline(yintercept=0)+geom_vline(yintercept=0)
> g<-g+ggtitle(paste("y =",a,"* x +",b))
> g
```

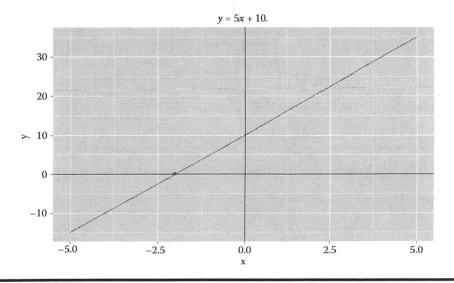

Figure 1.6 Function $y = 5x + 10$.

1.2.4.2 *The Quadratic Equations with One Unknown*

Given the quadratic equation with one unknown: $ax^2 + bx + c = 0$, let $a = 1$, $b = 5$, and $c = 6$, solve x.

```
> f2 <- function (x, a, b, c) a*x^2+b*x+c
> a<-1;b<-5;c<-6
> result <- uniroot(f2,c(0,-2),a=a,b=b,c=c,tol=0.0001)
> result$root
[1] -2
```

Passing the variables into the equation, we can solve one root of the equation with the function uniroot(). By changing the interval for calculation, we can get another root.

```
> result <- uniroot(f2,c(-4,-3),a=a,b=b,c=c,tol=0.0001)
> result$root
[1] -3
```

The equation has two roots, −2 and −3.

The function uniroot() solves only one root each time and requires that the f() values of end points of the interval input have signals opposite from each other. If we set the interval with (−10, 0), an error will be thrown from the function uniroot().

```
> result <- uniroot(f2,c(-10,0),a=a,b=b,c=c,tol=0.0001)
Error in uniroot(f2, c(-10, 0), a = a, b = b, c = c, tol = 1e-04):
The f() values of the end-points don't have opposite signals.
```

This should be the special design of uniroot() to statistically solve the polynomial equations with one unknown. In order to use uniroot(), we need to set different intervals to get the roots of the equation.

Show the function $y = x^2 + 5x + 6$ in graphics, as in Figure 1.7.

```
# Create the data points.
> x<-seq(-5,1,by=0.01)
```

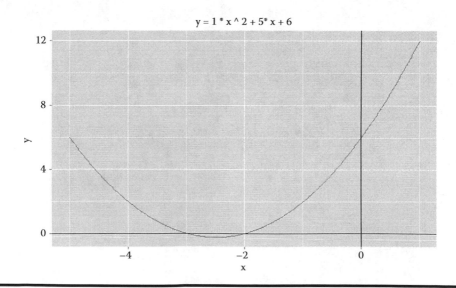

Figure 1.7 **Function** $y = x^2 + 5x + 6$.

```
> y<-f2(x,a,b,c)
> df<-data.frame(x,y)

# Draw the graphics using ggplot2.
> g<-ggplot(df,aes(x,y))
> g<-g+geom_line(col='red')
# The red curve.
> g<-g+geom_hline(yintercept=0)+geom_vline(yintercept=0)
# The axes.
> g<-g+ggtitle(paste("y =",a,"* x ^ 2 +",b,"* x +",c))
> g
```

From Figure 1.7, we can clearly see the value ranges of the two roots of x.

1.2.4.3 The Cubic Equations with One Unknown

Given the cubic equation with one unknown: $ax^3 + bx^2 + cx + d = 0$, let $a = 1$, $b = 5$, $c = 6$, and $d = -11$, solve x.

```
> f3 <- function (x, a, b, c,d) a*x^3+b*x^2+c*x+d
> a<-1;b<-5;c<-6;d<--11
> result <- uniroot(f3,c(-5,5),a=a,b=b,c=c,d=d,tol=0.0001)
> result$root
[1] 0.9461458
```

If we set the correct interval, we would easily solve the root of the equation.
Show the function $y = x^3 + 5x^2 + 6x - 11$ in graphics, as in Figure 1.8.

```
# Create the data points.
> x<-seq(-5,5,by=0.01)
```

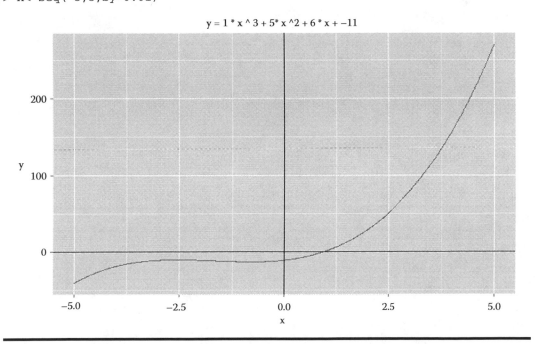

Figure 1.8 Function $y = x^3 + 5x^2 + 6x - 11$.

```
> y<-f3(x,a,b,c,d)
> df<-data.frame(x,y)

# Draw the graphics using ggplot2.
> g<-ggplot(df,aes(x,y))
# The cubic curve.
> g<-g+geom_line(col='red')
# The axes.
> g<-g+geom_hline(yintercept=0)+geom_vline(yintercept=0)
> g<-g+ggtitle(paste("y =",a,"* x ^ 3 +",b,"* x ^2 +",c,"* x + ",d))
> g
```

1.2.4.4 The Linear Equation Sets with Two Unknowns

R is able to solve the linear equation sets with two unknowns. In fact, the calculation is based on matrix operations.

Given the following equation set with two variables x_1 and x_2, solve x_1 and x_2:

$$\begin{cases} 3x_1 + 5x_2 = 4 \\ x_1 + 2x_2 = 1 \end{cases}$$

Create the equation set with the matrix form.

$$\begin{bmatrix} 3 & 5 \\ 1 & 2 \end{bmatrix} \cdot \begin{bmatrix} x_1 \\ x_2 \end{bmatrix} = \begin{bmatrix} 4 \\ 1 \end{bmatrix}$$

```
# The left matrix.
> lf<-matrix(c(3,5,1,2),nrow=2,byrow=TRUE)
# The right matrix.
> rf<-matrix(c(4,1),nrow=2)
# The result.
> result<-solve(lf,rf)
> result
      [,1]
[1,]    3
[2,]   -1
```

We have got the solutions of the equation set, $x_1 = 3$ and $x_2 = -1$.

Next, let us draw the graphics of the two linear equations, as shown in Figure 1.9. Let $y = x_2$ and $x = x_1$. We transform the original equation set into two functions.

```
# Define two functions.
> fy1<-function(x)  (-3*x+4)/5
> fy2<-function(x)  (-1*x+1)/2

# Define data.
> x<-seq(-1,4,by=0.01)
> y1<-fy1(x)
> y2<-fy2(x)
> dy1<-data.frame(x,y=y1,type=paste("y=(-3*x+4)/5"))
> dy2<-data.frame(x,y=y2,type=paste("y=(-1*x+1)/2"))
> df <- rbind(dy1,dy2)
```

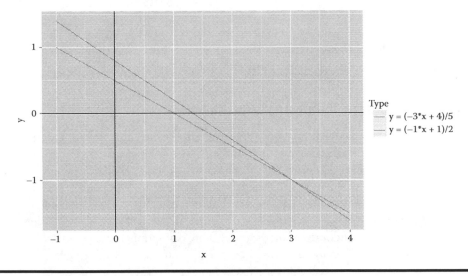

Figure 1.9   Set of linear equations with two unknowns.

```
# Draw the graphics using ggplot2.
> g<-ggplot(df,aes(x,y))
# Two lines.
> g<-g+geom_line(aes(colour=type,stat='identity'))
# The axes.
> g<-g+geom_hline(yintercept=0)+geom_vline(yintercept=0)
> g
```

We can see that the coordinates of the intersecting point of the two lines are the two roots. The solutions of linear equations with multiple unknowns can be evaluated through the same method.

We have implemented the various calculations in elementary mathematics with the R language. It is really very easy.

1.3 The Probability Basics and R

Question
How do we learn the probability theory with R?

The probability basics and R
http://blog.fens.me/r-probability/

Introduction

R is a language of statistics, whose foundation is the probability theory. So, we can imagine that the R language needs to provide complete, convenient, and easy-to-use functions for probability calculations. R can help us to effectively learn the basics of probability theory.

1.3.1 Introduction to Random Variables

Random variables are real-valued functions that represent possible outcomes of random phenomena. A random variable is defined in a sample space S, where the elements are results of random experiences, which are of randomness. So, the values of a random variable characterize randomness. A sample space is a set of all possible elementary outcomes of a random experience denoted by S. An element of the sample space, or a possible outcome, is called a sample point.

If the possible values of a random variable x are countable, we call x a discrete random variable. If the possible values of a random variable y can take any real number in a certain interval, or the values are continuous and uncountable, we call y as a continuous random variable.

The system environment used in this section:

- Win7 64bit
- R: 3.1.1 x86_64-w64-mingw32/x64 (64-bit)

R program: generate a discrete random variable x in sample space (1, 2, 3, 4, 5). The values of x can only be taken from 1, 2, 3, 4, or 5.

```
> S<-1:5
> x<-sample(S,1);x
[1] 2
```

Program: generate a continuous random variable y in sample space (0,1), taking 10 values.

```
> y<-runif(10,0,1);y
[1] 0.3819569 0.7609549 0.6692581 0.6314708 0.5552201 0.8225527 0.7633086
0.4667188 0.1883553 0.3741653
```

Probability distribution is used to describe the probability rule of the values of a random variable. For convenience, probability distributions have different forms according to the types that random variables belong to.

- Discrete distributions: two-point distribution, binomial distribution, Poisson's distribution, and so on
- Continuous distribution: uniform distribution, exponential distribution, normal distribution, gamma distribution, and so on

R implementations of the continuous distributions will be introduced in Section 1.4.

1.3.2 The Numerical Characteristics of Random Variables

In the following, I will introduce the numerical characteristics of random variables, which are frequently used concepts in probability theory, including: mathematical expectation, variance, standard deviation, quantiles (min, max, median, and quartile), covariance, correlation coefficient, and moments (moment about origin, moment about center, skewness, kurtosis, and covariance matrix.)

1.3.2.1 Mathematical Expectation

The mathematical expectation of a discrete random variable is defined as the sum of the products of each possible value x_i and its corresponding probability P_i of this discrete random variable, denoted by $E(X)$, as shown in formula (1.1). Mathematical expectation is one of the most basic numerical characteristics, measuring the mean of the possible values of a random variable. Usually, we also call mathematical expectation the mean.

$$E(X) = \sum_i x_i p_i. \tag{1.1}$$

The following R program calculates the mathematical expectation of data set (1,2,3,7,21):

```
> S<-c(1,2,3,7,21)
#Calculate the mathematical expectation of sample.
> mean(S)
[1] 6.8
```

Continuous random variables: if the distribution function $F(x)$ of a random variable X can be represented as the integration of a nonnegative integrable function $f(x)$, then we call X a continuous variable, $f(x)$ as the probability density function of X, its integration as the mathematical expectation of X, denoted by $E(X)$, as shown in formula (1.2).

$$E(X) = \int_{-\infty}^{+\infty} xf(x)dx. \tag{1.2}$$

1.3.2.2 Variance

The variance is the mean of the squares of the differences between each datum and all sample's mean, measuring the dispersion of a random variable from its mathematical expectation. Suppose that X is a random variable. If $E\{[X - E(X)]^2\}$ exists, then we call $E\{[X - E(X)]^2\}$ the variance of X, denoted by $\mathrm{Var}(X)$, as shown in formula (1.3).

$$\begin{aligned} \mathrm{Var}(X) &= E\{[X - E(X)]^2\} \\ \mathrm{Var}(X) &= E(X^2) - [E(X)]^2 \end{aligned} \tag{1.3}$$

The following R program calculates the variance of data set (1,2,3,7,21):

```
> S<-c(1,2,3,7,21)
#Calculate the variance of sample.
> var(S)
[1] 68.2
```

```
# Manually calculate the variance of samples.
> sum((S-mean(S))^2)/(length(S)-1)
[1] 68.2
```

1.3.2.3 Standard Deviation

The standard deviation is the arithmetic square root of variance, measuring the dispersion of a data set. Two data sets with same mean may have different standard deviations. The following R program calculates the standard deviation of data set (1,2,3,7,21):

```
> S<-c(1,2,3,7,21)
# Calculate the standard deviation of samples.
> sd(S)
[1] 8.258329
```

1.3.2.4 Quantile

A mode is the value that occurs most times in a data set. Sometimes, there are more than one mode in a data set. The following R program calculates the mode(s) of data set (1,2,3,3,3,7,7,7,7,9,10,21):

```
> S<-c(1,2,3,3,3,7,7,7,7,9,10,21)
# Calculate the mode(s).
>names(which.max(table(S)))
[1] "7"
```

A minimum is the smallest quantity or value that can be reached in a given scenario. The following R programs calculate the minimum of data set (2,3,3,3,7,7,7,7,9,10,21):

```
> S<-c(2,3,3,3,7,7,7,7,9,10,21)
# The minimum.
> min(S)
[1] 2
# The index of the minimum.
> which.min(S)
[1] 1
```

A maximum is the largest quantity or value that can be reached in a given scenario. The following R programs calculate the maximum of data set (2,3,3,3,7,7,7,7,9,10,21):

```
> S<-c(2,3,3,3,7,7,7,7,9,10,21)
# The maximum.
> max(S)
[1] 21
# The index of the maximum.
> which.max(S)
[1] 11
```

A median is the value at the middle position of values list sorted ascendant or descendant. The following R program calculates the median of data set (1,2,3,4,5):

```
> S<-c(1,2,3,4,5)
# The median.
> median(S)
[1] 3
```

Quartiles are used to describe the dispersion of any types of data sets, especially for skew distributional ones. Arrange all values from lowest to highest. The one that splits off the highest ¼ of data from the rest is called the upper quartile, while the one that splits off the lowest ¼ of data from the rest is called the lower quartile. The following R program calculates the quartiles of the data set (1,2,3,4,5,6,7,8,9):

```
> S<-c(1,2,3,4,5,6,7,8,9)
# The quartiles.
> quantile(S)
  0%  25%  50%  75% 100%
   1    3    5    7    9
# The five quantile numbers.
> fivenum(S)
[1] 1 3 5 7 9
```

R also provides a general statistical function, summary(), which encapsulated the above frequently used statistical measures. You can easily view the results by passing the data set into the function. The following R program calculates the statistical functions of data set (1,2,3,4,5,6,7,8,9):

```
> S<-c(1,2,3,4,5,6,7,8,9)
# The statistical function.
> summary(S)
   Min. 1st Qu.  Median    Mean 3rd Qu.    Max.
      1       3       5       5       7       9
```

1.3.2.5 Covariance

The covariance is used to measure the overall error of two variables, of which variance is a special situation, where the two variables are identical. Suppose X and Y are two random variables. Define $E\{[X - E(X)][Y - E(Y)]\}$ as the covariance of X and Y, denoted by $Cov(X,Y)$, as shown in formula (1.4).

$$Cov(X,Y) = E\{[X - E(X)][Y - E(Y)]\}. \tag{1.4}$$

The following R program calculates the covariance of X(1,2,3,4) and Y(5,6,7,8):

```
> X<-c(1,2,3,4)
> Y<-c(5,6,7,8)
# The covariance.
> cov(X,Y)
[1] 1.666667
```

1.3.2.6 Correlation Coefficient

The correlation coefficient is a statistical measure that describes how close the correlation is between two variables. The correlation coefficient is calculated according to the product-of-difference method, based on the deviations of two variables and their corresponding means. The correlation degree of two variables is reflected by multiplying the two deviations. If $\mathrm{Var}(X) > 0$ and $\mathrm{Var}(Y) > 0$, we define $\dfrac{\mathrm{Cov}(X,Y)}{\sqrt{\mathrm{Var}(X)\mathrm{Var}(Y)}}$ as the correlation coefficient of X and Y, as shown in formula (1.5).

$$\rho(X,Y) = \frac{\mathrm{Cov}(X,Y)}{\sqrt{\mathrm{Var}(X)\mathrm{Var}(Y)}}. \tag{1.5}$$

The following R program calculates the correlation coefficient of $X(1,2,3,4)$ and $Y(5,7,8,9)$:

```
> X<-c(1,2,3,4)
> Y<-c(5,7,8,9)
#The correlation coefficient.
> cor(X,Y)
[1] 0.9827076
```

1.3.2.7 Moments

Moments are a kind of numeric characteristic that has a wide range of applications. Mean and variance are first moment about origin and second moment about center.

Moments about origin: for positive integer k, if $E(X^k)$ exists, we call $ak = E(X^k)$ is the kth moment about origin of random variable X. The mathematical expectation of X is the first moment about origin of X, that is, $E(X)$, as shown in the formula (1.6).

$$\alpha_k = E(X^k) = \int_{-\infty}^{\infty} x^k \, \mathrm{d}F(x). \tag{1.6}$$

The following R program calculates the first moment about origin (mean) of $S(1,2,3,4,5)$:

```
> S<-c(1,2,3,4,5)
# The first moment about origin.
> mean(S)
[1] 3
```

Moments about center: for positive integer k, if both $E(X^k)$ and $E([X - E(X)]^k)$ exist, we call $E([X - E(X)]^k)$ the kth moment about center. The variance of X is the second moment about center of X, that is, $E([X - E(X)]^2)$, as shown in formula (1.7).

$$\mu_k = E\big([X - E(X)]^k\big) = \int_{-\infty}^{\infty} (x - E(X))^k \, \mathrm{d}F(x). \tag{1.7}$$

The following R program calculates the second moment about center (variance) of $S(1,2,3,4,5)$:

```
> S<-c(1,2,3,4,5)
# The second moment about center.
```

```
> var(S)
[1] 2.5
```

Skewness is a statistical measure that describes the direction and degrees how data distribution skews. It is a numeric characteristic that measures the symmetry of data distribution. Suppose that distribution function $F(x)$ has central moments $\mu_2 = E([X - E(X)]^2)$, $\mu_3 = E([X - E(X)]^3)$, then $C_s = \mu_3 / (\mu_2^{\frac{3}{2}})$ is the skewness coefficient, as shown in formula (1.8).

$$C_s = \mu_3 / \mu_2^{\frac{3}{2}}. \tag{1.8}$$

When $C_s > 0$, the probability distribution skews to the right of its mean, and vice versa.

The following R program calculates the skewness of a normal distribution data set that contains 10,000 sample points:

```
# Load PerformanceAnalytics package.
> library(PerformanceAnalytics)
# Generate  a normal distribution data set that contains 10,000 sample
points.
> S<-rnorm(10000)
# Calculate the skewness.
> skewness(S)
[1] -0.00178084

#Draw the histogram of this normal distribution, as shown in Fig. 1.10.
>hist(S,breaks=100)
```

Kurtosis, or the kurtosis coefficient, is a numeric characteristic that measures how much the distribution is peaked in the center. Kurtosis describes the degrees of centration or dispersion of different types of distributions. Suppose that a distribution function $F(x)$ has central moments

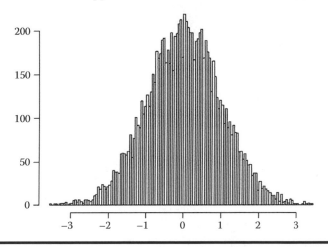

Figure 1.10 Histogram of the normal distribution data set that contains 10,000 sample points.

$\mu_2 = E([X - E(X)]^2)$ and $\mu_4 = E([X - E(X)]^4)$, then its kurtosis coefficient is $C_k = \mu_4/\mu_2{}^2 - 3$, as shown in formula (1.9).

$$C_k = \mu_4 / \mu_2^2 - 3. \tag{1.9}$$

The following R program calculates the kurtosis of sample with size of 10,000 (same as the data set in calculating the skewness):

```
#Calculate the kurtosis.
>kurtosis(S)
[1] -0.02443549
```

1.3.2.8 Covariance Matrix

The covariance matrix is a matrix such that each element is the covariance of two elements from two random vectors, respectively, which is the natural extension from a one-dimensional random variable to multidimensional random vectors. Suppose that $X = (X_1, X_2, \ldots, X_n)$ and $Y = (Y_1, Y_2, \ldots, Y_m)$ are two random vectors. $Cov(X,Y)$ is the covariance matrix of X and Y, as shown in formula (1.10).

$$Cov(X,Y) = (\sigma_{ij})_{n \times m}. \tag{1.10}$$

Calculate the covariance matrix in R as follows:

```
> x=as.data.frame(matrix(rnorm(10),ncol=2))
> x
           V1          V2
1 -2.11315384 -2.55189840
2 -0.96631271 -1.36148355
3 -0.02835058 -0.82328774
4 -1.86669567 -0.07201353
5  0.27324957 -2.23835218

# The covariance matrix.
> cov(x)
            V1          V2
V1  1.13470650 -0.09292042
V2 -0.09292042  1.03172261
```

1.3.3 Limit Theorems

1.3.3.1 Law of Large Numbers

Law of large numbers, or theorem of large numbers, is a theorem that judges whether the arithmetic mean of a random variable converges to a constant. It is one of the basic laws of probability theory and mathematical statistics.

Suppose that X_1, X_2, ..., X_k is a sequence of random variables and $E(X_k)(k = 1,2,3,...)$ exists. Let $Y_n = (X_1 + X_2 + \cdots + X_k)/n$. If formula (1.11) works for any $\varepsilon > 0$,

$$\lim_{n \to \infty} P\{|Y_n - E(Y_n)| \geq \varepsilon\} = 0,$$
$$\lim_{n \to \infty} P\{|Y_n - E(Y_n)| < \varepsilon\} = 1. \tag{1.11}$$

Then, we say that the random variable sequence $\{X_k\}$ follows the law of large numbers.

Suppose that the random variable X's mathematical expectation $E(X) = \mu$ and its $\mathrm{Var}(X) = \sigma^2$, then the law can be specialized as formula (1.12) through Chebyshev's inequality.

$$P\{|X - \mu| \geq \varepsilon\} \leq \frac{\sigma^2}{\varepsilon^2}. \tag{1.12}$$

Consider tossing a coin with the probability of 50% to get heads. Calculate the probability to get heads twice when tossing it four times. Calculate the probability to get heads 5000 times when tossing it 10,000 times.

Calculate the probability to get heads twice:

```
# The calculation of combination numbers: choose 2 from 4 objects.
> choose(4,2)/2^4
[1] 0.375
```

Calculate the probability to get heads for 5000 times:

```
# pbinom: the binomial distribution with 5,000 as quantile, generating
10,000 random numbers with probability of 0.5 for each.
> pbinom(5000, 10000, 0.5)
[1] 0.5039893
```

1.3.3.2 Central Limit Theorems

Central limit theorems are a kind of theorem that judge whether the distribution of the partial sum of a sequence of random variables approximates the normal distribution. Some phenomenon in the nature and production and scientific practices are affected by many random factors that are independent from each other. If the impact from each factor is small, then the overall impact can be considered to follow normal distribution. The central limit theorem proofs this rule mathematically. Choose a sample with size of n from any population that follows the distribution with μ as mean and σ^2 as variance. If n is large enough, the sampling distribution of the sample's mean approximately follows a normal distribution with μ as mean and σ^2/n as variance. The top two famous central limit theorems are Levy's theorem and Laplace's theorem.

Levy's theorem, or the central limit theorem for the sequence of independent random variables following the same distribution, states that the standardized normal distribution is the limit of the standardized sum of sequence of independent random variables with same distribution with finite mathematical expectation and variance.

Suppose that random variables X_1, X_2, ..., X_n, ... are independent from each other and follow the same distribution with $E(X_k) = \mu$ and $D(X_k) = \sigma^2 > 0$ ($k = 1,2, ...$), then for the standardized variable of the sum of the random variables, the distribution function $F(x)$ satisfies

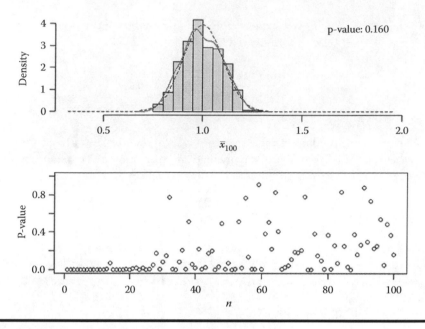

Figure 1.11   Animation simulation of central limit theorem.

$\lim_{n \to \infty} F_n(x) = \Phi(x)$, for any x, where $\Phi(x)$ is the distribution function of standardized normal distribution.

Laplace's theorem, or the central limit theorem for sequence of random variables that follows binomial distribution, states that the binomial distribution with parameters n and p has limit of a normal distribution with np as mean and $np(1-p)$ as variance.

Figure 1.11 illustrates the animation simulation of central limit theorem (from exponential distribution to normal distribution). The code is quoted from the help document of package animation.

```
> if (!require(animation)) install.packages("animation")
> library(animation)
> ani.options(interval = 0.1, nmax = 100)
> par(mar = c(4, 4, 1, 0.5))
> clt.ani()
```

After mastering the R language, we can easily perform various probability calculations with knowledge on probability. This is a big help for us to solve the problems in daily life.

1.4 Introduction to Frequently Used Continuous Distributions and Their R Implementations

Question

How do we draw the probability distribution function curves using the R language?

```
set.seed(1)

png("norm.png")

x <- seq(-5,5,length.out=100)
y <- dnorm(x,0,1)

plot(x,y,col="red",xlim=c(-5,5),ylim=
     xaxs="i", yaxs="i",ylab='density
     main="The Normal Density Distrib

lines(x,dnorm(x,0,0.5),col="green")
lines(x,dnorm(x,0,2),col="blue")
lines(x,dnorm(x,-2,1),col="orange")

legend("topright",legend=paste("m=",c
       col=c("red", "green","blue","c
```

$$f(x) = \frac{1}{\sqrt{2\pi}\sigma} \exp\left(-\frac{(x-\mu)^2}{2\sigma^2}\right)$$

Frequently used continuous distributions in R
http://blog.fens.me/r-density/

Introduction

We see the random variables anywhere in our daily life, such as daily weather, rising and declining of stock prices, winning of the lottery, and so on. These are unpredictable and even when we repeat the experience with the same conditions, the results may differ. Scientists abstract the rules and describe the values of random variables with probability distributions. Although we cannot predict stock prices, if we had known the probability distribution of its variation, then we are more likely to guess the answer.

This section will show you how to draw the curves of several continuous probability distributions, including normal, exponential, gamma, Weibull, F, T, beta, chi-square, and uniform distributions.

1.4.1 Uniform Distribution

The system environment used in this section:

- Win7 64bit
- R: 3.1.1 x86_64-w64-mingw32/x64 (64-bit)

The uniform distribution is a simple probability distribution that is uniform and not skew. The distribution has two types: discrete and continuous, of which the latter will be introduced in this section.

1.4.1.1 Probability Density Function

The probability density function of uniform distribution is given by formula (1.13).

$$f(x) = \begin{cases} \dfrac{1}{b-a} & \text{for } a \le x \le b \\ 0 & \text{elsewhere} \end{cases}. \tag{1.13}$$

The following R program draws the curve of the probability density function of uniform distribution:

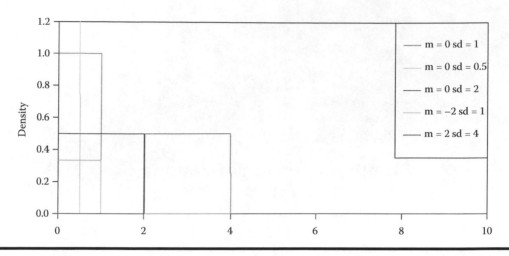

Figure 1.12 Illustration of probability density function of uniform distribution.

```
Set randomness seed.
> set.seed(1)
# Get 1,000 points with values from 0 to 10 in sequence.
> x<-seq(0,10,length.out=1000)
# Calculate the values of probability density function with 1000 points
that follow the uniform distribution of U(0,1).
> y<-dunif(x,0,1)

# Draw the curve of the probability density function, as shown in
Fig. 1.12.
> plot(x,y,col="red",xlim=c(0,10),ylim=c(0,1.2),type='l',xaxs="i", yaxs="
i",ylab='density',xlab='',main="The Uniform Density Distribution")
# Calculate the values that follows the uniform distribution U(0,0.5) and
add a curve.
> lines(x, dunif(x,0,0.5),col="green")
# Calculate the values that follows the uniform distribution U(0,2) and
add a curve.
> lines(x, dunif(x,0,2),col="blue")
# Calculate the values that follows the uniform distribution U(-2,1) and
add a curve.
> lines(x, dunif(x,-2,1),col="orange")
# Calculate the values that follows the uniform distribution U(2,4) and
add a curve.
> lines(x, dunif(x,2,4),col="purple")
# Add legend to the upper right corner.
> legend("topright",legend=paste("m=",c(0,0,0,-2,2)," sd=",
c(1,0.5,2,1,4)), lwd=1, col=c("red", "green","blue","orange","purple"))
```

1.4.1.2 Cumulative Distribution Function

The formula of cumulative distribution function of uniform distribution is (1.14)

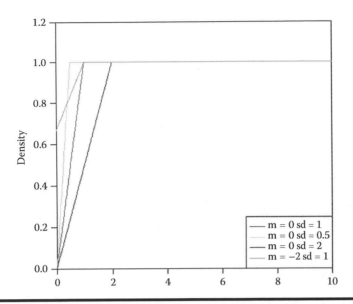

Figure 1.13   Illustration of cumulative probability functions of uniform distributions.

$$F(x) = \begin{cases} 0 & \text{for } x < a \\ \dfrac{x - a}{b - a} & \text{for } a \leq x < b. \\ 1 & \text{for } x \geq b \end{cases} \qquad (1.14)$$

The following R program draws the curve of the cumulative distribution function of uniform distribution:

```
# Set randomness seed.
> set.seed(1)
# Get 1,000 points with values from 0 to 10 in sequence.
> x<-seq(0,10,length.out=1000)
# Calculate the values of cumulative distribution function with
1,000 points that follow the uniform distribution U(0,1).
> y<-punif(x,0,1)
# Draw the curve of the cumulative probability function, as shown in
Fig. 1.13.
> plot(x,y,col="red",xlim=c(0,10),ylim=c(0,1.2),type='l',xaxs="i", yaxs="i"
,ylab='F(x)',xlab='',main="The Uniform Cumulative Distribution Function")
> lines(x,punif(x,0,0.5),col="green")
> lines(x,punif(x,0,2),col="blue")
> lines(x,punif(x,-2,1),col="orange")
> legend("bottomright",legend=paste("m=",c(0,0,0,-2)," sd=",
c(1,0.5,2,1)), lwd=1, col=c("red", "green","blue","orange","purple"))
```

1.4.1.3 Distribution Test

The Kolmogorov–Smirnov test for continuous distribution is a method that tests whether a single sample follows a presupposed specific distribution. Compare the cumulative frequency distribution with the specific theoretical distribution. If the difference is very small, then it deduces that the sample is taken from the specific distribution family. The null hypothesis is H_0: the data set follows the uniform distribution and the alternative hypothesis is H_1: the distribution of the population from which the sample is chosen does not follow the uniform distribution. Let $F_0(x)$ be the pre-suppose theoretical distribution and $F_n(x)$ be the cumulative probability (frequency) function. Let statistical measure $D = \max|F_0(x) - F_n(x)|$.

■ The less and nearer to 0 D is, the closer the sample data is to uniform distribution.
■ If the p-value is less than the significance level $\alpha(0.05)$, then H_0 is refused:

```
# Set randomness seed.
> set.seed(1)
# Generate 1,000 points that follow uniform distribution.
> S<-runif(1000)
# Kolmogorov-Smirnov test.
> ks.test(S, "punif")
    One-sample Kolmogorov-Smirnov test
data:  S
D = 0.0244, p-value = 0.5928
alternative hypothesis: two-sided
```

D is very small and p-value > 0.05, so we cannot refuse the null hypothesis. Therefore, the data set follows uniform distribution!

In the later text, the Kolmogorov–Smirnov test is also applied to the distribution test against other functions. So, the descriptions of the tests will be brief.

1.4.2 Normal Distribution

The normal distribution, or the Gaussian distribution, is a very important probability distribution in various fields such as mathematics, physics, and engineering. It has a huge impact on many aspects in statistics.

If a random variable follows a Gaussian distribution with position parameter μ and scale parameter σ^2, then the random variable is called a normal random variable. The distribution that this random follows is called normal distribution, denoted by $N(\mu, \sigma^2)s$. The mathematical expectation μ determines the position of the distribution and the standard deviation σ determines its scale. Because its curve is shaped like a bell, people often call it a bell-shaped curve. The usually motioned standard normal distribution is the one with $\mu = 0$ and $\sigma = 1$.

1.4.2.1 Probability Density Function

The probability density function of normal distribution is given by formula (1.15).

$$f(x|\mu, \sigma^2) = \frac{1}{\sqrt{2\sigma^2\pi}} e^{-\frac{(x-\mu)^2}{2\sigma^2}}. \tag{1.15}$$

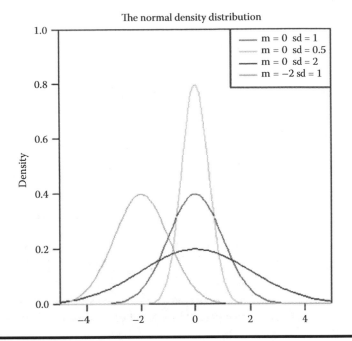

Figure 1.14   Illustration of probability density functions of normal distributions.

The following R program draws the curve of the probability density function of normal distribution:

```
> set.seed(1)
# Take 100 points whose values range from -5 to 5.
> x <- seq(-5,5,length.out=100)
# Calculate the values of probability density function with 1000 points
that follow the normal distribution of N(0,1).
> y <- dnorm(x,0,1)

# Draw the curve of the probability density function, as shown in
Fig. 1.14.
> plot(x,y,col="red",xlim=c(-5,5),ylim=c(0,1),type='l',xaxs="i", yaxs="i"
,ylab='density',xlab='',main="The Normal Density Distribution")
> lines(x,dnorm(x,0,0.5),col="green")
> lines(x,dnorm(x,0,2),col="blue")
> lines(x,dnorm(x,-2,1),col="orange")
> legend("topright",legend=paste("m=",c(0,0,0,-2)," sd=", c(1,0.5,2,1)),
lwd=1, col=c("red", "green","blue","orange"))
```

1.4.2.2 Cumulative Distribution Function

The formula of cumulative distribution function of normal distribution is (1.16)

$$F(x;\mu,\sigma) = \frac{1}{\sigma\sqrt{2\pi}} \int_{-\infty}^{x} \exp\left(-\frac{(t-\mu)^2}{2\sigma^2}\right) dt. \tag{1.16}$$

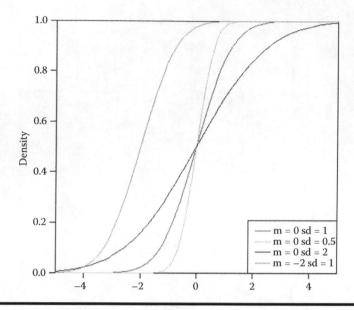

Figure 1.15   Illustration of cumulative probability functions of normal distributions.

The following R program draws the curve of the cumulative distribution function of normal distribution:

```
> set.seed(1)
> x <- seq(-5,5,length.out=100)
# Calculate the values of cumulative distribution function with 1,000
points that follow the normal distribution N(0,1).
> y <- pnorm(x,0,1)

# Draw the curve of the cumulative probability function, as shown in
Fig. 1.15.
> plot(x,y,col="red",xlim=c(-5,5),ylim=c(0,1),type='l',xaxs="i", yaxs="i"
,ylab='F(x)',xlab='',main="The Normal Cumulative Distribution")
> lines(x,pnorm(x,0,0.5),col="green")
> lines(x,pnorm(x,0,2),col="blue")
> lines(x,pnorm(x,-2,1),col="orange")
> legend("bottomright",legend=paste("m=",c(0,0,0,-2)," sd=",
c(1,0.5,2,1)), lwd=1,col=c("red", "green","blue","orange"))
```

1.4.2.3 Distribution Test

The Shapiro–Wilk normal distribution test is used to test whether data follow a normal distribution. The method tests the residual of the data points from the regression curves, just like linear regression. This method is suggested for use when the sample size is small, with a usual range from 3 to 5000. The Shapiro–Wilk normal distribution test's null hypothesis is H_0: the data set follows normal distribution. Let the statistical measure W be

$$W = \frac{\left(\sum_{i=1}^{n} a_i x_i\right)^2}{\sum_{i=1}^{n} (x_i - \bar{x})^2}. \tag{1.17}$$

- The maximum of W is 1. The closer it is to 1, the more the sample matches normal distribution.
- If the p-value is less than the significance level $\alpha(0.05)$, then H_0 is refused.

```
> set.seed(1)
# Generate 1,000 points that following normal distribution.
> S<-rnorm(1000)
# Shapiro-Wilk normal distribution test.
> shapiro.test(S)
    Shapiro-Wilk normality test
data:  S
W = 0.9988, p-value = 0.7256
```

Here, W is close to 1 and p-value > 0.05, so we cannot refuse the null hypothesis. Therefore, the data set follows normal distribution!

If we use Kolmogorov–Smirnov test for continuous distribution, the null hypothesis is H_0: the data set follows normal distribution and the alternative one is H_1: the distribution of the population from which the sample is chosen does not follow the normal distribution.

```
> set.seed(1)
> S<-rnorm(1000)
# Kolmogorov-Smirnov test.
> ks.test(S, "pnorm")
    One-sample Kolmogorov-Smirnov test
data:  S
D = 0.0211, p-value = 0.7673
alternative hypothesis: two-sided
```

Here, D is very small and p-value > 0.05, so we cannot refuse the null hypothesis. Therefore, the data set follows normal distribution!

1.4.3 Exponential Distribution

Exponential distribution describes the time intervals between independent random events, such as the time intervals of the events that passengers enter an airport, the time intervals of the emergence of new Chinese Wikipedia entries, and so on. The life cycle distributions of various electronic products follow the exponential distribution generally. The life cycle distribution of some systems can also be approximated by the exponential distribution. When the failure of products is by accident, the life cycle follows the exponential distribution. The failure ratio of exponential distribution is a constant that is unrelated to time t, so the distribution function is simple. It is one of the most frequently used distributions in reliability research.

1.4.3.1 Probability Density Function

The probability density function of exponential distribution is given by formula (1.18).

$$f(x; \lambda) = \begin{cases} \lambda e^{-\lambda x}, & x \geq 0, \\ 0, & x < 0. \end{cases} \tag{1.18}$$

Here, $\lambda > 0$ is a parameter of the distribution, usually known as rate parameter, that is, the times the event occurs each time unit. The value interval of exponential distribution is $[0,\infty)$. If a random variable X follows the exponential distribution, then we can denote it as $X \sim \text{Exponential}(\lambda)$.

The following R program draws the curve of the probability density function of exponential distribution:

```
> set.seed(1)
> x<-seq(-1,2,length.out=100)
# Calculate the values of probability density function with 100 points
that follow the exponential distribution of e(0.5).
> y<-dexp(x,0.5)

# Draw the curve of the probability density function, as shown in
Fig. 1.16.
> plot(x,y,col="red",xlim=c(0,2),ylim=c(0,5),type='l',xaxs="i", yaxs="i",
ylab='density',xlab='',main="The Exponential Density Distribution")
> lines(x,dexp(x,1),col="green")
> lines(x,dexp(x,2),col="blue")
> lines(x,dexp(x,5),col="orange")
```

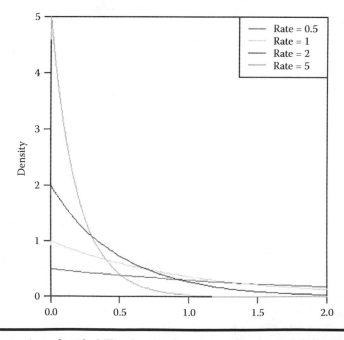

Figure 1.16   Illustration of probability density functions of exponential distributions.

```
> legend("topright",legend=paste("rate=",c(.5, 1, 2,5)),
lwd=1,col=c("red", "green","blue","orange"))
```

1.4.3.2 Cumulative Distribution Function

The formula of cumulative distribution function of exponential distribution is (1.19).

$$F(x;\lambda)=\begin{cases}1-e^{-\lambda x}, & x\ge 0,\\ 0, & x<0.\end{cases} \tag{1.19}$$

The following R program draws the curve of the cumulative distribution function of exponential distribution:

```
> set.seed(1)
> x<-seq(-1,2,length.out=100)
# Calculate the values of cumulative distribution function with
100 points that follow the exponential distribution e(0.5).
> y<-pexp(x,0.5)

# Draw the curve of the cumulative probability function, as shown in
Fig. 1.17.
> plot(x,y,col="red",xlim=c(0,2),ylim=c(0,1),type='l',xaxs="i", yaxs="i",yl
ab='F(x)',xlab='',main="The Exponential Cumulative Distribution Function")
> lines(x,pexp(x,1),col="green")
> lines(x,pexp(x,2),col="blue")
> lines(x,pexp(x,5),col="orange")
> legend("bottomright",legend=paste("rate=",c(.5, 1, 2,5)), lwd=1,
col=c("red", "green","blue","orange"))
```

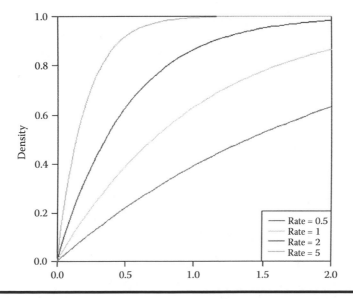

Figure 1.17 **Illustration of cumulative probability functions of exponential distributions.**

1.4.3.3 Distribution Test

For exponential distribution, the null hypothesis of the Kolmogorov–Smirnov continuous distribution test is H_0: the data set follows exponential distribution and the alternative one is H_1: the distribution of the population from which the sample is chosen does not follow the exponential distribution.

```
> set.seed(1)
> S<-rexp(1000)
# Kolmogorov-Smirnov test.
> ks.test(S, "pexp")
    One-sample Kolmogorov-Smirnov test
data:  S
D = 0.0387, p-value = 0.1001
alternative hypothesis: two-sided
```

Here, D is very small and p-value > 0.05, so we cannot refuse the null hypothesis. Therefore, the data set follows exponential distribution!

1.4.4 Gamma Distribution

Gamma distribution is an important member of the famous Pearson probability distribution function family, known as Pearson type III distribution. The curve has a peak but is not symmetric. Gamma function is the generalization of factorial function on the domain of real numbers. The formula is (1.20).

$$\Gamma(x) = \int_0^\infty t^{x-1} e^{-t} dt. \tag{1.20}$$

1.4.4.1 Probability Density Function

The probability density function of gamma distribution is given by formula (1.21).

$$f(x) = x^{k-1} \frac{\exp(-x/\theta)}{\Gamma(k)\theta^k}. \tag{1.21}$$

The following R program draws the curve of the probability density function of gamma distribution:

```
> set.seed(1)
> x<-seq(0,10,length.out=100)
# Calculate the values of probability density function with 100 points
  that follow the gamma distribution of Ga(1,2).
> y<-dgamma(x,1,2)

# Draw the curve of the probability density function, as shown in
Fig. 1.18.
> plot(x,y,col="red",xlim=c(0,10),ylim=c(0,2),type='l',xaxs="i", yaxs="i",
  ylab='density',xlab='',main="The Gamma Density Distribution")
> lines(x,dgamma(x,2,2),col="green")
> lines(x,dgamma(x,3,2),col="blue")
```

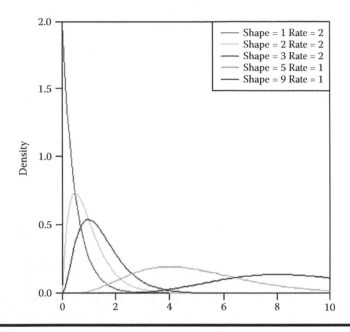

Figure 1.18 Illustration of probability density functions of gamma distributions.

```
> lines(x,dgamma(x,5,1),col="orange")
> lines(x,dgamma(x,9,1),col="black")
> legend("topright",legend=paste("shape=",c(1,2,3,5,9)," rate=",
c(2,2,2,1,1)), lwd-1, col-c("red", "green","blue","orange","black"))
```

1.4.4.2 Cumulative Distribution Function

The formula of cumulative distribution function of gamma distribution is (1.22).

$$f(x) = \frac{\gamma(k, x / \theta)}{\Gamma(k)}. \tag{1.22}$$

The following R program draws the curve of the cumulative distribution function of gamma distribution:

```
> set.seed(1)
> x<-seq(0,10,length.out=100)
# Calculate the values of cumulative distribution function with 100
points that follow the gamma distribution Ga(1,2).
> y<-pgamma(x,1,2)

# Draw the curve of the cumulative probability function, as shown in
Fig. 1.19.
> plot(x,y,col="red",xlim=c(0,10),ylim=c(0,1),type='l',xaxs="i", yaxs="i",
ylab='F(x)',xlab='',main="The Gamma Cumulative Distribution Function")
> lines(x,pgamma(x,2,2),col="green")
> lines(x,pgamma(x,3,2),col="blue")
> lines(x,pgamma(x,5,1),col="orange")
```

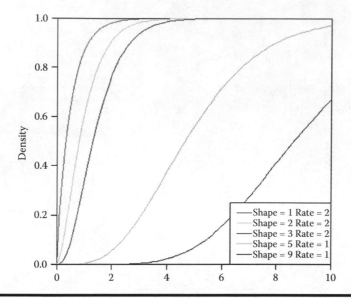

Figure 1.19   Illustration of cumulative probability functions of exponential distributions.

```
> lines(x,pgamma(x,9,1),col="black")
> legend("bottomright",legend=paste("shape=",c(1,2,3,5,9)," rate=",
c(2,2,2,1,1)), lwd=1, col=c("red", "green","blue","orange","black"))
```

1.4.4.3 Distribution Test

For gamma distribution, the null hypothesis of the Kolmogorov–Smirnov continuous distribution test is H_0: the data set follows gamma distribution and the alternative one is H_1: the distribution of the population from which the sample is chosen does not follow gamma distribution.

```
> set.seed(1)
> S<-rgamma(1000,1)
# Kolmogorov-Smirnov test.
> ks.test(S, "pgamma", 1)
    One-sample Kolmogorov-Smirnov test
data:  S
D = 0.0363, p-value = 0.1438
alternative hypothesis: two-sided
```

Here, D is very small and p-value > 0.05, so we cannot refuse the null hypothesis. Therefore, the data set follows gamma distribution!

In the following, let us try a failed test. If we the test parameter shape=2, then the result is not what we expected with the Kolmogorov–Smirnov test.

```
# Kolmogorov-Smirnov test.
> ks.test(S, "pgamma", 2)
    One-sample Kolmogorov-Smirnov test
data:  S
D = 0.3801, p-value < 2.2e-16
alternative hypothesis: two-sided
```

Here, *D* is not small enough and *p*-value < 0.05, so we refuse the null hypothesis. Therefore, the data set does not follow gamma distribution!

1.4.5 Weibull Distribution

Weibull distribution is the theoretical basics of reliability analysis and life cycle testing. The distribution can be used in different forms and determined by three parameters: shape, scale (range), and position, of which the parameter shape is the most important, which determines the basic shape of the distribution's density curve. The parameter scale can zoom the curve in and out, but does not affect the shape of distribution.

Weibull distribution is usually used in the field of failure analysis. In particular, it can mimic the distribution of the change of failure rate overtime. If the failure rate

- is a constant overtime, then $\alpha = 1$, suggesting the failure happens in random events.
- decreases overtime, then $\alpha < 1$, suggesting "infant mortality."
- increases overtime, then $\alpha > 1$, suggesting "wearing out"—the failure possibility increases as time proceeds.

1.4.5.1 Probability Density Function

The probability density function of Weibull distribution is given by formula (1.23).

$$f(x;\lambda,k) = \begin{cases} \dfrac{k}{\lambda}\left(\dfrac{x}{\lambda}\right)^{k-1} e^{-(x/\lambda)^k} & x \geq 0 \\ 0 & x < 0 \end{cases}. \tag{1.23}$$

The following R program draws the curve of the probability density function of Weibull distribution:

```
> set.seed(1)
> x<- seq(0, 2.5, length.out=1000)
# Calculate the values of probability density function with 1,000 points
that follow the Weibull distribution of W(0.5,1)
> y<- dweibull(x, 0.5)

# Draw the curve of the probability density function, as shown in
Fig. 1.20.
> plot(x, y, type="l", col="blue",xlim=c(0, 2.5),ylim=c(0, 6),xaxs="i",
yaxs="i",ylab='density',xlab='',main="The Weibulll Density
Distribution")
> lines(x, dweibull(x, 1), type="l", col="red")
> lines(x, dweibull(x, 1.5), type="l", col="magenta")
> lines(x, dweibull(x, 5), type="l", col="green")
> lines(x, dweibull(x, 15), type="l", col="purple")
> legend("topright", legend=paste("shape =", c(.5, 1, 1.5, 5, 15)),
lwd=1,col=c("blue", "red", "magenta", "green","purple"))
```

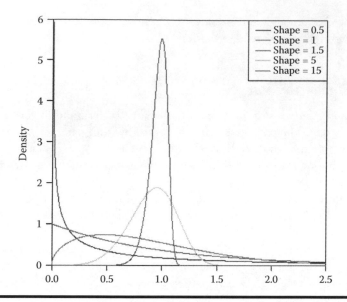

Figure 1.20   Illustration of probability density functions of Weibull distributions.

1.4.5.2 Cumulative Distribution Function

The formula of cumulative distribution function of Weibull distribution is (1.24).

$$f(x) = 1 - e^{-(x/\lambda)^k}. \tag{1.24}$$

The following R program draws the curve of the cumulative distribution function of Weibull distribution:

```
> set.seed(1)
> x<- seq(0, 2.5, length.out=1000)
# Calculate the values of cumulative distribution function with 1,000
points that follow the Weibull distribution W(0.5,1).
> y<- pWeibulll(x, 0.5)

# Draw the curve of the cumulative probability function, as shown in
Fig. 1.21.
> plot(x, y, type="l", col="blue",xlim=c(0, 2.5),ylim=c(0, 1.2),xaxs="i",
yaxs="i",ylab='F(x)',xlab='',main="The Weibulll Cumulative Distribution
Function")
> lines(x, pweibull(x, 1), type="l", col="red")
> lines(x, pweibull(x, 1.5), type="l", col="magenta")
> lines(x, pweibull(x, 5), type="l", col="green")
> lines(x, pweibull(x, 15), type="l", col="purple")
> legend("bottomright", legend=paste("shape =", c(.5, 1, 1.5, 5, 15)),
lwd=1, col=c("blue", "red", "magenta", "green","purple"))
```

1.4.5.3 Distribution Test

For Weibull distribution, the null hypothesis of the Kolmogorov–Smirnov continuous distribution test is H_0: the data set follows Weibull distribution and the alternative one is H_1: the

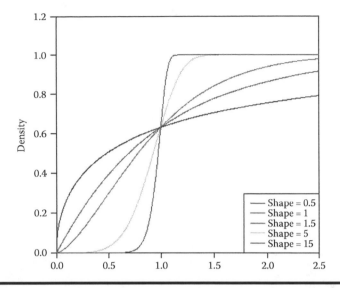

Figure 1.21 Illustration of cumulative probability functions of Weibull distributions.

distribution of the population from which the sample is chosen does not follow the Weibull distribution.

```
> set.seed(1)
> S<-rWeibull1(1000,1)
# Kolmogorov-Smirnov test.
> ks.test(S, "pWeibull",1)
     One-sample Kolmogorov-Smirnov test
data:  S
D = 0.0244, p-value = 0.5928
alternative hypothesis: two-sided
```

Here, D is very small and p-value > 0.05, so we cannot refuse the null hypothesis. Therefore, the data set follows Weibull distribution!

1.4.6 χ^2(Chi-Square) Distribution

If n independent random variables ξ_1, ξ_2, ..., ξ_n follow standard normal distribution (a.k.a. independent and identically distributed as standard normal distribution), then the sum of the squares of these n variables form a new random variable, whose distribution is called χ^2 (chi-square) distribution, where parameter n is the degree of freedom. A chi-square distribution with a different degree of freedom differs, just as a different normal distribution with a different mean or variance.

1.4.6.1 Probability Density Function

The probability density function of chi-square distribution is given by formula (1.25).

$$f_k(x) = \frac{(1/2)^{k/2}}{\Gamma(k/2)} x^{k/2-1} e^{-x/2}.$$

(1.25)

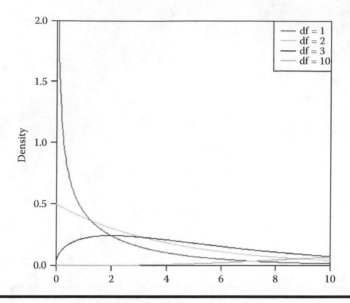

Figure 1.22   Illustration of probability density functions of chi-square distributions.

Here, the denominator γ is the gamma function.

The following R program draws the curve of the probability density function of chi-square distribution:

```
> set.seed(1)
> x<-seq(0,10,length.out=1000)
# Calculate the values of probability density function with 1,000 points
that follow the Chi-square distribution of X(1).
> y<-dchisq(x,1)

# Draw the curve of the probability density function, as shown in
Fig. 1.22.
> plot(x,y,col="red",xlim=c(0,5),ylim=c(0,2),type='l',xaxs="i", yaxs="i",
ylab='density',xlab='',main="The Chisq Density Distribution")
> lines(x,dchisq(x,2),col="green")
> lines(x,dchisq(x,3),col="blue")
> lines(x,dchisq(x,10),col="orange")
> legend("topright",legend=paste("df=",c(1,2,3,10)), lwd=1, col=c("red",
"green","blue","orange"))
```

1.4.6.2 Cumulative Distribution Function

The formula of cumulative distribution function of chi-square distribution is (1.26).

$$F_k(x) = \frac{\gamma(k/2, x/2)}{\Gamma(k/2)}.$$ (1.26)

Here, the denominator Γ is the gamma function and the numerator $\gamma(k, z)$ is the incomplete gamma function.

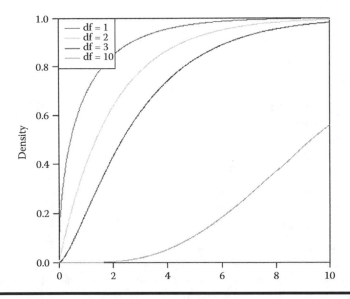

Figure 1.23 **Illustration of cumulative probability functions of chi-square distributions.**

The following R program draws the curve of the cumulative distribution function of chi-square distribution:

```
> set.seed(1)
> x<-seq(0,10,length.out=1000)
# Calculate the values of cumulative distribution function with 1,000
points that follow the Chi-square distribution X(1).
> y<-pchisq(x,1)

# Draw the curve of the cumulative probability function, as shown in
Fig. 1.23.
> plot(x,y,col="red",xlim=c(0,10),ylim=c(0,1),type='l',xaxs="i",
yaxs="i",ylab='F(x)',xlab='',main="The Chisq Cumulative Distribution
Function")
> lines(x,pchisq(x,2),col="green")
> lines(x,pchisq(x,3),col="blue")
> lines(x,pchisq(x,10),col="orange")
> legend("topleft",legend=paste("df=",c(1,2,3,10)), lwd=1, col=c("red",
"green","blue","orange"))
```

1.4.6.3 Distribution Test

For chi-square distribution, the null hypothesis of the Kolmogorov–Smirnov continuous distribution test is H_0: the data set follows chi-square distribution and the alternative one is H_1: the distribution of the population from which the sample is chosen does not follow the chi-square distribution.

```
> set.seed(1)
> S<-rchisq(1000,1)
# Kolmogorov-Smirnov test.
```

```
> ks.test(S, "pchisq",1)
    One-sample Kolmogorov-Smirnov test
data:  S
D = 0.0254, p-value = 0.5385
alternative hypothesis: two-sided
```

Here, D is very small and p-value > 0.05, so we cannot refuse the null hypothesis. Therefore, the data set follows chi-square distribution!

1.4.7 F Distribution

F distribution is a continuous probability distribution, widely used in likely rate tests, especially in ANOVA. The definition of F distribution: if X and Y are independent random variables, where X follows the chi-square distribution with degree of freedom of d_1 and Y follows the chi-square distribution with degree of freedom of d_2, then $F = \dfrac{X/d_1}{Y/d_2}$ follows the F distribution with degree of freedom d_1 and d_2.

F distribution is an asymmetric distribution with two degrees of freedom, d_1 and d_2. The corresponding distribution is denoted by $F(d_1, d_2)$. Usually, d_1 is known as the numerator degree of freedom and d_2 as the denominator degree of freedom. The F distributions form the distribution family with parameters d_1 and d_2, which determine the shape of F distribution.

1.4.7.1 Probability Density Function

The probability density function of F distribution is given by formula (1.27).

$$f(x, d_1, d_2) \frac{\Gamma\big((d_1 + d_2)/2\big)}{\Gamma(d_1/2)\,\Gamma(d_2/2)} \left(\frac{d_1}{d_2}\right)^{d_1/2} x^{\frac{d_1}{2}-1} \left(1 + \frac{d_1}{d_2} x\right)^{-(d_1+d_2)/2}. \tag{1.27}$$

Here, Γ is the gamma function.

The following R program draws the curve of the probability density function of F distribution:

```
> set.seed(1)
> x<-seq(0,5,length.out=1000)
# Calculate the values of probability density function with 1,000 points
that follow the F distribution of F(1,1,0).
> y<-df(x,1,1,0)

# Draw the curve of the probability density function, as shown in
Fig. 1.24.
> plot(x,y,col="red",xlim=c(0,5),ylim=c(0,1),type='l',xaxs="i", yaxs="i",
ylab='density',xlab='',main="The F Density Distribution")
> lines(x,df(x,1,1,2),col="green")
> lines(x,df(x,2,2,2),col="blue")
> lines(x,df(x,2,4,4),col="orange")
> legend("topright",legend=paste("df1=",c(1,1,2,2),"df2=",c(1,1,2,4),"
ncp=", c(0,2,2,4)), lwd=1, col=c("red", "green","blue","orange"))
```

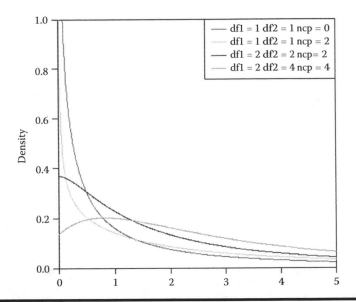

Figure 1.24   Illustration of probability density functions of F distributions.

1.4.7.2 Cumulative Distribution Function

The formula of cumulative distribution function of F distribution is (1.28).

$$F(x) = I_{d_1 x/(d_1 x + d_2)}\left(\frac{d_1}{2}, \frac{d_2}{2}\right). \tag{1.28}$$

Here, *I* is the incomplete beta function.

The following R program draws the curve of the cumulative distribution function of F distribution:

```
> set.seed(1)
> x<-seq(0,5,length.out=1000)
# Calculate the values of cumulative distribution function with 1,000
points that follow the F distribution F(1,1,0).
> y<-df(x,1,1,0)

# Draw the curve of the cumulative probability function, as shown in
Fig. 1.25.
> plot(x,y,col="red",xlim=c(0,5),ylim=c(0,1),type='l',xaxs="i", yaxs="i",
ylab='F(x)',xlab='',main="The F Cumulative Distribution Function")
> lines(x,pf(x,1,1,2),col="green")
> lines(x,pf(x,2,2,2),col="blue")
> lines(x,pf(x,2,4,4),col="orange")
> legend("topright",legend=paste("df1=",c(1,1,2,2),"df2=",c(1,1,2,4),"
ncp=", c(0,2,2,4)), lwd=1, col=c("red", "green","blue","orange"))
```

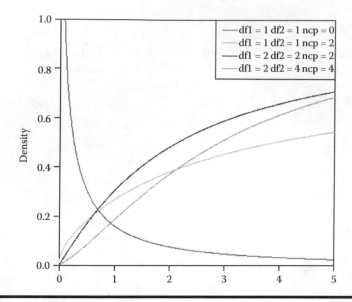

Figure 1.25 Illustration of cumulative probability functions of F distributions.

1.4.7.3 Distribution Test

For F distribution, the null hypothesis of the Kolmogorov–Smirnov continuous distribution test is H_0: the data set follows F distribution and the alternative one is H_1: the distribution of the population from which the sample is chosen does not follow the F distribution.

```
> set.seed(1)
> S<-rf(1000,1,1,2)
# Kolmogorov-Smirnov test.
> ks.test(S, "pf", 1,1,2)
    One-sample Kolmogorov-Smirnov test
data:  S
D = 0.0113, p-value = 0.9996
alternative hypothesis: two-sided
```

Here, D is very small and p-value > 0.05, so we cannot refuse the null hypothesis. Therefore, the data set follows F distribution!

1.4.8 t-Distribution

Student's t-distribution, short form t-distribution, is used to estimate the mean of a population that follows normal distribution. It is the basis of student's t-test that performs significance tests against the difference between the means of two samples. The student's t-test improves the Z-test, which has a precondition that the standard deviation is already known. While the Z-test can be used to approximate the value of the t-test when the sample size is large (>30), the error is significant when the sample size is small. In such a case, we must use the student's t-test to get a precise result.

1.4.8.1 Probability Density Function

The probability density function of t-distribution is given by formula (1.29).

$$f(x) = \frac{\Gamma((\nu+1)/2)}{\sqrt{\nu\pi}\,\Gamma(\nu/2)(1+x^2/\nu)(\nu+1)/2}. \tag{1.29}$$

Here, $\nu = n - 1$ is known as degree of freedom and Γ is the gamma function.

The following R program draws the curve of the probability density function of t-distribution:

```
>set.seed(1)
> x<-seq(-5,5,length.out=1000)
# Calculate the values of probability density function with 1,000 points
that follow the t-distribution of T(1,0).
> y<-dt(x,1,0)

# Draw the curve of the probability density function, as shown in
Fig. 1.26.
> plot(x,y,col="red",xlim=c(-5,5),ylim=c(0,0.5),type='l',xaxs="i", yaxs="
i",ylab='density',xlab='',main="The T Density Distribution")
>lines(x,dt(x,5,0),col="green")
>lines(x,dt(x,5,2),col="blue")
>lines(x,dt(x,50,4),col="orange")
> legend("topleft",legend=paste("df=",c(1,5,5,50)," ncp=", c(0,0,2,4)),
lwd=1, col=c("red", "green","blue","orange"))
```

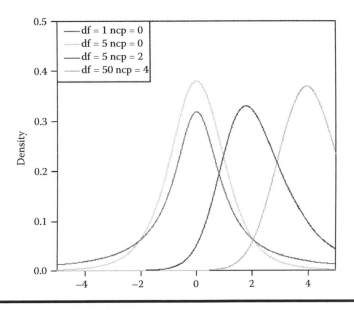

Figure 1.26   Illustration of probability density functions of t-distributions.

1.4.8.2 Cumulative Distribution Function

The formula of cumulative distribution function of t-distribution is (1.30).

$$f(x) = \frac{1}{2} + \frac{x\Gamma((\nu+1)/2)\,_2F_1\left(\frac{1}{2},(\nu+1)/2;\frac{3}{2};-\frac{x^2}{\nu}\right)}{\sqrt{\pi\nu}\Gamma(\nu/2)}.\qquad(1.30)$$

Here, $\nu = n - 1$ is known as degree of freedom and Γ is the gamma function.

The following R program draws the curve of the cumulative distribution function of t-distribution:

```
>set.seed(1)
> x<-seq(-5,5,length.out=1000)
# Calculate the values of cumulative distribution function with 1,000
points that follow the t-distribution T(1,0).
> y<-pt(x,1,0)

# Draw the curve of the cumulative probability function, as shown in
Fig. 1.27.
> plot(x,y,col="red",xlim=c(-5,5),ylim=c(0,0.5),type='l',xaxs="i", yaxs="
i",ylab='F(x)',xlab='',main="The T Cumulative Distribution Function")
>lines(x,pt(x,5,0),col="green")
>lines(x,pt(x,5,2),col="blue")
>lines(x,pt(x,50,4),col="orange")
> legend("topleft",legend=paste("df=",c(1,5,5,50)," ncp=", c(0,0,2,4)),
lwd=1, col=c("red", "green","blue","orange"))
```

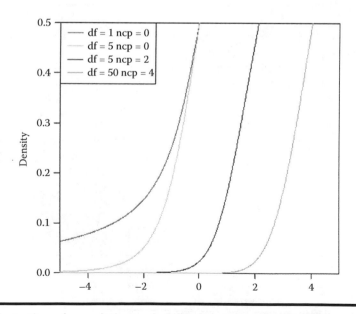

Figure 1.27   Illustration of cumulative probability functions of t-distributions.

1.4.8.3 Distribution Test

For t-distribution, the null hypothesis of the Kolmogorov–Smirnov continuous distribution test is H_0: the data set follows t-distribution and the alternative one is H_1: the distribution of the population from which the sample is chosen does not follow the t-distribution.

```
>set.seed(1)
> S<-rt(1000, 1,2)
# Kolmogorov-Smirnov test.
> ks.test(S, "pt", 1, 2)
     One-sample Kolmogorov-Smirnov test
data: S
D = 0.0253, p-value = 0.5461
alternative hypothesis: two-sided
```

Here, D is very small and p-value > 0.05, so we cannot refuse the null hypothesis. Therefore, the data set follows t-distribution!

1.4.9 Beta Distribution

Beta distribution is a family of continuous probability distributions defined on the interval [0,1]. The distribution has two parameters $\alpha, \beta > 0$, where α is the success times plus 1 and β is the failure time plus 1.

Usually, beta distribution is used to model the random behaviors limited to some finite interval [c,d]. Of course, if we let c be the origin and d-c be the unit length, then the interval can be converted to [0,1]. An important application of beta distribution is to be the conjugate prior probability distribution for the Bernoulli* distribution, which plays an important role in machine learning and mathematical statistics.

1.4.9.1 Probability Density Function

The probability density function of t-distribution is given by formula (1.31).

$$
\begin{aligned}
f(x; \alpha, \beta) &= \frac{x^{\alpha-1}(1-x)^{\beta-1}}{\displaystyle\int_0^1 u^{\alpha-1}(1-u)^{\beta-1}\,du} \\
&= \frac{\Gamma(\alpha+\beta)}{\Gamma(\alpha)\Gamma(\beta)} x^{\alpha-1}(1-x)^{\beta-1}. \\
&= \frac{1}{B(\alpha,\beta)} x^{\alpha-1}(1-x)^{\beta-1}
\end{aligned}
\tag{1.31}
$$

Here, the random variable X follows the beta distribution with parameters α and β, and Γ is the gamma function.

The following R program draws the curve of the probability density function of beta distribution:

```
>set.seed(1)
> x<-seq(-5,5,length.out=10000)
```

---

* https://en.wikipedia.org/wiki/Beta_distribution

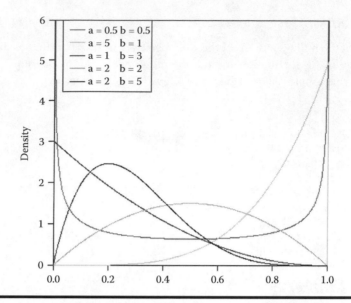

Figure 1.28 Illustration of probability density functions of beta distributions.

```
# Calculate the values of probability density function with 10,000 points
that follow the Beta distribution of B(0.5,0.5)
> y<-dbeta(x,0.5,0.5)

# Draw the curve of the probability density function, as shown in
Fig. 1.28.
> plot(x,y,col="red",xlim=c(0,1),ylim=c(0,6),type='l',xaxs="i", yaxs="i",
ylab='density',xlab='',main="The Beta Density Distribution")
>lines(x,dbeta(x,5,1),col="green")
>lines(x,dbeta(x,1,3),col="blue")
>lines(x,dbeta(x,2,2),col="orange")
>lines(x,dbeta(x,2,5),col="black")
> legend("top",legend=paste("a=",c(.5,5,1,2,2)," b=", c(.5,1,3,2,5)),
lwd=1,col=c("red", "green","blue","orange","black"))
```

1.4.9.2 Cumulative Distribution Function

The formula of cumulative distribution function of beta distribution is (1.32).

$$F(x; \alpha, \beta) = \frac{B_x(\alpha, \beta)}{B(\alpha, \beta)} = I_x(\alpha, \beta). \tag{1.32}$$

Here, I is the regularized incomplete beta function.

The following R program draws the curve of the cumulative distribution function of beta distribution:

```
>set.seed(1)
> x<-seq(-5,5,length.out=10000)
# Calculate the values of cumulative distribution function with 10,000
points that follow the Beta distribution B(0.5,0.5).
> y<-pbeta(x,0.5,0.5)
```

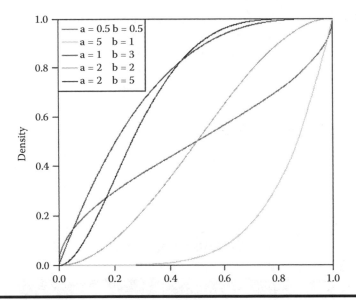

Figure 1.29   Illustration of cumulative probability functions of t-distributions.

```
# Draw the curve of the cumulative probability function, as shown in
Fig. 1.29.
> plot(x,y,col="red",xlim=c(0,1),ylim=c(0,1),type='l',xaxs="i", yaxs="i",
ylab='F(x)',xlab='',main="The Beta Cumulative Distribution Function")
>lines(x,pbeta(x,5,1),col="green")
>lines(x,pbeta(x,1,3),col="blue")
>lines(x,pbeta(x,2,2),col="orange")
>lines(x,pbeta(x,2,5),col="black")
> legend("topleft",legend=paste("a=",c(.5,5,1,2,2)," b=", c(.5,1,3,2,5)),
lwd=1,col=c("red", "green","blue","orange","black"))
```

1.4.9.3 Distribution Test

For beta distribution, the null hypothesis of the Kolmogorov–Smirnov continuous distribution test is H_0: the data set follows beta distribution and the alternative one is H_1: the distribution of the population from which the sample is chosen does not follow the beta distribution.

```
>set.seed(1)
> S<-rbeta(1000,1,2)
# Kolmogorov-Smirnov test.
> ks.test(S, "pbeta",1,2)
    One-sample Kolmogorov-Smirnov test
data: S
D = 0.0202, p-value = 0.807
alternative hypothesis: two-sided
```

Here, D is very small and p-value > 0.05, so we cannot refuse the null hypothesis. Therefore, the data set follows beta distribution!

After mastering the above frequently used continuous distributions, we are able to establish models based on them and understand many related algorithm models as well. Finally, please note that this book is not a textbook on statistics. The definitions for each distribution are quoted from the Internet. Please refer to the statistical textbooks for any differences.

1.5 The Calculation of Derivatives in R

Question

How do we perform the derivative calculations with R?

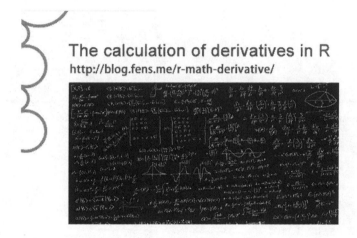

The calculation of derivatives in R
http://blog.fens.me/r-math-derivative/

Introduction

Advanced mathematics is a basic mathematical subject every undergraduate must study. Meanwhile it is the one that we forget most quickly after the examinations. I racked my brain when I studied the subject but never knew why I studied it. We can hardly see the applications of knowledge of advanced mathematics in daily life and work. Even in the fields of IT and finance, few aspects directly apply advanced mathematics. There is wide gap between academia and practice.

However, R opens a door for me to the applications of advanced mathematics, which is not only able to implement the calculations of advanced mathematics, but also apply the advanced mathematical formulas in a thesis to product practices. I re-studied the subject because of R. A life full of mathematics gets more interesting. Please note that the section is not a complete manual for advanced mathematical calculations. It only covers the R implementations of derivative and partial derivative calculations.

1.5.1 The Calculation of Derivative

Derivative is a basic concept in differential calculus for calculating the limits of functions. The definition of derivative is as following. If the function $y = f(x)$ has values in a certain neighborhood near x_0, and when the independent x gets increment Δx (the point $x_0 + \Delta x$ is still in the neighborhood), the function gets the corresponding increment $\Delta y = f(x_0 + \Delta x) - f(x_0)$. If the limit of the ratio of Δy to Δx exists when Δx approaches 0, we say that the function $y = f(x)$ is derivable at point x_0 and declare the limit as the derivative of function $y = f(x)$ as point x_0, denoted by $f'(x_0)$, that is,

$$f'(x_0) = \lim_{\Delta x \to 0} \frac{\Delta y}{\Delta x} = \lim_{\Delta x \to 0} \frac{f(x_0 + \Delta x) - f(x_0)}{\Delta x}. \tag{1.33}$$

It is also denoted by $y'\big|_{x=x_0}$, $dy/dx\big|_{x=x_0}$ or $df(x)/dx\big|_{x=x_0}$.

The calculations of derivative can be directly performed by using the function derive() in R. For instance, if we want to calculate the derivative of $y = x^3$, manually we transform the function as $y' = 3x^2$ according to the formula of derivative calculation. When $x = 1$, $y' = 3$; when $x = 2$, $y' = 3$.

The system environment used in this section:

- Win7 64bit
- R: 3.1.1 x86_64-w64-mingw32/x64 (64-bit)

The above derivative calculations can be implemented with R as following:

```
# Generate the derivative formula.
> dx <- deriv(y ~ x^3, "x") ; dx
expression({
    .value <- x^3
    .grad <- array(0, c(length(.value), 1L), list(NULL, c("x")))
    .grad[, "x"] <- 3 * x^2
    attr(.value, "gradient") <- .grad
    .value
})
# Check the type of variable dx.
> mode(dx)
[1] "expression"
# Assign the independent x with values.
> x<-1:2
# Calculate the derivative.
> eval(dx)
# The calculation result of the primitive function.
[1] 1 8
# The calculation result of the derivative function by using gradient
descent method.
>attr(,"gradient")
      x
# x=1,dx=3*1^2=3.
[1,]  3
# x=2,dx=3*2^2=12.
[2,] 12
```

The result is same as that we get manually. But the calculation processes are very different from each other. Manually, we complete the calculation after transforming the primitive function according to the given derivative formula, while the program calculation, which solves the first-order derivative using the gradient descent method, is an optimized approximating algorithm. If the function is too complex to apply the transformable formula, the manual calculation would be hardly to perform. However, as a general derivative calculating method, the program calculation is not affected by the difficulty of transformation.

Usually, the function deriv(expr, name) needs two parameters. The first one, expr, is the primitive function, with ~ to separate the two sides. The second one, name, is used to specify the independent variable. The function deriv() returns a variable with type of expression. The final result is obtained by running this expression using the function eval(), as shown in above code.

If we want to call the calculation formula in terms of function, we need to pass the third parameter, func, and assign it with TRUE. Refer the following implementation.

Calculate the derivative of the sine function $y = \sin x$. According to the derivative calculation formula, the transformation used in manually calculation is $y' = \cos x$. When $x = \pi$, $y' = -1$; when $x = 4\pi$, $y' = 1$.

```
# Generate the calling function of the derivative formula.
> dx <- deriv(y ~ sin(x), "x", func= TRUE) ; dx
function (x)
{
    .value <- sin(x)
    .grad <- array(0, c(length(.value), 1L), list(NULL, c("x")))
    .grad[, "x"] <- cos(x)
    attr(.value, "gradient") <- .grad
    .value
}
# Check the type of dx.
> mode(dx)
[1] "function"
# Call the function using parameter as independent variable.
> dx(c(pi,4*pi))
[1]   1.224606e-16 -4.898425e-16
attr(,"gradient")
# The calculation result of the derivative function.
     x
# x=pi,dx=cos(pi)=-1.
[1,] -1
# x=4*pi,dx=cos(4*pi)=1.
[2,]  1
```

1.5.2 *The Derivative Formulas of Elementary Functions*

We can manually calculate the derivatives of the basic elementary functions by using the derivative formulas. The following is the derivative formula of one-variable elementary functions, where y is the primitive function, x is the independent variable of function y, y' is the derivative function of function y, C, n, a are constant numbers, and ln is the logarithmic function with base of the natural constant e.

Function	Primitive	Derivative
linear	$y = C + ax$	$y' = a$
power	$y = x^n$	$y' = nx^{n-1}$
exponential	$y = a^x$	$y' = a^x \ln(a)$
logarithmic	$y = \log_a x$	$y' = \dfrac{1}{x \ln(a)}, \; (a > 0 \,\&\, a \neq 1, x > 0)$
sine	$y = \sin x$	$y' = \cos x$
cosine	$y = \cos x$	$y' = -\sin x$
tangent	$y = \tan x$	$y' = \sec^2 x = \dfrac{1}{\cos^2 x}$
cotangent	$y = \cot x$	$y' = -\csc^2 x = \dfrac{1}{\sin^2 x}$
secant	$y = \sec x$	$y' = \sec x \tan x$
cosecant	$y = \csc x$	$y' = -\csc x \cot x$
arc-sine	$y = \arcsin x$	$y' = \dfrac{1}{\sqrt{1 - x^2}}$
arc-cosine	$y = \arccos x$	$y' = -\dfrac{1}{\sqrt{1 - x^2}}$
arc-tangent	$y = \arctan x$	$y' = \dfrac{1}{\sqrt{1 + x^2}}$
arc-cotangent	$y = \operatorname{arccot} x$	$y' = -\dfrac{1}{\sqrt{1 + x^2}}$
arc-secant	$y = \operatorname{arcsec} x$	$y' = \dfrac{1}{|x|(x^2 - 1)}$
arc-cosecant	$y = \operatorname{arccsc} x$	$y' = -\dfrac{1}{|x|(x^2 - 1)}$

In the following, let us calculate the first-order derivative of the above one-variable elementary functions.

1.5.2.1 The Linear Function

Calculate the derivative of function $y = 3 + 10x$. According to the derivative formulas, in manual calculation, the transformation is $y' = 0 + 10$, where the derivative of the constant item 3 is 0. When $x = 1$, $y' = 10$.

```
# Generate the derivative expression with the form of function.
> dx<-deriv(y~ 3+10*x,"x",func = TRUE)
# Input the independent and do calculation.
> dx(1)
# The calculation result of the primitive function is y=3+10*1=13.
[1] 13
attr(,"gradient")
     x
# The calculation result of the derivative function is y'=10.
 [1,] 10
```

1.5.2.2 The Power Function

Calculate the derivative of function $y = x^4$. According to the derivative formulas, in manual calculation, the transformation is $y' = 4x^3$. When $x = 2$, $y' = 32$.

```
> dx<-deriv(y~x^4,"x",func = TRUE)
> dx(2)
[1] 16
attr(,"gradient")
     x
# The calculation result of the derivative function is y'=4*x^3=4*2^3=32.
 [1,] 32
```

1.5.2.3 The Exponential Function

Calculate the derivative of function $y = 4^x$. According to the derivative formulas, in manual calculation, the transformation is $y' = 4^x \ln 4$. When $x = 2$, $y' = 22.18071$.

```
> dx<-deriv(y~4^x ,"x",func = TRUE)
> dx(2)
[1] 16
attr(,"gradient")
         x
# The calculation result of the derivative function is
y'=4^x*log(4)=4*2^3=22.18071.
 [1,] 22.18071
```

Calculate the derivative of function $y = e^x$. According to the derivative formulas, in manual calculation, the transformation is $y' = e^x$. When $x = 2$, $y' = 7.389056$.

```
> dx<-deriv(y~exp(1)^x ,"x",func = TRUE)
> dx(2)
[1] 7.389056
attr(,"gradient")
           x
# The calculation result of the derivative function is
y'=exp(1)^x=exp(1)^2=7.389056.
 [1,] 7.389056
```

1.5.2.4 The Logarithmic Function

Calculate the derivative of function $y = \ln x$. According to the derivative formulas, in manual calculation, the transformation is $y' = \dfrac{1}{x}$. When $x = 2$, $y' = 0.5$.

```
> dx<-deriv(y~log(x),"x",func = TRUE)
> dx(2)
[1] 0.6931472
attr(,"gradient")
       x
# The calculation result of the derivative function is y'=1/x=1/2=0.5.
  [1,] 0.5
```

Calculate the derivative of function $y = \log_2 x$. According to the derivative formulas, in manual calculation, the transformation is $y' = \dfrac{1}{x} \ln 2$. When $x = 3$, $y' = 0.4808983$. However, the R language can only calculate the derivative of logarithmic function with base of the natural constant. The function whose base is not the natural constant must be transformed to the one with base of the natural constant before calculating its derivative. According to the base-changing formula, the logarithm to base 2 is changed to the logarithm with the natural constant as its base: $y = \log_2 x = \dfrac{\ln x}{\ln 2}$.

```
> dx<-deriv(y~log(x)/log(2),"x",func = TRUE)
> dx(3)
[1] 1.584963
attr(,"gradient")
       x
# The calculation result of the derivative function is y'=1/(x*log(2)=1/
(3*log(2)=0.4808983.
  [1,] 0.4808983
```

1.5.2.5 The Sine Function

Calculate the derivative of function $y = \sin x$. According to the derivative formulas, in manual calculation, the transformation is $y' = \cos x$. When $x = \pi$, $y' = -1$.

```
> dx<-deriv(y~sin(x),"x",func = TRUE)
> dx(pi)
[1] 1.224606e-16
attr(,"gradient")
       x
# The calculation result of the derivative function is
y'=cos(x)=cos(pi)=-1.
  [1,] -1
```

1.5.2.6 The Cosine Function

Calculate the derivative of function $y = \cos x$. According to the derivative formulas, in manual calculation, the transformation $y' = -\sin x$. When $x = \pi/2$, $y' = -1$.

```
> dx<-deriv(y~cos(x),"x",func = TRUE)
> dx(pi/2)
[1] 6.123032e-17
attr(,"gradient")
     x
# The calculation result of the derivative function is
y'=-sin(x)=-sin(pi/2)=-1.
 [1,] -1
```

1.5.2.7 The Tangent Function

Calculate the derivative of function $y = \tan x$. According to the derivative formulas, in manual calculation, the transformation is $y' = \sec^2 x = 1/\cos^2 x$. When $x = \pi/6$, $y' = 1.333333$.

```
> dx<-deriv(y~tan(x),"x",func = TRUE)
> dx(pi/6)
[1] 0.5773503
attr(,"gradient")
       x
# The calculation result of the derivative function is y'=1/cos(x)^2=1/
cos(pi/6)^2=1.333333
 [1,] 1.333333
```

1.5.2.8 The Cotangent Function

Calculate the derivative of function $y = \cot x$. Since there is no such a function cot(), we need to transform the primitive function as $y = \cot x = 1/\tan x$ before calculation. According to the derivative formulas, in manual calculation, the transformation is $y' = -\csc^2 x = -1/\sin^2 x$. When $x = \pi/6$, $y' = -4$.

```
> dx<-deriv(y~1/tan(x),"x",func = TRUE)
> dx(pi/6)
[1] 1.732051
attr(,"gradient")
      x
# The calculation result of the derivative function is y'=-1/sin(x)^2=-1/
sin(pi/6)^2=-4.
 [1,] -4
```

1.5.2.9 The Arc-Sine Function

Calculate the derivative of function $y = \arcsin x$. According to the derivative formulas, in manual calculation, the transformation is $y' = 1/\sqrt{1-x^2}$. When $x = \pi/6$, $y' = 1.173757$.

```
> dx<-deriv(y~asin(x),"x",func = TRUE)
> dx(pi/6)
[1] 0.5510696
attr(,"gradient")
        x
# The calculation result of the derivative function is y'=1/
sqrt(1-x^2)=1/sqrt(1-(pi/6)^2)=1.173757
 [1,] 1.173757
```

1.5.2.10 The Arc-Cosine Function

Calculate the derivative of function $y = \arccos x$. According to the derivative formulas, in manual calculation, the transformation is $y' = -1/\sqrt{1-x^2}$. When $x = \pi/8$, $y' = -1.08735$.

```
> dx<-deriv(y~acos(x),"x",func = TRUE)
> dx(pi/8)
[1] 1.167232
attr(,"gradient")
         x
# The calculation result of the derivative function is y'=-1/
sqrt(1-x^2)=-1/sqrt(1-(pi/8)^2)=-1.08735
 [1,] -1.08735
```

1.5.2.11 The Arc-Tangent Function

Calculate the derivative of function $y = \arctan x$. According to the derivative formulas, in manual calculation, the transformation is $y' = 1/\sqrt{1+x^2}$. When $x = \pi/6$, $y' = 0.7848335$.

```
> dx<-deriv(y~atan(x),"x",func = TRUE)
> dx(pi/6)
[1] 0.4823479
attr(,"gradient")
           x
# The calculation result of the derivative function is y'= 1/(1+x^2) = 1/
(1+(pi/6)^2)=0.7848335
 [1,] 0.7848335
```

1.5.3 The Calculation of Second Derivative

When we do the derivative calculations against a function for multiple time, we can get the higher-order derivatives. Generally speaking, the derivative function of the function $y = f(x)$, $y' = f'(x)$, is still a function of x. We call the derivative of $y' = f'(x)$ the second derivative of function $y = f(x)$, denoted by y'', that is,

$$y'' = \left(y'\right)' \text{ or } \frac{d^2 y}{dx^2} = \frac{d}{dx}\left(\frac{dy}{dx}\right). \tag{1.34}$$

The derivative of a first derivative is called the second derivative. The derivative of a second derivative is called the third derivative. The derivative of a $(N-1)$th derivative is called the Nth derivative. We used to call the derivatives with order more than two as the higher-order derivatives.

Next, we calculate the second derivative y'' of the function $y = \sin ax$, where a is a constant. According to the derivative formulas, in manual calculation, the transformation is the first derivative $y' = a \cos x$. We transform y' as $y'' = -a^2 \sin ax$ to calculate its derivative.

Implement the program using R:

```
# Set the value of a.
> a<-2
# Generate the expression of the first derivative.
> dx<-deriv(y~sin(a*x),"x",func = TRUE)
# Calculate the first derivative.
> dx(pi/3)
[1] 0.8660254
attr(,"gradient")
      x
```

```
# The calculation result of the derivative function is y'=
a*cos(a*x)=2*cos(2*pi/3)=-1.
 [1,] -1
```

```
# Calculate the derivative of the first derivative.
> dx<-deriv(y~a*cos(a*x),"x",func = TRUE)
> dx(pi/3)
[1] -1
attr(,"gradient")
        x
# The calculation result of the derivative function is y'=
-a^2*sin(a*x)=-2^2*sin(2*pi/3)=-3.464102.
 [1,] -3.464102
```

In the above calculation process, we manually split the second derivative calculation into two steps. In fact using the function deriv3(), we can complete the process just in one step.

```
# Generate the expression of the second derivative.
> dx<-deriv3(y~sin(a*x),"x",func = TRUE)
# Calculate the derivative.
> dx(pi/3)
[1] 0.8660254
attr(,"gradient")
      x
# The result of the first derivative.
 [1,] -1
attr(,"hessian")
, , x

          x
# The result of the second derivative.
 [1,] -3.464102
```

Now, let us calculate another second derivative. Calculate the second derivative of function $y = ax^4 + bx^3 + x^2 + x + c$, where $a = 2$, $b = 1$, and $c = 3$ are constants. According to the derivative formulas, in manual calculation, the transformation is the first derivative $y' = (2x^4 + x^3 + x^2 + x + 3)' = 8x^3 + 3x^2 + 2x + 1$. When $x = 2$, $y' = 81$. We transform y' as $y'' = 24x^2 + 6x + 2$ to calculate its derivative.

```
# Specify the values of the constants through the parameter func.
> dx<-deriv3(y~a*x^4+b*x^3+x^2+x+c,"x",func=function(x,a=2,b=1,c=3){})
> dx(2)
[1] 49
attr(,"gradient")
      x
```

```
# The result of the first derivative.
 [1,] 81
attr(,"hessian")
, , x

      x
# The result of the second derivative.
 [1,] 110
```

Therefore, we complete the calculation for the second derivative. In the R language, the second derivatives can be calculated directly. If you want to calculate the derivatives with higher orders, other mathematical toolkits should be employed.

1.5.4 The Calculation of Partial Derivative

We have understood that the derivative of a one-variable function is its rate of change. For two-variable functions, we still need to study its "rate of change." However, the one more variable makes the situation more complex. In mathematics, a partial derivative of a function of several variables is its derivative with respect to one of those variables, with the others held constant (as opposed to the total derivative, in which all variables are allowed to vary).* The operator symbol of partial derivative is ∂. It is denoted by $\dfrac{\partial f}{\partial x}$, or f'_x. Partial derivative measures the rate of change of the function toward the positive direction of the axis. Partial derivatives are used in vector calculus and differential geometry.*

In plane xOy, with the moving point changing its position from P(x_0, y_0) toward a different direction, the changing speeds of function $f(x,y)$ are different generally speaking. So, we need to study the rates of change toward different directions of function $f(x,y)$ at point(x_0, y_0). In this section, we just study the rates of change toward two specific directions which are parallel to the x and y axis, respectively, of function $f(x,y)$ in plane xOy.

The partial derivative toward x direction: let $z = f(x,y)$ be the two-variable function. Point(x_0, y_0) is in the domain of the function's definition. Fix y to y_0 and let x have increment Δx at x_0. The function $z = f(x,y)$ has increment (called the partial increment with respect to x) $\Delta z = f(x_0 + \Delta x, y_0) - f(x_0, y_0)$. If the limit of the ratio of Δz to Δx exists when $\Delta x \to 0$, then the limit is called the partial derivative of function $z = f(x,y)$ with respect to x at point (x_0, y_0), denoted by $f'x(x_0, y_0)$.

The partial derivative toward x direction: the derivative of function $z = f(x,y)$ with respect to x at point (x_0, y_0), is in fact the derivative of a one-variable $z = f(x, y_0)$ at point x_0 when fixing y to y_0 (seen as a constant). Likely, we fix x to x_0 and let y have increment Δy. If the limit exists, then it is called the partial derivative of function $z = f(x,y)$ with respect to y at point (x_0, y_0), denoted by $f'y(x_0, y_0)$.

Similarly, the calculation of partial derivatives can be performed using the function derive() in R. In the following, we calculate the partial derivative of function $f(x,y) = 2x^2 + y + 3xy^2$. It is difficult to describe the derivative of this function because there are infinite tangent lines at each point on the surface of the two-variable function. If we fix the values of a variable y to a constant, then we can calculate the partial derivative with respect to another variable x, saying $\partial f / \partial x$.

Let us calculate the partial derivatives with respect to the independent variables x and y, respectively. Let y be constant and calculate the partial derivative with respect to x $\dfrac{\partial f}{\partial x} = 4x + 3y^2$. When $x = 1$ and $y = 1$, the partial derivative $\dfrac{\partial f}{\partial x} = 4x + 3y^2 = 7$. Let x be constant and calculate

---

* https://en.wikipedia.org/wiki/Partial_derivative

the partial derivative with respect to y $\frac{\partial f}{\partial y} = 1 + 6xy$. When $x = 1$ and $y = 1$, the partial derivative $\frac{\partial f}{\partial y} = 1 + 6xy = 7$. The R implementation is as following:

```
# The expression of the two-variable function.
> fxy = expression(2*x^2+y+3*x*y^2)
> dxy = deriv(fxy, c("x", "y"), func = TRUE)
> dxy
function (x, y)
{
    .expr4 <- 3 * x
    .expr5 <- y^2
    .value <- 2 * x^2 + y + .expr4 * .expr5
    .grad <- array(0, c(length(.value), 2L), list(NULL, c("x","y")))
    .grad[, "x"] <- 2 * (2 * x) + 3 * .expr5
    .grad[, "y"] <- 1 + .expr4 * (2 * y)
    attr(.value, "gradient") <- .grad
    .value
}
# Specify the independent variables.
> dxy(1,1)
[1] 6
attr(,"gradient")
     x y
# The calculation result. The partial derivative to x is 7 and the
partial derivative to y is 7
[1,] 7 7
```

The programming calculation result of the partial calculation is same as the one by manual.

Let us calculate the partial derivative of a complex function. Calculate the partial derivative of the two-variable function $f(x, y) = x^y + e^{xy} + x^2 + 2xy + y^3 + \sin xy$ at points (1,3) and (0,0). The R implementation is as following:

```
> fxy = expression(x^y + exp(x * y) + x^2 - 2 * x * y + y^3 + sin(x*y))
> dxy = deriv(fxy, c("x", "y"), func = TRUE)
# Specify the independent variables.
> dxy(1,3)
[1] 43.22666
attr(,"gradient")
            x          y
# The calculation result. The partial derivative to x is 56.28663 and the
partial derivative to y is 44.09554
 [1,] 56.28663 44.09554
> dxy(0,0)
[1] 2
attr(,"gradient")
     x   y
# The calculation result. The partial derivative to x has no meaning and
the partial derivative to y is negative infinite
 [1,] NaN -Inf
```

The people that doubt the results may have a manual try.

Through this section, we have mastered the calculation of derivatives in advanced mathematics. With this ease of use, we have more motivation to learn advanced mathematics.

Chapter 2

The Algorithm Implementations in R

Four algorithm cases are implemented with R in this chapter including the collaborative filtering, the PageRank, the moving average (MA) model, and the genetic algorithm.

2.1 Rewriting the Collaborative Filtering Algorithm (UserCF) on Mahout with R

Question

How do we implement the recommendation algorithm with R?

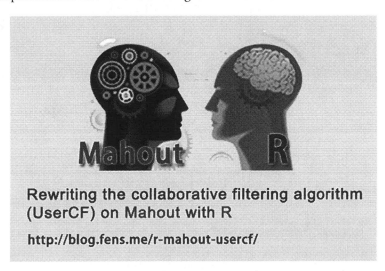

Rewriting the collaborative filtering algorithm (UserCF) on Mahout with R

http://blog.fens.me/r-mahout-usercf/

Introduction

We often see recommendation systems in Internet applications. For example, Amazon recommends books for you and Douban recommends a movie list for you. Mahout is a distributed computing framework in the Hadoop family that is used in machine learning. The framework mainly contains three categories of algorithms: recommendation algorithms, clustering algorithms, and classification algorithms. This section rewrites the user-based collaborative filtering (UserCF) algorithm in R in the recommendation category. The rewriting is implemented completely following the logics and designs of Mahout. The result is compared with that of Mahout.

2.1.1 The Recommendation Algorithm Model of Mahout

First of all, let us understand how the collaborative filtering algorithm is implemented in Mahout. The recommendation algorithms in Mahout define a set of standard processes for model building and executing. Take the UserCF algorithm, for instance, as shown in Figure 2.1.

From Figure 2.1, we can see that the UserCF is modularized. The universal method calling is performed through four modules. First, build the data model and then define user similarity algorithm. Second, define user neighborhood algorithm, and finally call the recommender to complete the computation. The item-based collaborative filtering (ItemCF) algorithm is similar, just removing the user neighborhood algorithm in step 3.

This section will rewrite the algorithm in Mahout version 0.5. Following is the version definition of Mahout in Maven.

```
<dependency>
<groupId>org.apache.mahout</groupId>
<artifactId>mahout-core</artifactId>
<version>0.5</version>
</dependency>
```

A simple test data set testCF.csv is chosen. The data set has three columns: user ID, item ID, and the item score rated by user. There are 21 rows in total.

```
1,101,5.0
1,102,3.0
1,103,2.5
2,101,2.0
2,102,2.5
```

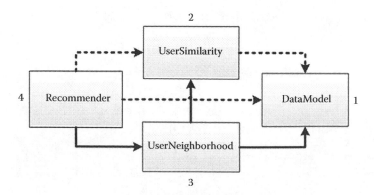

Figure 2.1   The Mahout recommendation algorithm model (referencing *Mahout In Action*).

```
2,103,5.0
2,104,2.0
3,101,2.5
3,104,4.0
3,105,4.5
3,107,5.0
4,101,5.0
4,103,3.0
4,104,4.5
4,106,4.0
5,101,4.0
5,102,3.0
5,103,2.0
5,104,4.0
5,105,3.5
5,106,4.0
```

Let us review how Java calls the APIs in Mahout Library to implement UserCF algorithm.

```java
/**
 * The test against UserCF algorithm. This is an implementation for single machine.
 */
public class UserCF {

    //choose 2 nearest neighbors
    final static int NEIGHBORHOOD_NUM = 2;
    //reserve 3 recommendations
    final static int RECOMMENDER_NUM = 3;

    public static void main(String[] args) throws IOException, TasteException {
        //read the data set
        String file = "item.csv";
        //load data to memory object
        DataModel model = new FileDataModel(new File(file));
        //define user similarity distance
        UserSimilarity user = new EuclideanDistanceSimilarity(model);
        //defined the neighborhood
        NearestNUserNeighborhood neighbor = new NearestNUserNeighborhood(NEIGHBORH
OOD_NUM, user, model);
        //build the recommendation model
        Recommender r = new GenericUserBasedRecommender(model, neighbor, user);
        //get the user list
        LongPrimitiveIterator iter = model.getUserIDs();

        // iterate the user list to calculate the recommendation result for each user
        while (iter.hasNext()) {
            long uid = iter.nextLong();
            List<RecommendedItem> list = r.recommend(uid, RECOMMENDER_NUM);
            System.out.printf("uid:%s", uid);
            for (RecommendedItem ritem : list) {
                //print the recommendation result
                System.out.printf("(%s,%f)", ritem.getItemID(), ritem.getValue());
            }
            System.out.println();
        }
    }
}
Run the Java program and output the recommendation result.
```

```
uid:1(104,4.250000)(106,4.000000)
uid:2(105,3.956999)
uid:3(103,3.185407)(102,2.802432)
uid:4(102,3.000000)
uid:5
```

The following explains the result:

- For user with uid=1, recommend the 2 items with highest scores, 104 and 106.
- For user with uid=2, recommend the 1 item with highest score, 105.
- For user with uid=3, recommend the 2 items with highest scores, 103 and 102.
- For user with uid=4, recommend the 1 item with highest score, 102.
- For user with uid=5, no recommendation is given.

2.1.2 The Model Implementations in R

In the following, let us rewrite the Mahout implementations using R. The R code implementations follow the design logics of Mahout's source code. To keep consistency with Java program's logics, the R code uses loop statements in implementations. Here, the performance of R program is not considered.

We will follow the five steps in building the algorithm models using R language:

1. Creating the data model
2. Implementing the Euclidean distance similarity algorithm
3. Implementing the nearest-neighborhood algorithm
4. Implementing the recommendation algorithm
5. Running the program

2.1.2.1 Creating the Data Model

The function to create the data model, FileDataModel(), is mainly used to read data from CSV file and load the data to memory with the type of matrix in R.

```
# The data model function.
FileDataModel<-function(file){
# Read csv file to memory.
data<-read.csv(file,header=FALSE)
  # Add column names.
  names(data)<-c("uid","iid","pref")
  # Calculate the number of users.
  user <- unique(data$uid)
  # Calculate the number of items.
  item <- unique(sort(data$iid))
  uidx <- match(data$uid, user)
  iidx <- match(data$iid, item)
  # Define the storing matrix.
  M <- matrix(0, length(user), length(item))
  i <- cbind(uidx, iidx, pref=data$pref)
```

```
  # Assign the matrix.
  for(n in 1:nrow(i)){
    M[i[n,][1],i[n,][2]]<-i[n,][3]
  }

  dimnames(M)[[2]]<-item
  # Return the matrix.
  M
}
```

2.1.2.2 *Implementing Euclidean Distance Similarity Algorithm*

There are various algorithms for us to calculate the user similarity, such as Euclidean distance similarity algorithm, Pearson similarity algorithm, the cosine similarity algorithm, Spearman rank correlative coefficient similarity algorithm, logarithmic likelihood similarity algorithm, and so on. Many of them have the corresponding R functions which can be directly called. The Mahout implementation makes some optimization to the basic algorithms, so we need to rewrite the algorithm completely from the lower level.

The following code creates the function EuclideanDistanceSimilarity() for calculating the Euclidean distance similarity, loading user-item matrix data and calculating user similarity with Euclidean distance.

```
EuclideanDistanceSimilarity<-function(M){
  row<-nrow(M)
  # The similarity matrix.
  s<-matrix(0, row, row)

  for(z1 in 1:row){
    for(z2 in 1:row){
      if(z1<z2){
        # The calculable columns              .
        num<-intersect(which(M[z1,]!=0),which(M[z2,]!=0))

        sum<-0
        for(z3 in num){
          sum<-sum+(M[z1,][z3]-M[z2,][z3])^2
        }

        s[z2,z1]<-length(num)/(1+sqrt(sum))

        # Limit the thresholds of the algorithm.
        if(s[z2,z1]>1)  s[z2,z1]<-1
        if(s[z2,z1]<-1)  s[z2,z1]<- -1
      }
    }
  }
  # Supplement the triangle matrix.
  ts<-t(s)
  w<-which(upper.tri(ts))
  s[w]<-ts[w]
  # Return the user similarity matrix.
  s
}
```

2.1.2.3 *Implementing the Nearest-Neighborhood Algorithm*

There are two algorithms for us to choose to calculate the user's nearest neighbors. One is based on the number, choosing the top nearest neighbors. The other is based on percentage, choosing the top percentages of nearest neighbors. The following code implements the number-based algorithm which chooses the top N nearest neighbors. Create the function NearestNUserNeighborhood() for calculating nearest, pass the user similarity matrix and the number, and we can get the nearest neighbors for users.

```
NearestNUserNeighborhood<-function(S,n){
    row<-nrow(S)
    neighbor<-matrix(0, row, n)
    for(z1 in 1:row){
        for(z2 in 1:n){
            m<-which.max(S[,z1])
            neighbor[z1,][z2]<-m
            S[,z1][m]=0
        }
    }
    # Return the top n nearest neighbor.
    neighbor
}
```

2.1.2.4 *Implementing the Recommendation Algorithm*

When calculating the recommendations, we also have several algorithms to choose, such as user based, item based, slopeOne, itemKNN, SVD, TreeCluster, and so on. These algorithms need to match the defined data models and similarity algorithms for use.

In the following, we create the user-based recommendation algorithm function UserBasedRecommender(), calculating the recommendations using user-item data matrix, user similarity matrix, user nearest neighborhood as input and implementing the UserCF.

```
UserBasedRecommender<-function(uid,n,M,S,N){
    row<-ncol(N)
    col<-ncol(M)
    r<-matrix(0, row, col)
    N1<-N[uid,]
    for(z1 in 1:length(N1)){
        # The calculable columns.
        num<-intersect(which(M[uid,]==0),which(M[N1[z1],]!=0))
        for(z2 in num){
            r[z1,z2]=M[N1[z1],z2]*S[uid,N1[z1]]
        }
    }

    # Print the recommendation matrix for each user.
    # print(r).
    sum<-colSums(r)
    s2<-matrix(0, 2, col)
    for(z1 in 1:length(N1)){
        num<-intersect(which(colSums(r)!=0),which(M[N1[z1],]!=0))
```

```
        for(z2 in num){
        s2[1,][z2]<-s2[1,][z2]+S[uid,N1[z1]]
        s2[2,][z2]<-s2[2,][z2]+1
        }
    }

    s2[,which(s2[2,]==1)]=10000
    s2<-s2[-2,]

    r2<-matrix(0, n, 2)
    rr<-sum/s2
    item <-dimnames(M)[[2]]
    for(z1 in 1:n){
        w<-which.max(rr)
        if(rr[w]>0.5){
            r2[z1,1]<-item[which.max(rr)]
            r2[z1,2]<-as.double(rr[w])
            rr[w]=0
        }
    }
    r2
}
```

Here, we have implemented the UserCF algorithms using R. Then we run these functions in the order of calling relationship.

2.1.2.5 Running the Program

```
# The data file.
> FILE<-"item.csv"

# Choose 2 nearest neighbors
> NEIGHBORHOOD_NUM<-2
//reserve 3 recommendations
> RECOMMENDER_NUM<-3

# Load data to memory converting the data file to matrix.
> M<-FileDataModel(FILE)
# Calculate the user similarity matrix.
> S<-EuclideanDistanceSimilarity(M)
# Calculate the user nearest neighbors.
> N<-NearestNUserNeighborhood(S,NEIGHBORHOOD_NUM)

# View the recommendation result for user=1.
> R1<-UserBasedRecommender(1,RECOMMENDER_NUM,M,S,N);R1
      [,1]   [,2]
[1,] "104"  "4.25"
[2,] "106"  "4"
[3,] "0"    "0"
# View the recommendation result for user=2.
> R2<-UserBasedRecommender(2,RECOMMENDER_NUM,M,S,N);R2
      [,1]   [,2]
```

```
[1,]  "105"  "3.95699903407931"
[2,]  "0"    "0"
[3,]  "0"    "0"
# View the recommendation result for user=3.
> R3<-UserBasedRecommender(3,RECOMMENDER_NUM,M,S,N);R3
     [,1]   [,2]
[1,]  "103"  "3.18540697329411"
[2,]  "102"  "2.80243217111765"
[3,]  "0"    "0"
# View the recommendation result for user=4.
> R4<-UserBasedRecommender(4,RECOMMENDER_NUM,M,S,N);R4
     [,1]   [,2]
[1,]  "102"  "3"
[2,]  "0"    "0"
[3,]  "0"    "0"
# View the recommendation result for user=5.
> R5<-UserBasedRecommender(5,RECOMMENDER_NUM,M,S,N);R5
     [,1]  [,2]
[1,]   0    0
[2,]   0    0
[3,]   0    0
```

Finally, we see that the calculating results are the same as that of the Java program calling Mahout's APIs.

2.1.3 The Principle of the Algorithm Implementation

What is the principle of the algorithm implementation? In fact, the collaborative filtering algorithm is the result of matrix transformations! In the following, please pay attention to the matrix transformation in each step. Let us start from the original data file, as shown below.

```
1,101,5.0
1,102,3.0
1,103,2.5
2,101,2.0
2,102,2.5
2,103,5.0
2,104,2.0
3,101,2.5
3,104,4.0
3,105,4.5
3,107,5.0
4,101,5.0
4,103,3.0
4,104,4.5
4,106,4.0
5,101,4.0
5,102,3.0
5,103,2.0
5,104,4.0
5,105,3.5
5,106,4.0
```

The matrix transformation in the first step is to output the user-item matrix through the function FileDataModel().

```
     101 102 103 104 105 106 107
[1,] 5.0 3.0 2.5 0.0 0.0   0   0
[2,] 2.0 2.5 5.0 2.0 0.0   0   0
[3,] 2.5 0.0 0.0 4.0 4.5   0   5
[4,] 5.0 0.0 3.0 4.5 0.0   4   0
[5,] 4.0 3.0 2.0 4.0 3.5   4   0
```

The second step performs another matrix transformation using the Euclidean similarity algorithm. Calling the function EuclideanDistanceSimilarity() gets the following result.

```
          [,1]         [,2]         [,3]         [,4]         [,5]
[1,] 0.0000000 0.6076560 0.2857143 1.0000000 1.0000000
[2,] 0.6076560 0.0000000 0.6532633 0.5568464 0.7761999
[3,] 0.2857143 0.6532633 0.0000000 0.5634581 1.0000000
[4,] 1.0000000 0.5568464 0.5634581 0.0000000 1.0000000
[5,] 1.0000000 0.7761999 1.0000000 1.0000000 0.0000000
```

The third step calculates users nearest neighbors through the user similarity matrix. Calling the function NearestNUserNeighborhood() gets the following result.

```
     top1 top2
[1,]    4    5
[2,]    5    3
[3,]    5    2
[4,]    1    5
[5,]    1    3
```

The fourth step combines the above matrixes to produce the recommendation matrix for each user through user-based recommendation algorithm. Take the user with uid=1, for example. Calling the function UserBasedRecommender() gets the following result.

```
    101   102   103   104   105   106   107
4     0     0     0   4.5   0.0     4     0
5     0     0     0   4.0   3.5     4     0
```

Uncomment the line print(r) in the function UserBasedRecommender().

The fifth step is to filter the result of recommendation matrix, returning the second highest scored items. Take the user with uid=1, for example. The result is shown as follows.

```
     Recommendation  score
[1,] "104"     "4.25"
[2,] "106"     "4"
```

Through the matrix transformations in the five steps, we can clearly get the nature of user-based collaborative algorithm. Of course, the algorithms provided by Mahout are based on matrix and able to process a huge amount of data. If the data amount is small, we can optimize and improve the above algorithms to reduce the difficulty of matrix calculation.

2.1.4 Summary of the Algorithm

This section only rewrote Mahout's collaborative filtering algorithm that is based on the user, uses Euclidean distance similarity, and the number of nearest-neighborhood algorithms. The processes for rewriting other algorithms are similar. Besides, during reading Mahout's source code, I found that Mahout has its own optimizations when implementing the various algorithms, instead of copying the algorithm formulas in textbooks. For instance, Mahout uses not only the standard Euclidean distance algorithm but also the improved algorithm when calculating user similarity based on Euclidean distance.

```
# The standard Euclidean distance algorithm.
similar = 1/(1+sqrt( (a-b)^2 + (a-c)^2 ))
# The improved Euclidean distance algorithm.
similar = n/(1+sqrt( (a-b)^2 + (a-c)^2 ))
```

Explanations:

- a,b are items that are rated by both users. Can be 1 or more items.
- n is the number of items that are rated by both users.
- If similar >1, then similar = 1.
- If similar <–1, then similar = –1.

Mahout is able to give more precise recommendation results by optimizing the algorithms. So the guys who want to use Mahout as the recommendation engine should hold yourself to a higher standard. You need to not only master Java and Mahout, but also understand the underlying algorithms and the improvements made by Mahout to the algorithms. By doing the above, Mahout plays real power in your hand! For more articles about Mahout, please review my blog http://blog.fens.me/hadoop-mahout-roadmap/.

2.2 The R Implementation of the PageRank Algorithm

Question

How do we implement the PageRank algorithm?

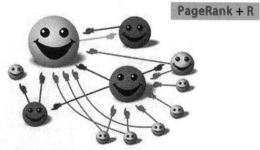

The R implementation of the PageRank algorithm

http://blog.fens.me/algorithm-pagerank-r/

Introduction

The Google searching service has long been my essential daily tool. Time and time again, I am astonished by the preciseness of the search result it provides. In the meanwhile, I am working on Google SEO to promote my blog. After tries over several months, the PR of my blog rises to 2 and there are tens of thousands of backward links. When summarizing, I am still touched by the amazingness of PageRank. The writer thinks that PageRank is an algorithm that changes the Internet!

2.2.1 Introduction to the PageRank Algorithm

PageRank is an algorithm that is exclusively owned by Google. It is used to measure the specific page's importance to other pages in the search engine index. The algorithm was innovated by Larry Page and Sergey Brin in the late 1990s. PageRank makes it true to consider the linking value concept into ranking factors.

PageRank makes links to "vote." The "votes" of a page is determined by the importance of the pages that links to it. One hyperlink to a page means a vote for it. The PageRank of a page is evaluated iteratively using the importance of all pages that refers it (in-link pages). A page with more link-in pages has higher rank, while if a page has no in-link pages, then it has no rank. In one word, the page that is linked by many high-quality pages is also high quality.

The calculation of PageRank is based on the two basic assumptions.

- Quantity assumption: a page is more important if the page node receives more links from other pages.
- Quality assumption: different in-link pages referring page A have different quality. The pages with high quality would propagate more weight to other pages through links. So the more the pages with high-quality refer page A, the more important page A is.

To raise PageRank, there are three important points, which need to be considered when we perform SEO:

- The number of backward links: the more the backward links are, the higher the weight of the referred page is.
- Whether the backward link is from the page with higher PageRank: the higher the weight of backward linking page is, the higher the weight assigned to the referred page is.
- The number of links on the source page of the backward link: the more the links are on the source page of the backward link, the higher the weight is of the source page of the backward link, which in turn explains the previous two points.

2.2.2 The Principle of PageRank Algorithm

In the initial stage, pages construct a directed graph through linkages, with equal PageRank value for each page. After rounds of iterative calculations, the final PageRank value of each page is obtained. The current PageRank value of a page gets updated every round of the calculations.

During a round of updating pages' PageRank scores, every page assigns its current PageRank value equally to all the out-links on it. Therefore, every out-link gets the corresponding weight. And then, every page sums all the weights that are passed through all the in-links that refer it to get its new PageRank score. When all the pages get updated PageRank value, the round of PageRank calculation completes.

2.2.2.1 Principle of the Algorithm

PageRank algorithm is established on the model random surfer. The basic idea is that the importance order of pages is determined by the linking relationship. The algorithm evaluates the rank and importance of each page according to the linking structure among pages. The PageRank value of a page is not only related to the number of linking pages that refer to it, but also related to the importance of other pages that refer to it.

PageRank has two properties. One is the transitivity of PR value, meaning when page A refers to page B, the PR value of page A will be partially transferred to page B. The other is the transitivity of importance, meaning the weight transferred by a more important page is more than that by a less important page.

2.2.2.2 The Calculation Formula

The calculation formula of PageRank is

$$PR(p_i) = \frac{1-d}{n} + d \sum_{p_j \in M(i)} \frac{PR(p_j)}{L(j)}. \tag{2.1}$$

Here, $PR(p_i)$ is the PageRank value of page p_i; n the number of all pages; p_i the different pages such as p_1, p_2, and p_3; M(i) the collection of all in-link pages of pi; L(j) the collection of all the out-link pages of p_i; d the damping coefficient, meaning the probability at which the user lands on the specific page and continues to navigate. According to experiences, Google set d = 0.85, which means 1−d = 0.15 representing the probability for user to stop clicking and randomly picking a new URL. The value range of d is $0 < d \leq 1$.

2.2.2.3 Constructing an Instance with Data of Four Pages

Let us construct a simple PageRank instance model with data of four pages, as shown in Figure 2.2.

In Figure 2.2, the page with ID = 1 links to pages 2, 3, and 4, so the probabilities at which a user jumps from page with ID = 1 to pages 2, 3, and 4 are equally 1/3. The page with ID = 2 links

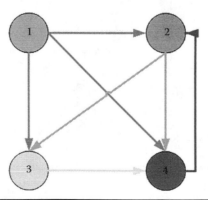

Figure 2.2   A PageRank instance.

to pages 3 and 4 so the probabilities at which a user jumps from page with ID = 2 to pages 3 and 4 are equally 1/2. The page with ID = 3 links to page 4 so the probability at which a user jumps from page with ID = 3 to page 4 is 1. The page with ID = 4 links to page 2 so the probability at which a user jumps from page with ID = 4 to page 2 is 1.

In the following, we construct PageRank's data model and transition matrix. In step 1, we construct the adjacent table using the linking relationship between pages.

```
Linking source page linking target page
      1                   2,3,4
      2                   3,4
      3                   4
      4                   2
```

In step 2, we construct adjacent matrix (a square matrix) with adjacent table, where the columns represent the source pages and the rows represent the target pages:

```
      [,1] [,2] [,3] [,4]
[1,]    0    0    0    0
[2,]    1    0    0    1
[3,]    1    1    0    0
[4,]    1    1    1    0
```

In step 3, we convert the adjacent matrix to probability matrix (the transition matrix):

```
      [,1]   [,2]  [,3]  [,4]
[1,]    0      0    0     0
[2,]   1/3     0    0     1
[3,]   1/3    1/2   0     0
[4,]   1/3    1/2   1     0
```

Thus, we construct the probability matrix with the linking relationship between pages.

2.2.3 R Implementation of the Stand-Alone Algorithm

In the following, we implement the construction process from adjacent table to transition matrix with R language. First, we create a data file page.csv, representing the adjacent table of pages.

```
1,2
1,3
1,4
2,3
2,4
3,4
4,2
```

We implement the PageRank model with the following three scenarios: without consideration of the damping coefficient, with consideration of the damping coefficient, and calculation directly using the matrix eigenvalue function in R.

2.2.3.1 Scenario without Consideration of the Damping Coefficient

The R implementation is as follows:

```
# Construct the adjacent matrix.
> adjacencyMatrix<-function(pages){
+   n<-max(apply(pages,2,max))
+   A <- matrix(0,n,n)
+   for(i in 1:nrow(pages)) A[pages[i,]$dist,pages[i,]$src]<-1
+   A
+}

# Convert to the probability matrix.
> probabilityMatrix<-function(G){
+   cs <- colSums(G)
+   cs[cs==0] <- 1
+   n <- nrow(G)
+   A <- matrix(0,nrow(G),ncol(G))
+   for (i in 1:n) A[i,] <- A[i,] + G[i,]/cs
+   A
+}

# Calculate the eigenvalues of matrix iteratively.
> eigenMatrix<-function(G,iter=100){
+   iter<-10
+   n<-nrow(G)
+   x <- rep(1,n)
+   for (i in 1:iter) x <- G %*% x
+   x/sum(x)
+}

# Run the program.
# Read the data file into memory.
> pages<-read.table(file="page.csv",header=FALSE,sep=",")
# Set the header of data.
> names(pages)<-c("src","dist");pages
  src dist
1   1    2
2   1    3
3   1    4
4   2    3
5   2    4
6   3    4
7   4    2

# Construct the adjacent matrix.
> A<-adjacencyMatrix(pages);A
     [,1] [,2] [,3] [,4]
[1,]    0    0    0    0
[2,]    1    0    0    1
[3,]    1    1    0    0
[4,]    1    1    1    0

# The probability matrix.
> G<-probabilityMatrix(A);G
```

```
          [,1] [,2] [,3] [,4]
[1,]  0.0000000  0.0    0    0
[2,]  0.3333333  0.0    0    1
[3,]  0.3333333  0.5    0    0
[4,]  0.3333333  0.5    1    0

# The PageRank value.
> q<-eigenMatrix(G,100);q
          [,1]
[1,]  0.0000000
[2,]  0.4036458
[3,]  0.1979167
[4,]  0.3984375
```

The result is explained as follows:

- For the page with ID = 1, the PR value is 0, since no pages refer to the page with ID1 = 1.
- For the page with ID = 2, the PR value is 0.4, since both pages 1 and 4 refer to page 2, and the weight of page 4 is higher and there is only one link on page 4 that refers to page 2, so there is no loss for weight transferring.
- For the page with ID = 3, the PR value is 0.19. Although both pages 1 and 2 refer to page 3, they also refer to other pages, the weight is divided, so the PR value for page with ID = 3 is not high.
- For the page with ID = 4, the PR value is 0.39, which is very high. This is because the page is referred by all other pages 1, 2, and 3.

For the above result we can see that for the page with ID = 1, the PR value is 0. Therefore, the page with ID = 1 cannot output the weight to other pages, which is unreasonable. So we need to add d the damping coefficient to correct the pages in order to make sure the minimum PR value >0.

2.2.3.2 Scenario with Consideration of the Damping Coefficient

Add a function dProbabilityMatrix:

```
# Convert to probability matrix with consideration of the damping coefficient d.
> dProbabilityMatrix<-function(G,d=0.85){
+   cs <- colSums(G)
+   cs[cs==0] <- 1
+   n <- nrow(G)
+   delta <- (1-d)/n
+   A <- matrix(delta,nrow(G),ncol(G))
+   for (i in 1:n) A[i,] <- A[i,] + d*G[i,]/cs
+   A
+}

# Run the program.
# Read the data file to memory.
> pages<-read.table(file="page.csv",header=FALSE,sep=",")
# Set the head of data.
> names(pages)<-c("src","dist")
```

```
# Construct the adjacent matrix.
> A<-adjacencyMatrix(pages);A
     [,1] [,2] [,3] [,4]
[1,]    0    0    0    0
[2,]    1    0    0    1
[3,]    1    1    0    0
[4,]    1    1    1    0
```

```
# The probability transition matrix, with consideration of the damping
  coefficient d.
> G<-dProbabilityMatrix(A);G
           [,1]      [,2]     [,3]     [,4]
[1,]  0.0375000 0.0375 0.0375 0.0375
[2,]  0.3208333 0.0375 0.0375 0.8875
[3,]  0.3208333 0.4625 0.0375 0.0375
[4,]  0.3208333 0.4625 0.8875 0.0375
```

```
# The PageRank value.
> q<-eigenMatrix(G,100);q
           [,1]
[1,]  0.0375000
[2,]  0.3738930
[3,]  0.2063759
[4,]  0.3822311
```

After adding the damping coefficient d, the page with ID = 1 has PR(1)=(1−d)/n=(1−0.85)/4 = 0.0375, meaning the minimum value of pages without out-links.

2.2.3.3 Calculation Directly Using the Matrix Eigenvalue Function in R

In the above implementations, we use the iteration method to calculate the eigenvalues of matrix through loops. In fact, R has provided functions that directly calculate the eigenvalues of matrix. We can call the function directly to reduce the complexity of code.

Add a function that calculates the eigenvalues.

```
# Directly calculate the eigenvalues of matrix.
> calcEigenMatrix<-function(G){
+     x <- Re(eigen(G)$vectors[,1])
+     x/sum(x)
+ }
```

```
#Run the program.
# Read the data file to memory.
> pages<-read.table(file="page.csv",header=FALSE,sep=",")
# Set the header of data.
> names(pages)<-c("src","dist")
# Construct the adjacent matrix.
> A<-adjacencyMatrix(pages);A
     [,1] [,2] [,3] [,4]
[1,]    0    0    0    0
[2,]    1    0    0    1
[3,]    1    1    0    0
[4,]    1    1    1    0
```

```
# The probability matrix.
> G<-dProbabilityMatrix(A);G
          [,1]    [,2]   [,3]   [,4]
[1,]  0.0375000 0.0375 0.0375 0.0375
[2,]  0.3208333 0.0375 0.0375 0.8875
[3,]  0.3208333 0.4625 0.0375 0.0375
[4,]  0.3208333 0.4625 0.8875 0.0375

# Calculate the PageRank values through the eigenvalue function.
# The PageRank value.
> q<-calcEigenMatrix(G);q
[1] 0.0375000 0.3732476 0.2067552 0.3824972
```

Directly calculating the eigenvalues of matrix can reduce the loop operations effectively and improve the performance of program.

2.2.4 R Implementation of the Distributed Algorithm

Most of the scenarios involving PageRank are based on massive data, where the programs usually employ the Hadoop solutions, which calculate every page's scores through distributed parallel algorithms. We can build the prototype of the distributed algorithm with R language and make sure the result of the distributed algorithm is the same as what the stand-alone algorithm yields. And then we extract the calculation formula of the distributed algorithm, which is then rewritten using Hadoop's MapReduce to meet the requirements of the product environment to go live.

2.2.4.1 The Principle of PageRank's Distributed Algorithm

Simply speaking, the principle of PageRank's distributed algorithm is to perform a parallel calculation through matrix computation. Let us start from the adjacent matrix. Hadoop needs to store the columns of the matrix as rows, while R can represent the matrix directly using matrix object.

```
          [,1]    [,2]   [,3]   [,4]
[1,]  0.0375000 0.0375 0.0375 0.0375
[2,]  0.3208333 0.0375 0.0375 0.8875
[3,]  0.3208333 0.4625 0.0375 0.0375
[4,]  0.3208333 0.4625 0.8875 0.0375
```

The next step is to calculate the matrix eigenvalues iteratively. According to the design idea of MapReduce, we separate the R algorithm as map process and reduce process. The process of map is mainly used to data separation. The process of reduce is mainly used to data calculation. We iterate the map and reduce to calculate the matrix eigenvalues. The MapReduce calculation of Hadoop is shown as Figure 2.3.

The input of the map process of Hadoop is the adjacent matrix and PR value. In the output, key is the row number of PR and value is the multiplication-sum formula of the adjacent matrix and PR value. For the reduce process of Hadoop, in the input key is the row number of PR and value is the multiplication-sum formula of the adjacent matrix and PR value; in the output, k is the row number of PR and value is the calculation result, that is, the value of PR.

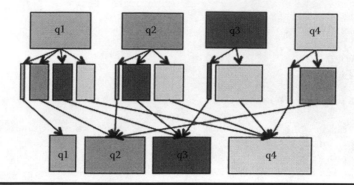

Figure 2.3 The MapReduce calculation of PageRank.

2.2.4.2 R Program Simulation

Let us implement the program in R and separate the matrix calculation in two processes: map and reduce. In the process of map, we split the data into blocks following a certain rule to simulate the distribution of massive data on multiple storage nodes. The data set used in this section contains only four pages, so we split data in two groups, as to store them on two Hadoop nodes.

```
# The function of map process.
> map <- function(S0, node = "a") {
+     S <- apply(S0, 2, function(x) x/sum(x))
+     if (node == "a")
+         S[, 1:2] else S[, 3:4]
+ }
```

In the process of reduce, the data from multiply map nodes are aggregated into the reduce nodes for the combination calculation of data. In the R program, we complete the iterative calculation directly inside the function reduce(). However for completely distributed MapReduce, each iteration should contain the complete map process and reduce process.

```
# The function of reduce process.
> reduce <- function(A, B, a = 0.85, niter = 100) {
+     n <- nrow(A)
+     q <- rep(1, n)
+     Ga <- a * A + (1 - a)/n * (A[A != 1] = 1)
+     Gb <- a * B + (1 - a)/n * (B[B != 1] = 1)
+     # Calculate iteratively.
+     for (i in 1:niter) {
+         qa <- as.matrix(q[1:ncol(A)])
+         qb <- as.matrix(q[(ncol(A) + 1):n])
+         q <- Ga %*% qa + Gb %*% qb
+     }
+     # Standardize the PR value.
+     as.vector(q/sum(q))
+ }
```

Run the program and iterate for 100 times to calculate the matrix eigenvalues (the PR values). Check the result and we can see that it is the same as that of the stand-alone calculation.

```
# The initial data matrix.
> S0 <- t(matrix(c(0, 0, 0, 0, 1, 0, 0, 1, 1, 1, 0, 0, 1, 1, 1, 0),nrow = 4))
# Matrix data separation.
> A <- map(S0, "a")
> B <- map(S0, "b")
# Matrix data calculation.
> reduce(A, B)
[1] 0.0375000 0.3732476 0.2067552 0.3824972
```

2.2.4.3 The Matrix Calculation Process

We can print each iteration of calculating the PR values to clearly view the result of each iteration. Make small modifications to the code of the function reduce(), adding output of original and standardized PR values.

```
> reduce <- function(A, B, a = 0.85, niter = 100) {
+     n <- nrow(A)
+     q <- rep(1, n)
+     Ga <- a * A + (1 - a)/n * (A[A != 1] = 1)
+     Gb <- a * B + (1 - a)/n * (B[B != 1] = 1)
+     for (i in 1:niter) {
+         qa <- as.matrix(q[1:ncol(A)])
+         qb <- as.matrix(q[(ncol(A) + 1):n])
+         q <- Ga %*% qa + Gb %*% qb
+     }
+     # The original PR values.
+     print(q)
+     # The standardized PR values.
+     print(as.vector(q/sum(q)))
+ }
```

The calculation result of the first iteration.

```
# The original PR values.
> reduce(A, B, niter=1)
          [,1]
[1,] 0.1500000
[2,] 1.2833333
[3,] 0.8583333
[4,] 1.7083333
# The standardized PR values.
[1] 0.0375000 0.3208333 0.2145833 0.4270833
```

Following is the matrix calculation formula for the original PR values in the first iteration.

```
0.0375000 0.0375 0.0375 0.0375        1      0.150000
0.3208333 0.0375 0.0375 0.8875    *   1  =   1.283333
0.3208333 0.4625 0.0375 0.0375        1      0.858333
0.3208333 0.4625 0.8875 0.0375        1      1.708333
```

The calculation result of the second iteration.

```
> reduce(A, B, niter=2)
          [,1]
[1,] 0.1500000
[2,] 1.6445833
```

```
[3,]  0.7379167
[4,]  1.4675000
[1]  0.0375000 0.4111458 0.1844792 0.3668750
```

Following is the matrix calculation formula for the original PR values in the second iteration.

```
0.0375000 0.0375 0.0375 0.0375       0.150000       0.150000
0.3208333 0.0375 0.0375 0.8875   *   1.283333   =   1.6445833
0.3208333 0.4625 0.0375 0.0375       0.858333       0.7379167
0.3208333 0.4625 0.8875 0.0375       1.708333       1.4675000
```

The calculation result of the 10th iteration.

```
> reduce(A, B, niter=10)
          [,1]
[1,]  0.1500000
[2,]  1.4955721
[3,]  0.8255034
[4,]  1.5289245
[1]  0.0375000 0.3738930 0.2063759 0.3822311
```

Following is the matrix calculation formula for the standardized PR values in the 10th iteration.

```
0.150000                                                    0.0375000
1.4955721   / (0.15+1.4955721+0.8255034+1.5289245)  =       0.3738930
0.8255034                                                   0.2063759
1.5289245                                                   0.3822311
```

It is very easy to construct the PageRank model in R after understanding the principle of PageRank. In practical applications, we are willing to first model and validate the algorithm and then implement the enterprise applications using other lower level programming languages.

There are some advanced contents in my blog introducing how to implement the PageRank model using distributed MapReduce algorithm. Please reference the article, the parallel implementation of the PageRank algorithm, http://blog.fens.me/algorithm-pagerank-mapreduce/

2.3 Surfing the Stock Market with MA Model

Question

How do we code financial models in R?

Introduction

MA is one of the most frequently used technical analysis tools. It is used to find effective trading signals in fluctuating phases during big trends. MA is simple and effective to guide market operations magically. According to financial staff, the MA model runs over most subjective strategies and is the essential tool for investments in stocks and futures. This section will study in depth how the MA model plays its roles in the stock market.

2.3.1 Moving Average

MAs are a technical analysis tool theoretically based on the concept of "average cost" from Dow Johns. According to the principle of "moving average" in statistics, the model chains all the averages of stock prices in periods to form a curve to show the historical fluctuation of stock prices, which in turn reflects the trends of stock indices. The MA algorithm is to calculate the arithmetic average of closing prices of continuous days, where number of days is the parameter of MA algorithm.

The formula is MA = $(C_1 + C_2 + C_3 + C_4 + C_5 + \cdots + C_n)/n$, where C is closing price and n the period of moving average. For example, the calculation of 5-day MA is:

```
MA5 = (closing price of the 4th day before today + closing price of the
3rd day before today + closing price of the day before yesterday + closing
price of yesterday + closing price of the day before today)/5
```

According to the period involved in the formula, MA lines can be divided into three kinds: short-term, mid-term, and long-term. Short-term MA takes 5 or 10 days as calculation period, mid-term 30 or 60 days, and long-term 100 or 200 days.

Depending on the processing methods, MA can be divided into three categories:

- Simple moving average (SMA), also called arithmetic MA, means to simply average the closing prices in certain periods. The usually mentioned MA is indeed SMA. The algorithm model introduced in this section is SMA as well.
- Weighted moving average (WMA) performs weighted calculation in terms of time. The closer the time, the larger the price weight is. Based on the number of days, the calculation elevates the weight of every previous date. Each price is multiplied by a weight. The latest price has the largest weight. The previous prices have weights decreasing with days backward. WMA is an improvement to MA.
- Exponential moving average (EMA) is an MA that assigns weights decreasing exponentially. The weighted influence goes smaller exponentially and with time goes back. The more recent the datum is, the heavier the weighted influence goes. But the older data are also given certain weights.

2.3.2 MA Model

In a daily candlestick chart, except for the standard k-line, there are generally also four lines, that is, white, yellow, purple, and green lines representing 5-day, 10-day, 20-day, and 60-day MAs, respectively. The crosses of the four lines and k-line form different MA models. Let us take the daily candlestick chart of LeTV300104, for example. The stock trading data from August 2012 to July 2014 are shown in Figure 2.4.

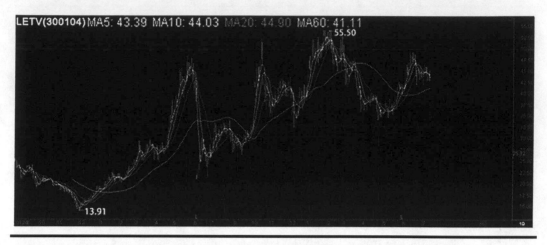

Figure 2.4 The daily candlestick chart of LeTV stock.

Figure 2.4 shows us the lowest price of LeTV's stock is 13.91 RMB, appearing in December 2012. The highest is 55.50 in January 2014. During the period, the stock price raised with fluctuations all the way. The green line with smallest ripples is 60-day MA smoothing price line, showing obvious trends.

Considering the characteristics of MA smoothing, we can find that MA lines cross with the price k-line, and that crosses also exist between MA lines. We can judge the trading signals through the crossing points.

■ Golden cross is the cross where 10-day MA goes through 30-day MA from blow, with 10-day MA over 30-day MA. The golden cross means the long position. When the golden cross occurs, the market will surely have space for price rising. It is the best time to buy.
■ Death cross is the cross where the 30-day MA goes through the 10-day MA from blow, with 30-day MA over 10-day MA. The death cross means the signal is short where the market will fall. It is the best time to sell.

Considerable profit would be obtained if we could apply the theory of MA well and get the real market trends.

However, the theory has the following constraints:

■ MA is generated after the stock price gets consolidated so it is a slow reflection to market and therefore just fit to inter-day trading.
■ MA cannot reflect the changes of stock price affected by trading volume, so there would be certain error in day trading.
■ MA is a trending model which is not suitable where the stock price has no trends and just fluctuates frequently.

2.3.3 Implementing MA Model in R

In the following, let's implement an instance of MA model by operating stock data using R.

2.3.3.1 Download Data from the Internet

By itself, R provided rich financial function toolkits including quantmod. But quantmod needs to be used together with zoo (a time-series package), xts (an extensible time-series package), TTR (an indicator calculation package), and ggplot2 (a data visualization package). For the detailed usage description about zoo and xts, please refer to Sections 2.1 and 2.2 of *R Programmers: Mastering the Tools*.

First of all let us use quantmod to download the stock data and save the data to local disk.

```
# Load the packages.
> library(plyr)
> library(quantmod)
> library(TTR)
> library(ggplot2)
> library(scales)

# Download data and save locally.
> download<-function(stock,from="2010-01-01"){
+ # Download data.
+ df<-getSymbols(stock,from=from,env=environment(),auto.assign=FALSE)
+ names(df)<-c("Open","High","Low","Close","Volume","Adjusted")
+ # Save to local file.
+ write.zoo(df,file=paste(stock,".csv",sep=""),sep=",",quote=FALSE)
+}
# Read data from local file.
> read<-function(stock){
+   as.xts(read.zoo(file=paste(stock,".csv",sep=""),header = TRUE,sep=",",
format="%Y-%m-%d"))
+}

# Download the stock market data of IBM.
> stock<-"IBM"
> download(stock,from='2010-01-01')
# Load data to memory.
> IBM<-read(stock)

# View the data types.
> class(IBM)
[1] "xts" "zoo"

# View the top 6 rows of data.
> head(IBM)
             Open    High     Low   Close   Volume  Adjusted
2010-01-04 131.18  132.97  130.85  132.45  6155300    121.91
2010-01-05 131.68  131.85  130.10  130.85  6841400    120.44
2010-01-06 130.68  131.49  129.81  130.00  5605300    119.66
2010-01-07 129.87  130.25  128.91  129.55  5840600    119.24
2010-01-08 129.07  130.92  129.05  130.85  4197200    120.44
2010-01-11 131.06  131.06  128.67  129.48  5730400    119.18
```

The function getSymbols() in package quantmod downloads data through the open APIs of Yahoo finance. We choose IBM's stock market data, the inter-day trading data for more than

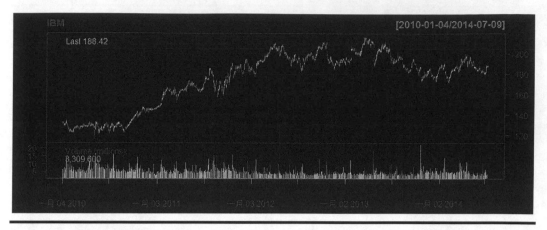

Figure 2.5 A candlestick chart.

four years from 2010-01-01 to 2014-07-09, with data type as the time-series type in xts. The data contain seven columns, indexed by date column. The other six columns are open, high, low, close, volume, and adjusted.

2.3.3.2 Implementing the Simple Candlestick Chart

We can draw a good-looking candlestick chart directly using the function chartSeries() in package quantmod. A simple candlestick chart is shown in Figure 2.5.

```
> chartSeries(IBM)
```

It is convenient to add some technical indicators on the candlestick chart, by directly passing the indicator function to chartSeries() as parameters. Figure 2.6 shows a candlestick chart with indicators such as SMA, MACD, ROC, and so on.

```
# Draw a candlestick chart with indicators.
> chartSeries(IBM,TA = "addVo(); addSMA(); addEnvelope();addMACD();
addROC()")
```

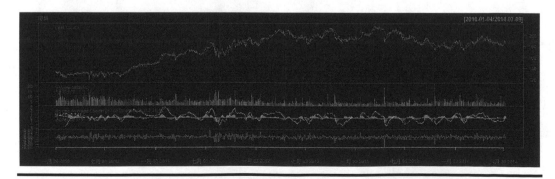

Figure 2.6 A candlestick chart with indicators.

As we can see, the visualization of stock data can be easily implemented by the two simple functions. Of course, the functionality is a general function that is already encapsulated. If we want to customize a strategy model, for example, a customized SVM classifier, we need to write code by ourselves. This section does not involve models that are too complex. It just implements the MA models.

2.3.3.3 Customized MA Chart

By customization, we can play freely without the constraints of package quantmod. We need to customize some MA indicators:

- Set the date time series as index.
- Set the close price as the price indicator.
- Ignore the volume and the other dimensional columns.
- Retrieve the market data from 2010-01-01 to 2012-01-01.
- Draw the price curve and 5-day MA, 20-day MA, and 60-day MA.

The following R code implements the above ideas.

```
# The calculation function for moving average.
> ma<-function(cdata,mas=c(5,20,60)){
+     ldata<-cdata
+     for(m in mas){
+         ldata<-merge(ldata,SMA(cdata,m))
+     }
+     ldata<-na.locf(ldata, fromLast=TRUE)
+     names(ldata)<-c('Value',paste('ma',mas,sep=''))
+     return(ldata)
+ }

# Draw MA chart.
>drawLine<-function(ldata,title="Stock_MA",sDate=min(index(ldata)),eDate=
max(index(ldata)),out=FALSE){
+     g<-ggplot(aes(x=Index, y=Value),data=fortify(ldata[,1],melt=TRUE))
+     g<-g+geom_line()
+     g<-g+geom_line(aes(colour=Series),data=fortify(ld
ata[,-1],melt=TRUE))
+     g<-g+scale_x_date(labels=date_format("%Y-%m"),breaks=date_breaks("2
months"),limits = c(sDate,eDate))
+     g<-g+xlab("") + ylab("Price")+ggtitle(title)
+
+     if(out) ggsave(g,file=paste(title,".png",sep=""))
+     else g
+ }
# Run the application.
# Retrieve the close price.
> cdata<-IBM['2010/2012']$Close

# The chart title.
> title<-"Stock_IBM"
```

```
# The starting date.
> sDate<-as.Date("2010-1-1")
# The ending date.
> eDate<-as.Date("2012-1-1")
# Choose moving average indicators.
> ldata<-ma(cdata,c(5,20,60))
# Draw the chart, shown as in Fig. 2.7
> drawLine(ldata,title,sDate,eDate)
```

Using the customized MA function and data visualization function, we realized the visualization output combining the daily candlestick chart and multiple MA lines, which is similar to commonly used trading software.

2.3.3.4 The Trading Strategy with One MA

On the basis of the MA function, we are able to design our own trading strategy model. The idea of model design is as follows:

1. Make judgment for trading signals based on the crosses of price line and 20-day MA.
2. Buy when price line goes up through 20-day MA and sell when price line goes down through 20-day MA.

Draw the lines of stock price and 20-day MA, as shown in Figure 2.8.

```
# Choose the moving average indicator.
> ldata<-ma(cdata,c(20))
# Draw the chart.
> drawLine(ldata,title,sDate,eDate)
```

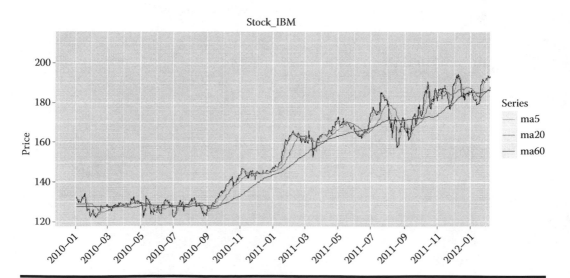

Figure 2.7   An MA chart.

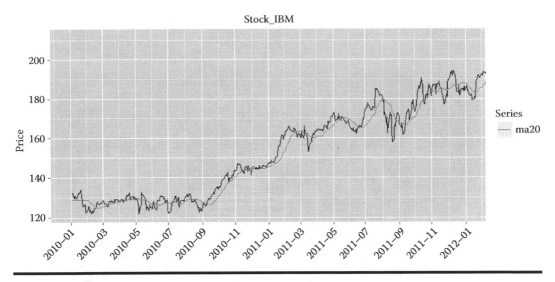

Figure 2.8   20-day MA chart.

Draw scatters covering 20-day MA, where the red points mean to buy and hold while the blue ones mean to sell and short, as shown in Figure 2.9.

```
# MA chart + scatters.
> drawPoint<-function(ldata,pdata,titie,sDate,eDate){
+   g<-ggplot(aes(x=Index, y=Value),data=fortify(ldata[,1],melt=TRUE))
+   g<-g+geom_line()
+   g<-g+geom_line(aes(colour=Series),data=fortify(ldata[,-1],melt=TRUE))
+   g<-g+geom_point(aes(x=Index,y=Value,colour=Series),data=fortify(pdata
    ,melt=TRUE))
```

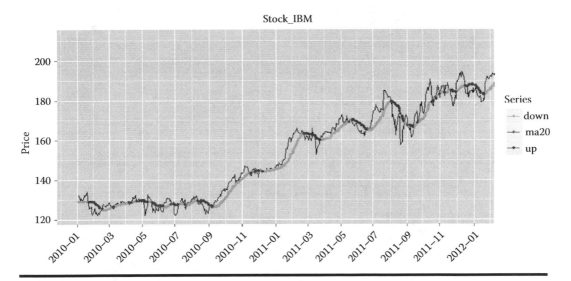

Figure 2.9   The trading signal chart with 20-day MA.

```
+    g<-g+scale_x_date(labels=date_format("%Y-%m"),breaks=date_breaks("2
     months"),limits = c(sDate,eDate))
+    g<-g+xlab("") + ylab("Price")+ggtitle(title)
+    g
+ }

# The scattering data.
> pdata<-merge(ldata$ma20[which(ldata$Value-ldata$ma20>0)],ldata$ma20[whi
ch(ldata$Value-ldata$ma20<0)])
> names(pdata)<-c("down","up")
> pdata<-fortify(pdata,melt=TRUE)
> pdata<-pdata[-which(is.na(pdata$Value)),]

> head(pdata)
       Index Series    Value
1 2010-01-04   down 128.7955
2 2010-01-05   down 128.7955
3 2010-01-06   down 128.7955
4 2010-01-07   down 128.7955
5 2010-01-08   down 128.7955
6 2010-01-11   down 128.7955

# Draw the chart as shown in Fig. 2-9
> drawPoint(ldata,pdata,title,sDate,eDate)
```

Compare the stock price with 20-day MA price, assigning blue to the points where the stock price > the 20-day MA price and red to the points where the stock price < the 20-day MA price. From the figure we can see that the blue and red points appear alternatively. We can buy the stock where the first red point appears and sell the stock where the blue point appears in each period. Thus the trading signals are constituted. It looks intuitively good.

We need to find these trading signal points and perform quantitative statistics to check whether the strategy is able to make money and how much.

```
# The trading signal.
> Signal<-function(cdata,pdata){
+    tmp<-''
+    tdata<-ddply(pdata[order(pdata$Index),],.(Index,Series),function(row){
+        if(row$Series==tmp) return(NULL)
+        tmp<<-row$Series
+    })
+    tdata<-data.frame(cdata[tdata$Index],op=ifelse(tdata$Series=='down'
     ,'B','S'))
+    names(tdata)<-c("Value","op")
+    return(tdata)
+ }

> tdata<-Signal(cdata,pdata)
> tdata<-tdata[which(as.Date(row.names(tdata))<eDate),]
> head(tdata)
            Value op
2010-01-04 132.45  B
2010-01-22 125.50  S
2010-02-17 126.33  B
```

```
2010-03-09 125.55  S
2010-03-11 127.60  B
2010-04-08 127.61  S

# The row number of trading records.
> nrow(tdata)
[1] 72
```

The statistics show that there are 72 trading records; buying and selling is half-by-half.

In the following let's do trading simulation with the trading signal data. Specify the trading parameters with capital of $100,000, full position, and fee of 0.

```
# The trading simulation.
# Trading signal, capital, position ratio and fee ratio.
> trade<-function(tdata,capital=100000,position=1,fee=0.00003){
+       # The amount of stock.
+       amount<-0
+       # Cash.
+       cash<-capital
+
+       ticks<-data.frame()
+       for(i in 1:nrow(tdata)){
+           row<-tdata[i,]
+           if(row$op=='B'){
+               amount<-floor(cash/row$Value)
+               cash<-cash-amount*row$Value
+           }
+
+           if(row$op=='S'){
+               cash<-cash+amount*row$Value
+               amount<-0
+           }
+
+           # Cash.
+           row$cash<-cash
+           # Stock amount held.
+           row$amount<-amount
+           # The asset total amount.
+           row$asset<-cash+amount*row$Value
+           ticks<-rbind(ticks,row)
+       }
+
+       # The difference of the asset total amount.
+       ticks$diff<-c(0,diff(ticks$asset))
+       # The operations making money.
+       rise<-ticks[c(which(ticks$diff>0)-1,which(ticks$diff>0)),]
+       rise<-rise[order(row.names(rise)),]
+
+       # The operations losing money.
+       fall<-ticks[c(which(ticks$diff<0)-1,which(ticks$diff<0)),]
+       fall<-fall[order(row.names(fall)),]
+
+       return(list(
```

```
+              ticks=ticks,
+              rise=rise,
+              fall=fall
+      ))
+ }

> result1<-trade(tdata,100000)

# View each tick.
> head(result1$ticks)
             Value op      cash amount     asset     diff
2010-01-04 132.45  B      0.25    755 100000.00     0.00
2010-01-22 125.50  S 94752.75      0  94752.75 -5247.25
2010-02-17 126.33  B      5.25    750  94752.75     0.00
2010-03-09 125.55  S 94167.75      0  94167.75  -585.00
2010-03-11 127.60  B    126.55    737  94167.75     0.00
2010-04-08 127.61  S 94175.12      0  94175.12     7.37

# The ticks that make profit.
> head(result1$rise)
             Value op      cash amount     asset     diff
2010-03-11 127.60  B    126.55    737 94167.75     0.00
2010-04-08 127.61  S 94175.12      0 94175.12     7.37
2010-07-22 127.47  B    108.79    633 80797.30     0.00
2010-08-12 128.30  S 81322.69      0 81322.69   525.39
2010-09-09 126.36  B    120.40    632 79979.92     0.00
2010-11-16 142.24  S 90016.08      0 90016.08 10036.16

# The ticks that make loss.
> head(result1$fall)
             Value op      cash amount     asset     diff
2010-01-04 132.45  B      0.25    755 100000.00     0.00
2010-01-22 125.50  S 94752.75      0  94752.75 -5247.25
2010-02-17 126.33  B      5.25    750  94752.75     0.00
2010-03-09 125.55  S 94167.75      0  94167.75  -585.00
2010-04-09 128.76  B     51.56    731  94175.12     0.00
2010-04-12 128.36  S 93882.72      0  93882.72  -292.40
```

By simulating the trades, we can precisely calculate the profit and loss of each tick. Can you believe that in fact 56 ticks lost money and only 16 ticks made money? Check the final result of the capital.

```
> tail(result1$ticks,1)
             Value op      cash amount     asset     diff
2011-12-21 181.47  S 96363.76      0 96363.76 -3063.87
```

The balance of the capital is $96,363.76, meaning that we lost $3,636.24. Why did we finally lose money? We should have made enough money in the middle phases with big trends. We can find the reasons by examining the capital curve. Draw the capital curve as shown in Figure 2.10.

```
# Stock price + cash flow.
> drawCash<-function(ldata,adata){
+    g<-ggplot(aes(x=Index, y=Value),data=fortify(ldata[,1],melt=TRUE))
```

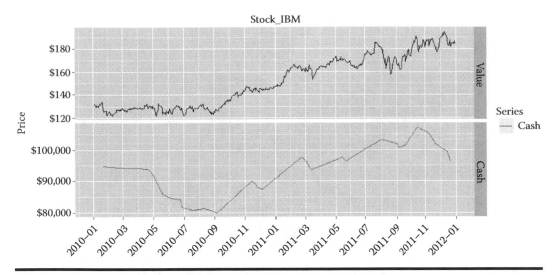

Figure 2.10   The cash curve under 20-day MA strategy.

```
+    g<-g+geom_line()
+    g<-g+geom_line(aes(x=as.Date(Index), y=Value,colour=Series),data=fort
ify(adata,melt=TRUE))
+    g<-g+facet_grid(Series ~ .,scales = "free_y")
+    g<-g+scale_y_continuous(labels = dollar)
+    g<-g+scale_x_date(labels=date_format("%Y-%m"),breaks=date_breaks("2
months"),limits = c(sDate,eDate))
+    g<-g+xlab("") + ylab("Price")+ggtitle(title)
+    g
+ }

# Cash flow.
> adata<-as.xts(result1$ticks[which(result1$ticks$op=='S'),]['cash'])
# Draw the chart.
> drawCash(ldata,adata)
```

We placed the stock price and cash flow side by side and see that the MA strategy started making money from September 2010 and went to a peak in October 2011, which was more than the capital. From then on, the cash curve went down and lost $3,859.86 till June 2012. That is because we put the profit into investment together with the capital, which enlarged the position and caused the big loss in late 2011 when the fluctuation made the model fail.

Here, we have completed the trading strategy model with one 20-day MA and done the test using IBM stock.

2.3.3.5 The Trading Strategy with Two MAs

The one-MA model can make money stably during big trends. But the mode is very sensitive to fluctuation. Frequent fluctuations can increase the times of trading and make the model invalid. Therefore, the two-MA strategy model was developed which can reduce the sensitivity to fluctuations.

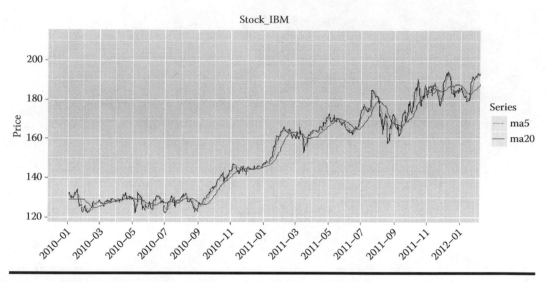

Figure 2.11   The chart of 5-day MA and 20-day MA.

Similar to the idea of one-MA strategy model, the two-MA strategy model replaces stock price with 5-day MA price and makes trading according to the crosses of 5-day MA and 20-day MA. First, let us draw the chart of stock price, 5-day MA, and 20-day MA, as shown in Figure 2.11.

```
# Choose the moving average indicators.
> ldata<-ma(cdata,c(5,20))
# Draw the chart.
> drawLine(ldata,title,sDate,eDate)
```

Draw scatters covering 20-day MA, where the red points mean to buy and hold while the purple ones mean to sell and short, as shown in Figure 2.12.

```
# The scatter data.
> pdata<-merge(ldata$ma20[which(ldata$ma5-ldata$ma20>0)],ldata$ma20[which
(ldata$ma5-ldata$ma20<0)])
> names(pdata)<-c("down","up")
> pdata<-fortify(pdata,melt=TRUE)
> pdata<-pdata[-which(is.na(pdata$Value)),]

> head(pdata)
       Index Series     Value
1 2010-01-04   down 128.7955
2 2010-01-05   down 128.7955
3 2010-01-06   down 128.7955
4 2010-01-07   down 128.7955
5 2010-01-08   down 128.7955
6 2010-01-11   down 128.7955
# Draw the chart as shown in Fig. 2-12.
> drawPoint(ldata,pdata,title,sDate,eDate)
```

Compare the 5-day MA price with the 20-day MA price, assigning purple to the points where the 5-day MA price >20-day MA price and red to the points where the 5-day MA <20-day MA

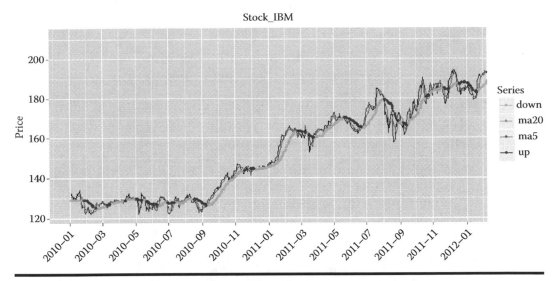

Figure 2.12   The trading signal chart with 5-day MA and 20-day MA.

price. From the figure we can see that the purple and red points appear alternatively. It is likely we can buy the stock where the first red point appears and sell the stock where the purple point appears in each period. The model looks same as the one-MA model and can make money as well.

We need to find these trading signal points and perform quantitative statistics to check whether the strategy is able to make money.

```
> tdata<-Signal(cdata,pdata)
> tdata<-tdata[which(as.Date(row.names(tdata))<eDate),]
> head(tdata)
            Value op
2010-01-04 132.45  B
2010-01-26 125.75  S
2010-02-18 127.81  B
2010-03-10 125.62  S
2010-03-16 128.67  B
2010-04-12 128.36  S

# The trading records.
> nrow(tdata)
[1] 36
```

There are 36 trading records, with half of buying and half of selling, less than the one-MA model for 36 trading records.

```
# The trading simulation.
> result2<-trade(tdata,100000)

# View each ticks.
> head(result2$ticks)
            Value op   cash amount      asset    diff
2010-01-04 132.45  B   0.25    755 100000.00    0.00
```

```
2010-01-26  125.75  S 94941.50      0  94941.50 -5058.50
2010-02-18  127.81  B    106.48    742  94941.50     0.00
2010-03-10  125.62  S 93316.52      0  93316.52 -1624.98
2010-03-16  128.67  B     30.77    725  93316.52     0.00
2010-04-12  128.36  S 93091.77      0  93091.77  -224.75
```

```
# The ticks that make profit.
> head(result2$rise)
            Value op      cash amount       asset     diff
2010-09-10  127.99  B     75.34    649  83140.85     0.00
2010-11-18  144.36  S 93764.98      0  93764.98 10624.13
2010-12-07  144.02  B      2.66    638  91887.42     0.00
2011-02-23  160.18  S 102197.50     0 102197.50 10310.08
2011-03-28  161.37  B    124.70    582  94042.04     0.00
2011-05-20  170.16  S  99157.82      0  99157.82  5115.78
```

```
# The ticks that make loss.
> head(result2$fall)
            Value op      cash amount       asset     diff
2010-01-04  132.45  B      0.25    755 100000.00     0.00
2010-01-26  125.75  S 94941.50      0  94941.50 -5058.50
2010-02-18  127.81  B    106.48    742  94941.50     0.00
2010-03-10  125.62  S 93316.52      0  93316.52 -1624.98
2010-03-16  128.67  B     30.77    725  93316.52     0.00
2010-04-12  128.36  S 93091.77      0  93091.77  -224.75
```

By simulating the trades, we can precisely calculate the profit and loss of each tick precisely. Twenty-six ticks lost money and 10 ticks made money.

Check the final result of the capital.

```
> tail(result2$ticks,1)
            Value op    cash amount   asset      diff
2011-12-19  182.89  S 96828.9      0 96828.9 -3581.33
```

We simulated trading with trading signal data. We set parameters with $100,000 as capital, position ratio as 1, and fee as 0. Finally, the balance is $96,828.9, with $3,171.1 lost. Check the capital curve as shown in Figure 2.13.

```
> adata<-as.xts(result2$ticks[which(result2$ticks$op=='S'),]['cash'])
> drawCash(ldata,adata)
```

We can find that although the capital lost $3,171.1, a little less than in the one-MA strategy mode, there are three times when cash > capital and the worst scenario is better than the worst one in the one-MA model. If we put the profit out of our account and keep the original capital trading, then we can lock the profit.

2.3.3.6 Compare the Profiting Scenarios in the Two Models

Let us furthermore compare the profiting scenarios in the two models. Find all the trades in two models that made profit.

```
# Trades that made profit.
> rise<-merge(as.xts(result1$rise[1]),as.xts(result2$rise[1]))
> names(rise)<-c("plan1","plan2")
```

```
# View the data.
> rise
            plan1  plan2
2010-03-11 127.60     NA
2010-04-08 127.61     NA
2010-07-22 127.47     NA
2010-08-12 128.30     NA
2010-09-09 126.36     NA
2010-09-10     NA 127.99
2010-11-16 142.24     NA
2010-11-18     NA 144.36
2010-12-07     NA 144.02
2010-12-08 144.98     NA
2011-02-22 161.95     NA
2011-02-23     NA 160.18
2011-03-25 162.18     NA
2011-03-28     NA 161.37
2011-05-16 168.86     NA
2011-05-20     NA 170.16
2011-06-21 166.22     NA
2011-06-23     NA 166.12
2011-08-02 178.05     NA
2011-08-04     NA 171.48
2011-09-14 167.24     NA
2011-09-16     NA 172.99
2011-09-22 168.62     NA
2011-09-23 169.34     NA
2011-10-18 178.90     NA
2011-10-21     NA 181.63
```

Plan1 is the one-MA model and plan2 the two-MA model. Plan1 has six trades more than plan2, which, however, were caused by the sensitivity to fluctuations and reduced the benefits obtained by market trends.

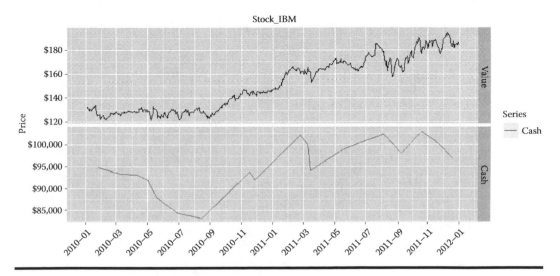

Figure 2.13   The cash curve under 5-day MA and 20-day MA strategy.

Finally, let us draw the trading ranges for the profiting parts of the two models.

```
# MA chart + trading ranges.
> drawRange<-function(ldata,plan,titie="Stock_2014",sDate=min(index(ldata
)),eDate=max(index(ldata)),out=FALSE){
+    g<-ggplot(aes(x=Index, y=Value),data=fortify(ldata[,1],melt=TRUE))
+    g<-g+geom_line()
+    g<-g+geom_line(aes(colour=Series),data=fortify(ldata[,-1],melt=TRUE))
+    g<-g+geom_rect(aes(NULL, NULL,xmin=start,xmax=end,fill=plan),ymin =
yrng[1], ymax = yrng[2],data=plan)
+    g<-g+scale_fill_manual(values =alpha(c("blue", "red"), 0.2))
+    g<-g+scale_x_date(labels=date_format("%Y-%m"),breaks=date_breaks("2
months"),limits = c(sDate,eDate))
+    g<-g+xlab("")  + ylab("Price")+ggtitle(title)
+
+    if(out) ggsave(g,file=paste(titie,".png",sep=""))
+    else g
+ }


# The profiting ranges.
> yrng <-range(ldata$Value)
> plan1<-as.xts(result1$rise[c(1,2)])
> plan1<-data.frame(start=as.Date(index(plan1)
[which(plan1$op=='B')]),end=as.Date(index(plan1)[which(plan1$op=='S')]),p
lan='plan1')
> plan2<-as.xts(result2$rise[c(1,2)])
> plan2<-data.frame(start=as.Date(index(plan2)
[which(plan2$op=='B')]),end=as.Date(index(plan2)[which(plan2$op=='S')]),p
lan='plan2')


#The profiting ranges of plan1.
> plan<-rbind(plan1)
# Draw.
> drawRange(ldata,plan,title,sDate,eDate)
```

The profiting ranges of plan1 are shown in Figure 2.14.

```
# Combine the profiting ranges of plan1 and plan2.
> plan<-rbind(plan1,plan2)
# Draw.
> drawRange(ldata,plan,title,sDate,eDate)
```

The profiting ranges existing in both plan1 and plan2 are shown in Figure 2.15.

The above profiting ranges prove the sensitivity of the one-MA model to fluctuations. The two-MA model is an optimization of the one-MA model. Thus, we have finished the development of a complete instance of MA models.

2.3.3.7 The Optimization to Models

From the perspective of trading, we cannot say the above model is completed, since there are still many trades that lose money. The model needs more improvement to reduce the maximum

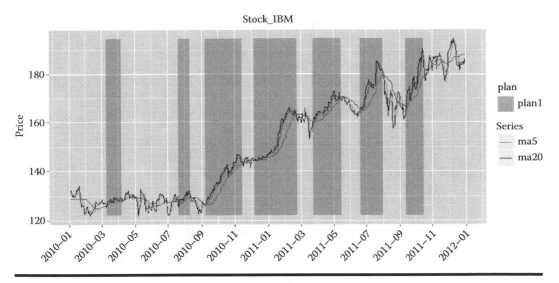

Figure 2.14 **The profiting ranges of one-MA strategy.**

drawdown, to go long at more certain opportunities, and to go short reversely. The problems about the model optimizations will be discussed in detail in *R for Programmers: Quantitative Investments*.

It seems that the MA model is so easy. But to win the two-MA (optimized) model in a trending market within practice trading is really not an easy thing. We cannot say that the two-MA model always wins but at least it wins most of the time.

The friends who have made money following the model I introduced, please treat me with a cup of tea and share your experience; while the ones who didn't need more hard work to reach the target of financial freedom someday.

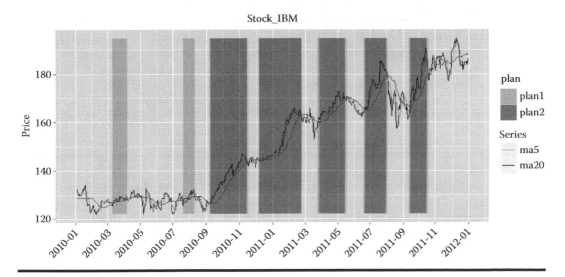

Figure 2.15 **The profiting ranges of two-MA strategy.**

2.4 Genetic Algorithms in R*

Question

How to perform the calculation of genetic algorithms using R language?

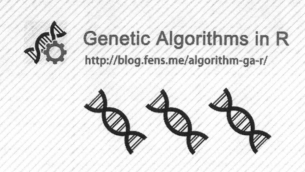

Introduction

People tend to explore the laws in their lives, summarize them as experiences, and then pass them over to descendants, who in turn find more laws. Each delivery of knowledge is a process of evolution, all of which eventually constitutes the intelligence of human beings. Humans survive according to the natural principle of "survival of the fittest." Such a biological evolution law was refined by smart scientists as genetic algorithms and expanded to various domains. This chapter leads you to the world of genetic algorithms.

2.4.1 Introducing Genetic Algorithms

A genetic algorithm is a kind of searching algorithms for solving optimization problems and belongs to the family of evolutionary algorithms. Originally, the algorithm was influenced by Darwin's evolutionism and Mendel's genetic theory, and extracted from some phenomena of biological evolution such as inheritance, mutation, natural selection and crossover, etc. It builds randomly global search and optimization methods by mimicking biological evolution in the natural world. Genetic algorithms are highly efficient, parallel and globally searching algorithms, which are able to automatically retrieve and consolidate the knowledge related to search space during the search process, and control the process of solution searching in order to produce the global optimum.

The calculation of genetic algorithms employs the principle of "survival of the fittest." It generates approximately optimized solutions from potential populations in iteration processes. From each generation, individuals are selected according to the fitness in the problem domain and by the reproduction methods which are inspired from natural genetics to produce a new approximate solution. The process results in the evolution of individuals in the population. The newly generated individuals fit the environment better than the last generation, just as in natural selection.

Let us understand genetic algorithms from the perspective of biological evolution. Certain populations with a number of individuals produce a new generation by selection and reproduction, including crossover and mutation. By the principle of "survival of the fittest," the selection

---

* Genetic algorithm on Wikipedia (https://en.wikipedia.org/wiki/Genetic_algorithm).

process filters individuals according to their fitness. However, the process is not controlled by just fitness. This is because simply selecting individuals with higher fitness may produce a population with local optima instead of the global optimum, which would not evolve and is therefore called a "premature" one. In the next step, the reproduction process generates new generation individuals by crossing pairs of individuals, which passes the good genes from parents to children and replaces the bad genes with genes from the spouse. Finally, the population also makes mutational evolution by producing a next generation with a genetic mutation with small probability.

By such a series of processes of selection, crossover, and mutation, the new individuals are different from their parent generation and evolve toward increasing the fitness of entire population. The best individuals have more opportunities to be selected to reproduce the next generation while less fit ones are discarded as time passes. Each individual is evaluated to calculate its fitness, pairs of which crossover each other, and then mutate and reproduce the next generation. This process iterates time and again until a certain termination condition is reached.

There are some limitations that need attention while working with genetic algorithms:

■ Genetic algorithms may converge toward local optima instead of the global optimum when inappropriate fitness function is employed.

■ The number of individuals in the initial population matters. If the number is too large, the algorithm would consume a lot of system resources; while if the number is too small, the algorithm may probably ignore the optimum.

■ Encoding each solution practically helps writing mutation function and fitness function.

■ In encoded genetic algorithms, the code length of each mutation affects the performance. Too large a length will limit the diversity of mutation while too small a length significantly decreases the performance of genetic algorithms. An appropriate mutation code length is the key to performance.

■ Mutation rate is an important parameter.

■ It is difficult for GAs to solving the optimum on dynamic data sets, as genomes begin to converge early on toward solutions which may no longer be valid for later data. To remedy this, several methods have been proposed by increasing genetic diversity and preventing early convergence. One is called *triggered hypermutation*, which increases the mutation probability when the genome population quality drops (difference from each other decreases). Another is called random immigrants, which increases genetic diversity by occasionally introducing entirely new, randomly generated elements into the gene pool.

■ Scientists agree on the importance of the selection process but differ on whether crossover or mutation is more important. One opinion considers crossover more important than mutation since the latter only guarantees the calculation and does not lose certain possible solutions. However, another believes that crossover is just a kind of update as a result of evacuation of the mutation across population. For population in early stages, crossover is almost equivalent to mutation with a very large rate which may affect evolution progression.

■ Genetic algorithms can find good solutions efficiently even in a very complex solution space.

■ Genetic algorithms are not always the best strategy for optimization problems, which need practical analysis. Try to use other algorithms together with or instead of the genetic algorithms.

■ Genetic algorithms cannot solve the "needle in haystack" problems, where there is no specific fitness function to represent the quality of individuals and therefore can mislead the evolution of algorithms.

■ Adjusting parameters such as number of individuals, crossover rate, and mutation rate may help to converge more efficiently and with higher quality for any specific optimization problems. For example, a mutation rate that is too high may lead to loss of the optimum while too small rates may converge to local optima in an unexpected early stage. There are no practical bounds for these parameters for now.

■ Fitness function is also important for the speed and result of genetic algorithms.

Genetic algorithms can be applied to domains including computer automatic design, production scheduling, circuit design, game design, robot learning, fuzzy control, timetabling, neural network training, etc. However, I am going to apply genetic algorithms to parameterize optimization solutions for financial domains such as the backtest system, which I will introduce in *R for Programmers: Quantitative Investment*.

2.4.2 Principle of Genetic Algorithms

In genetic algorithms, a candidate solution of an optimization problem is called an individual. It is represented as a sequence of variables, called a chromosome, or genotype. Typically, the chromosome is expressed as a simple character or number string, or in another form. The process is called encoding. First of all, we need to create a population. An algorithm usually generates a certain number of individuals randomly. Sometime the process can be interfered by manual operation to increase the quality of the population. Every individual in a generation is evaluated and given a fitness by calling fitness function. The individuals in the population are sorted by their fitness from higher to lower.

Second, generate the population of next generation individuals by processes of selection and reproduction.

Selection is performed according to the fitness of new individuals. But it does not mean that the process is directed completely by the value of fitness. This is because simply selecting individuals with higher fitness may lead the algorithm rapidly converging to local optima instead of the global optimum, which is called "premature." As a compromise, genetic algorithms comply to the rule where the higher the fitness, the higher the opportunity to be selected, and vice versa. The initial data can constitute a relatively optimized population by such a process of selection.

Reproduction means the selected individuals enter the process of crossing each other, including crossover and mutation. The crossover process corresponds to the crossover operation in algorithms. Generally, a genetic algorithm has a crossover rate, with a typical range of 0.6–1, which reflects the probability of crossover happening between the selected individuals.

For instance, a crossover rate of 0.8 means 80% of "parent" individuals would reproduce children. Every two individuals generate two new ones by crossover replacing the "old" ones, while others without crossover keep unchanged. In the process of crossover, the chromosomes of father and mother are swapped to generate two "child" chromosomes. For the first child, the starting half segment is from the father and the ending half is from the mother. The situation is reversed for the second child. Here, "half segment" does not mean actual half. The position is called the crossover point, which is calculated randomly and can be in any position on a chromosome.

The mutation process means to generate next generation individuals by mutating. A usual genetic algorithm has a fixed mutation constant, which is also called mutation probability and is 0.1 or less, representing the probability at which mutation takes place. According to the probability, the chromosomes of new individuals mutate randomly, usually by changing one bit on the chromosome.

Genetic algorithms implementation iterates the following process: each individual is evaluated to calculate its fitness, pairs of which crossover each other, and then mutate and reproduce the next generation. It stops when the terminating conditions are satisfied. Common terminating conditions are

- The constraint of evolutionary iterations.
- The constraint of computational resources, such as computing time, consumed CPU, and memory.
- The individual has satisfied the optimization criteria, that is, the best solution is found.
- The fitness has reached a plateau such that successive iterations no longer produce better results.
- Manual interference.

The idea of implementing a genetic algorithm is shown in Figure 2.16.

1. Create the initial population.
2. Loop: produce next generation.
3. Evaluate individual fitness in population.
4. Define fitness function for selection.
5. Alter the population (crossover and mutation).
6. Go back to step 2.
7. Terminate when terminating conditions are satisfied.

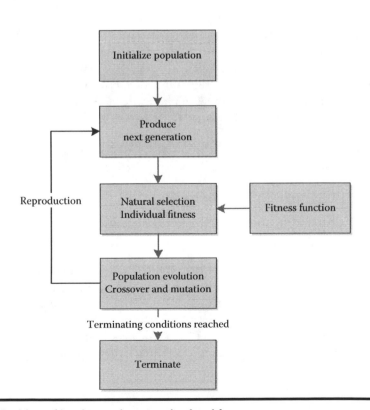

Figure 2.16   The idea of implementing genetic algorithm.

2.4.3 Genetic Algorithms in R

System environment used in this section:

- Win7 64bit
- R: 3.1.1x86_64-w64-mingw32/x64 (64-bit)

A typical genetic algorithm requires a solution domain represented by genes and a fitness function to evaluate candidate solutions.

The genetic algorithm parameters are usually as follows:

- Population size, the number of chromosomes in the population.
- Size of genes in the chromosome, the number of variables.
- The probability of crossover, used to control the probability of using crossover calculation. Crossover is able to speed up the convergence to make the solution reach the optimization area. Thus, the value is assigned higher. But very large values may lead to premature convergence.
- The probability of mutation, used to control the probability of using mutation calculation, determining the ability to perform local search of the genetic algorithm.
- Terminating condition, the signal of terminating the iterations.

Some third-part packages in R have implemented genetic algorithms. We can use them directly.

- Package mcga, a multivariable genetic algorithm for solving the minimum of multidimensional functions. It is highly efficient for large-scale search spaces.
- Package genalg, a multivariable genetic algorithm for solving the minimum of multidimensional functions, providing the functionality of data visualization.
- Package rgenoud, a complex genetic algorithm combining a genetic algorithm and quasi-Newton algorithm, able to solve complex optimization problems.
- Package gafit, for solving the minimum of one-dimensional functions. Does not support R 3.1.1.
- Package GALGO, for solving the optimization problem of one-dimensional functions. Does not support R 3.1.1.

Mcga and genalg packages are introduced in this section.

2.4.3.1 Package mcga

We can implement the multivariable genetic algorithm with mcga() function in package mcga.

Package mcga is a toolkit of genetic algorithms, mainly for solving real-valued optimization problems. It uses variables rather than binaries to represent gene sequence and thus there is no need for the encoding–decoding process. Mcag has implemented the operations of crossover and mutation of genetic algorithms and can perform the calculation of search spaces with large-scale and high precision. Using a 256-unary alphabet is the main disadvantage of this algorithm.

First, install the package mcga.

```
> install.packages("mcga")
> library(mcga)
```

View the definition of function mcga().

```
> mcga
function (popsize, chsize, crossprob = 1, mutateprob = 0.01, elitism = 1,
minval, maxval, maxiter = 10, evalFunc)
```

- popsize, number of individuals or chromosomes
- chsize, number of genes, or parameters
- cross prob, crossover probability. By default it is 1.0
- mutateprob, mutation probability. By default it is 0.01
- elitsm, number of elites. Number of chromosomes to be copied directly into next generation. By default it is 1
- minval, the lower bound of the randomized initial population
- maxval, the upper bound of the randomized initial population
- maxiter, the maximum number of generations. By default it is 10
- evalFunc, fitness function for evaluating individuals

Next, let us give an optimization problem and solve the best solution using mcga().

Problem 1: let $f(x) = (x_1 - 5)^2 + (x_2 - 55)^2 + (x_3 - 555)^2 + (x_4 - 5555)^2 + (x_5 - 55555)^2$. Calculate the minimum of $f(x)$, where x_1, x_2, x_3, x_4, x_5 are five different variables.

Obviously, for solving the minimum of $f(x)$, actually when $x_1 = 5$, $x_2 = 55$, $x_3 = 555$, $x_4 = 5555$, $x_5 = 55555$, $f(x) = 0$ is the minimum. If we use the method of exhaustion, it would be very costly to find the values of five variables using loops, so I won't do the test. Let us take a look at the execution of the genetic algorithm.

```
# Define the fitness function.
> f<-function(x) {
  return ((x[1]-5)^2 + (x[2]-55)^2 +(x[3]-555)^2 +(x[4]-5555)^2
+(x[5]-55555)^2)
}
```

```
# Execute the genetic algorithm.
> m <- mcga( popsize=200,
+            chsize=5,
+            minval=0.0,
+            maxval=999999,
+            maxiter=2500,
+            crossprob=1.0,
+            mutateprob=0.01,
+            evalFunc=f)
```

```
# The optimized individual.
> print(m$population[1,])
[1]   5.000317    54.997099   554.999873  5555.003120 55554.218695
```

```
# Execution time.
> m$costs[1]
[1] 3.6104556
```

Here, we have got the optimized result: $x_1 = 5.000317$, $x_2 = 54.997099$, $x_3 = 554.999873$, $x_4 = 5555.003120$, $x_5 = 55554.218695$, which is very approximate to the expected result and takes only 3.6 s. The result is very satisfactory. However if we use the method of exhaustion, the time complexity is $O(n^5)$ and the result would not be retrieved within five minutes.

Of course, the execution time and precision of the algorithm can be configured with arguments. If we enlarge the number of individuals or iterations, on the one hand the execution time would be increased and on the other hand the results would be more precise. Therefore, we need to adjust these arguments using our experience when using the algorithm in action.

2.4.3.2 Package Genalg

The function rbga() in package genalg can also be used to implement a multivariable genetic algorithm.

Package genalg not only implements the genetic algorithm, but also provides data visualization functionalities for a genetic algorithm, allowing for users to understand the algorithm more visually. The package generates the default graph including optimized evaluation values for each iteration, which represent the calculation progression of the genetic algorithm. The histogram shows the frequency of gene choice, that is, the number of genes being selected in current individuals. The variable graph represents the evaluation function and variable values, which allows for checking the relationship between evaluation function and variables.

First, install the package genalg.

```
> install.packages("genalg")
> library(genalg)
View the definition of fuction rbga().
> rbga(stringMin=c(), stringMax=c(),
    suggestions=NULL,
    popSize=200, iters=100,
    mutationChance=NA,
    elitism=NA,
    monitorFunc=NULL, evalFunc=NULL,
    showSettings=FALSE, verbose=FALSE)
```

Arguments:

- stringMin, vector with minimum values for each gene.
- stringMax, vector with maximum values for each gene.
- suggestions, optional list of suggested chromosomes.
- popSize, the population size, or numbers of chromosomes. By default it is 200.
- iters, the number of iterations. By default it is 100.
- mutationChance, the chance that a gene in the chromosome mutates. By default 1/(size+1). It affects the convergence rate and the probing of search space: a low chance results in quicker convergence, while a high chance increases the span of the search space.
- elitism, the number of chromosomes that are kept into the next generation. By default this is about 20% of the population size.
- monitor, method run after each generation to allow monitoring of the optimization.
- evalFunc, fitness function for evaluating individuals.
- showSettings, print settings. By default it is False.
- verbose, print algorithm's execution log. By default it is False.

Next, let us give an optimization problem and solve the best solution using rbga().

Problem 2: let $f(x) = \left| x_1 - \sqrt{e} \right| + \left| x_2 - \ln \pi \right|$. Calculate the minimum of $f(x)$, where x_1, x_2 are two different variables.

Obviously, for solving the minimum of $f(x)$, actually when $x_1 = \sqrt{e} = 1{,}648721$, $x_2 = \ln \pi = 1.14473$, $f(x) = 0$ is the minimum. If we use the method of exhaustion, it would be very costly to find the values of five variables using loops, so I won't do the test. Let us take a look at the execution of the genetic algorithm rbga().

```
# Define fitness function.
> f<-function(x){
   return (abs(x[1]-sqrt(exp(1)))+abs(x[2]-log(pi)))
}

# Define monitor fuction.
> monitor <- function(obj){
   xlim = c(obj$stringMin[1], obj$stringMax[1]);
   ylim = c(obj$stringMin[2], obj$stringMax[2]);
   plot(obj$population, xlim=xlim, ylim=ylim, xlab="sqrt(exp(1))",
ylab="log(pi)");
}

# Execute the genetic algorithm.
> m2 = rbga(c(1,1),
+           c(3,3),
+           popSize=100,
+           iters=1000,
+           evalFunc=f,
+           mutationChance=0.01,
+           verbose=TRUE,
+           monitorFunc=monitor
+           )

Testing the sanity of parameters…
Not showing GA settings...
Starting with random values in the given domains...
Starting iteration 1
Calucating evaluation values... ................ done.
Sending current state to rgba.monitor()...
Creating next generation...
   sorting results...
   applying elitism...
   applying crossover...
   applying mutations... 2 mutations applied
Starting iteration 2
Calucating evaluation values... ............. done.
Sending current state to rgba.monitor()...
Creating next generation...
   sorting results...
   applying elitism...
   applying crossover...
   applying mutations... 4 mutations applied
Starting iteration 3
Calucating evaluation values... ............... done.

# Omit the rest output.
```

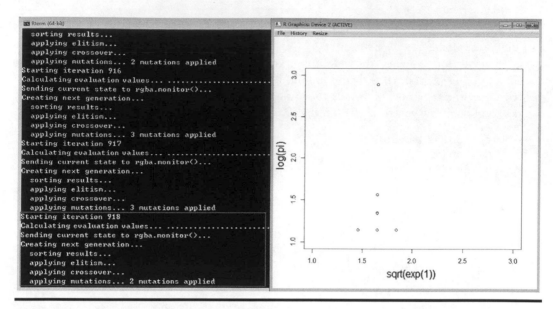

Figure 2.17   The execution of algorithm rbga.

Screenshot of execution, shown as Figure 2.17.

Notice that the program needs to be executed in cmd. The following error would be shown if the program was run in RStudio.

```
Error in get(name, envir = asNamespace(pkg), inherits = FALSE):
  object 'rversion' not found
Graphics error: Error in get(name, envir = asNamespace(pkg), inherits =
FALSE):
  object 'rversion' not found
```

Let us check the result after doing 1000 iterations.

```
# Calculation result.
> m2$population[1,]
[1] 1.650571 1.145784
```

The optimized result is: $x_1 = 1.650571$, $x_2 = 1.145784$, which is very approximate to the expected result. Besides, we can check each iteration result with the visualization functionality of the package genalg. The following screenshots correspond to the calculation results for iteration 1, iteration 10, iteration 200, and iteration 1000. From Figure 2.18, we can see that the optimized result sets are getting less and more precise, with the iteration number increasing.

Other than visually tracing the algorithm, genalg also provides three visualization graphs (the default graph, the histogram, and the parameter graphs) to analyze the algorithm effects. The default graph is the output by R shown in Figure 2.19.

```
> plot(m2)
```

The default graph is used to demonstrate the progression of the genetic process. The X axis is the iteration number and Y is evaluation. The closer the evaluation value to 0 the better the result is. The precise result is almost found after 1000 iterations.

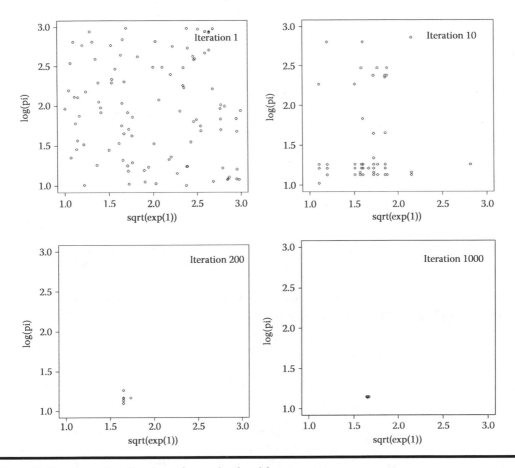

Figure 2.18   Data visualization of genetic algorithm.

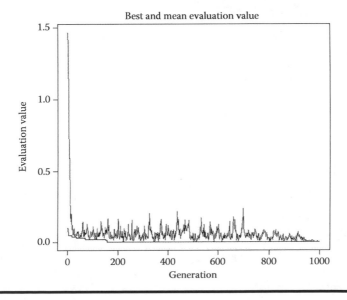

Figure 2.19   Output of the default graph.

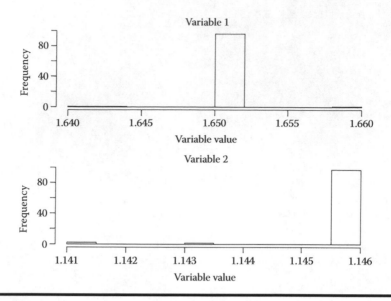

Figure 2.20   Output of the histogram.

The histogram is the output by R, shown in Figure 2.20.

```
> plot(m2,type='hist')
```

The histogram is used to demonstrate the frequency of choosing for chromosome genes, or the numbers for a gene being selected in a chromosome. When variable x_1 is in interval 1.65, it is selected for more than 80 times. When variable x_2 is in interval 1.146, it is selected for more than 80 times. Through the histogram, we can understand that the best genes are passed on to the descendants.

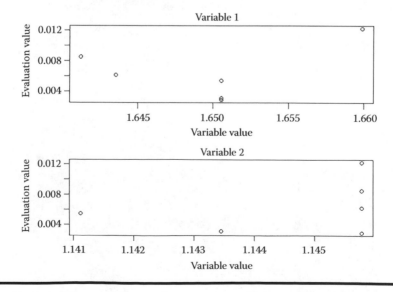

Figure 2.21   Output of the variable graph.

The variable graph is output by R, shown as in Figure 2.21.

```
> plot(m2,type='vars')
```

The variable graph is used to demonstrate the relationship between evaluation function and variable values. For x_1, the smaller the evaluation value is, the more precise the variable is. The relationship is rough. While for x_2, we cannot figure out the relationship.

Comparing mcga and genalg, we can conclude that mcga is suitable for calculating the optimum solution in large-scale-valued space while genalg fits less. However, on the other hand, genalg provides visualization tools, which enable us to see the convergence process of the genetic algorithm. It is very helpful in understanding and tuning the genetic algorithm.

After mastering genetic algorithms, we can solve some optimization problems quickly. I will introduce the parameterized optimization solution for the financial backtest system. Let us get rid of the method of exhaustion and cherish every second of CPU.

R PROGRAMMING IN DEPTH

Chapter 3

The Kernel Programming in R

The chapter introduces the programming related to the kernel of R. By studying pryr package we will understand the features of the lower level of R, learn how to define the environments in R and use the function environments, master the management of file system with R, and interpret the new features of R version 3.1. This chapter will help readers learn the underlying knowledge.

3.1 The Advanced Toolkit pryr Which Levers the R Kernel

Question

How do we understand the internal execution principle of R?

Introduction

As working with R in more depth, we need to learn more about the lower level of R, such as object-orienting data types S3, S4, and RC, and the mechanism of function calling. The pryr package is such a tool that helps us understand the execution mechanism of R. With the pryr package, we can touch the core of R more easily.

3.1.1 Introducing pryr

The pryr package is a tool that helps us to understand the execution principle of R and to get closer to the core of R. We need to master R more deeply in order to develop more advanced application of R. The APIs of the pryr package mainly include functions such as internal implementation utilities, OO (object-oriented) checking utilities, assisted programming functions, code simplification utilities, and so on, as listed in the following.

Internal implementation utilities:

- Utilities for promise objects: uneval(),is_promise()
- To query the environment variables: where(),rls(),parenv()
- To view the enclosure function variables: unenclose()
- The calling relationship of functions: call_tree()
- To view the underlying C language types of objects: address(), refs(),typename()
- To check whether an object is modified: track_copy()

Object-orientation checking utilities:

- To check the object type: otype()
- To check the function type: ftype()

Auxiliary programming functions:

- To create function through arguments: make_function(), f()
- To substitute the variable expression: substitute_q(), subs()
- To modify objects in batch: modify_lang()
- To create list object quickly: dots(), named_dots()
- To create call to anonymous function: partial()
- To find eligible functions: find_funs()

Code simplification utilities:

- To create delayed or direct binding: %<d-%, %<a-%
- To create constant binding: %<c-%
- To create re-binding: rebind, <<-

3.1.2 Installing pryr

The system environment used in this section:

- Linux: Ubuntu Server 12.04.2 LTS 64bit
- R: 3.1.1 x86_64-pc-linux-gnu (64-bit)

Note: pryr supports both Win7 and Linux.

When installing the pryr package, we can download directly from CRAN or install through Github. Let's download it directly from CRAN and install. The version of the pryr package used in this sections is v0.1. The source code of the corresponding Gitbhub project is available at https://github.com/hadley/pryr/releases/tag/v0.1.

```
# Launch the R application.
~ R
# Install the pryr package.
> install.packages(pryr)
> library(pryr)
```

3.1.3 Using pryr

In the following, we will see various examples of using functions in the pryr package.

3.1.3.1 Creating Anonymous Function : f()

We can create an anonymous function using function f() to complete the definition, the calling, and the calculation of function in just one line.

```
# Create an anonymous function.
> f(x + y)
function (x, y)
x + y

# Create an anonymous function and do calculation with assigned
arguments.
> f(x + y)(1, 10)
[1] 11

# Create an anonymous function and specify the arguments x and assign the
argument y with default valuc of 2.
> f(x, y = 2, x + y)
function (x, y = 2)
x + y

# Create an anonymous function and specify the arguments x and assign the
argument y with default value of 2 and do calculation with assigned
arguments.
> f(x, y = 2, x + y)(1)
[1] 3

# Create an anonymous function that has multi-line operation and do
calculation with assigned arguments.
>  f({y <- runif(1); x + y})(3)
[1] 3.7483
```

3.1.3.2 Creating Function through Arguments: make_function()

Unlike the original syntax with assignments using function(), we can dynamically create functions by passing different arguments to make_function(). The advantage is that we can dynamically generate a function at runtime and bind it with the environment, and reduce the amount of static code.

Let's view the definition of the function make_function(). It has three arguments.

```
make_function(args, body, env = parent.frame())
```

In the definition, the argument args, with the type of list, is used to generate the arguments part of the function; body, a syntax expression, with the type of call, is used to generate the expression part of the function; env is used to generate the system environment part of the function, with the type of environment, defaulted as the parent of current environment.

```
# Create a standard function.
> f <- function(x) x + 3
> f
function(x) x + 3
# Run the application.
> f(12)
[1] 15

# Create a function through arguments.
> g <- make_function(alist(x = ), quote(x + 3))
> g
function (x)
x + 3
# Run the function.
> g(12)
[1] 15
```

3.1.3.3 Creating Anonymous Function Call: Partial()

Using the partial() function can reduce the definition of arguments to make it easy to call anonymous functions. The code is as follows:

```
# Define an ordinary function.
> compact1 <- function(x) Filter(Negate(is.null), x)
> compact1
function(x) Filter(Negate(is.null), x)

# Define an anonymous function through partial().
> compact2 <- partial(Filter, Negate(is.null))
> compact2
function (...)
Filter(Negate(is.null), ...)
```

As we see, of the two function definitions, one is ordinary and the other is defining an anonymous function using partial().

Here is another example: output the result of a uniform distribution runif().

```
# Implement using an ordinary function.
> f1 <- function(){runif(rpois(1, 5))}
> f1()
[1] 0.09654228 0.93089395 0.85530142 0.33021067 0.16728877 0.79099825
> f1()
[1] 0.6166580 0.2100876 0.3125176

# Implement using an anonymous function call through partial().
> f2 <- partial(runif, n = rpois(1, 5))
> f2()
```

```
[1] 0.25955143 0.12858459 0.04994997 0.11505708 0.10509429
> f2()
[1] 0.9710866 0.1469317
```

3.1.3.4 Substituting the Variable Expression: substitute_q(), subs()

The function substitute_q() can directly perform argument substitution for expression call.

```
# Define an expression call.
>  x <- quote(a + b)
> class(x)
[1] "call"

# Perform argument substitution for x but there is no effect.
>  substitute(x, list(a = 1, b = 2))

# Perform argument substitution for direct variables.
3> substitute(a+b, list(a = 1, b = 2))
1 + 2
# Perform argument substitution for x using substitute_q().
>  substitute_q(x, list(a = 1, b = 2))
1 + 2
# Perform argument call.
> eval(substitute_q(x, list(a = 1, b = 2)))
[1] 3
```

The function subs() can perform direct substitution for variable expressions.

```
> a <- 1
> b <- 2

# Perform substitution for variable expression but there is no effect.
> substitute(a + b)
a + b
# Perform substitution for variable using subs.
> subs(a + b)
1 + 2
```

3.1.3.5 Object-Orientation Type Checking: otype(),ftype()

The functions from base package are not so good to check the type of objects. The function otype() from pryr package can easily distinguish the object types such as S3, S4, RC, and so on, which is much more efficient than the built-in type checking.

```
# The primitive types.
> otype(1:10)
[1] "primitive"
> otype(c('a','d'))
[1] "primitive"
> otype(list(c('a'),data.frame()))
[1] "primitive"

# S3 type.
> otype(data.frame())
[1] "S3"
```

```
# The customized S3 type.
> x <- 1
> attr(x,'class')<-'foo'
> is.object(x)
[1] TRUE
> otype(x)
[1] "S3"

# S4 type.
> setClass("Person",slots=list(name="character",age="numeric"))
> alice<-new("Person",name="Alice",age=40)
> isS4(alice)
[1] TRUE
> otype(alice)
[1] "S4"

# RC type.
> Account<-setRefClass("Account")
> a<-Account$new()
> class(a)
[1] "Account"
attr(,"package")
[1] ".GlobalEnv"

> is.object(a)
[1] TRUE
> isS4(a)
[1] TRUE
> otype(a)
[1] "RC"
```

The function ftype() can easily distinguish the functions with types of function, primitive, S3, S4, and internal. It is much more efficient than the built-in function type checking.

```
# The Standard function.
> ftype(`%in%`)
[1] "function"

# The primitive function.
> ftype(sum)
[1] "primitive" "generic"

# The internal function.
> ftype(writeLines)
[1] "internal"
> ftype(unlist)
[1] "internal" "generic"

# The S3 function.
>  ftype(t.data.frame)
[1] "s3"       "method"
> ftype(t.test)
[1] "s3"        "generic"
```

```
# The S4 function.
> setGeneric("union")
[1] "union"
> setMethod("union",c(x="data.frame",y="data.frame"),function(x, y)
  {unique(rbind (x, y))})
[1] "union"
> ftype(union)
[1] "s4"        "generic"

# The RC function.
> Account<-setRefClass("Account",
+    fields=list(balance="numeric"),
+    methods=list(
+    withdraw=function(x){balance<<-balance-x},
+    deposit=function(x){balance<<-balance+x}))
> a<-Account$new(balance=100)
> a$deposit(100)
> ftype(a$deposit)
[1] "rc"        "method"
```

3.1.3.6 Viewing the Underlying C Types of Objects: address(), refs(), typename()

We can view the underlying C types of R objects through address(), refs(), and typename(). The function typename() returns the C type name; the function address() returns the memory address; the function refs() returns the pointer index. The following code views the C type of variables.

```
# Define a variable x.
>   x <- 1:10
# Print the C type name.
> typename(x)
[1] "INTSXP"
# Return the pointer.
> refs(x)
[1] 1
# Print the memory address.
> address(x)
[1] "0x365f560"

# Define a list object.
>   z <- list(1:10)
# Print the C type name.
>   typename(z)
[1] "VECSXP"

# The delayed assignment.
> delayedAssign("a", 1 + 2)
# Print the C type name.
> typename(a)
[1] "PROMSXP"

# Print the variable a.
> a
[1]  3
```

```
> typename(a)
[1] "PROMSXP"

# Print the variable b to compare with a.
> b<-3
> typename(b)
[1] "REALSXP"
```

3.1.3.7 Checking whether an Object Is Modified: track_copy()

Using the function track_copy(), we can track an object and check whether it is modified through its memory address.

```
# Define a variable.
> a<-1:3;a
[1] 1 2 3
# View the memory address of the variable.
> address(a)
[1] "0x2ad77f0"

# Track the variable.
> track_a <- track_copy(a)
# Check whether the variable is modified. No change.
> track_a()

# Assign the variable.
. a[3] <- 3L
# View the memory address of the variable. No change.
> address(a)
[1] "0x2ad77f0"
# Check whether the variable is modified. No.
>  track_a()

# Assign the variable again.
> a[3]<-3
# View the memory address of the variable. Changed.
> address(a)
[1] "0x37f8580"
# Check whether the variable is modified. Modified as a copy.
>  track_a()
a copied
```

3.1.3.8 View the Enclosure Function Variable: unenclose()

Assign the enclosure function variable using unenclose().

```
# Define an embedded function power.
>  power <- function(exp) {
+       function(x) x ∧ exp
+  }

# Call the enclosure function.
>  square <- power(2)
>  cube <- power(3)
```

```
# View the square function, showing the result with exp unassigned.
> square
function(x) x ^ exp
<environment: 0x4055f28>

# View the square function, showing the result with exp assigned.
> unenclose(square)
function (x)
x ^ 2

# Call the square function.
> square(3)
[1] 9
```

3.1.3.9 Modifying Objects in Batch: modify_lang()

This is a magic function which can conveniently replace the variable definitions in list object, expression, or function. In the following, let's try to replace the variable a in list object with variable b.

```
# Define a list object and its internal data.
> examples <- list(
+       quote(a <- 5),
+       alist(a = 1, c = a),
+       function(a = 1) a * 10,
+       expression(a <- 1, a, f(a), f(a = a))
+       )

# View the data of the object.
> examples
[[1]]
a <- 5
[[2]]
[[2]]$a
[1] 1
[[2]]$c
a
[[3]]
function (a = 1)
a * 10
[[4]]
expression(a <- 1, a, f(a), f(a = a))

# Define the conversion function a_to_b.
>  a_to_b <- function(x) {
+       if (is.name(x) && identical(x, quote(a))) return(quote(b))
+       x
+  }

# Modify the object in batch. Replace all the occurrences of variable a
with variable b.
> modify_lang(examples, a_to_b)
[[1]]
b <- 5
```

```
[[2]]
[[2]]$a
[1] 1
[[2]]$c
b
[[3]]
function (a = 1)
b * 10
[[4]]
expression(b <- 1, b, f(b), f(a = b))
```

3.1.3.10 Creating List Object Quickly: dots(), named dots()

Using dots(), we can quickly create a list object, setting the name-values of data in list.

```
# Initialize an object.
> y <- 2
# Create a list object.
> dots(x = 1, y, z = )
$x
[1] 1
[[2]]
Y
$z

# View the object type.
> class(dots(x = 1, y, z = ))
[1] "list"

#View the internal structure of the object.
> str(dots(x = 1, y, z = ))
List of 3
 $ x: num 1
 $  : symbol y
 $ z: symbol
```

We can also use named_dots() to create list object quickly, setting the name-value of the list data. The difference from dots() lies that the argument variable itself is the name of list datum. Although the variable y is not assigned, it is still used as the name of list datum and the value of the datum can be obtained by $y.

```
# Create a list object.
> named_dots(x = 1, y, z =)
$x
[1] 1
$y
y
$z

# View the object type.
> class(named_dots(x = 1, y, z =))
[1] "list"
```

```
#View the internal structure of the object.
> str(named_dots(x = 1, y, z =))
List of 3
 $ x: num 1
 $ y: symbol y
 $ z: symbol
```

3.1.3.11 Finding the Eligible Functions: find_funs()

We can use find_funs() to quick find the eligible functions match criteria. The following statement searches the base package to find the function names that match the string "match.fun" in their calls.

```
> find_funs("package:base", fun_calls, "match.fun", fixed = TRUE)
Using environment package:base
 [1] "apply"  "eapply" "Find"   "lapply" "Map"    "mapply" "Negate" "outer"
 [9] "Reduce" "sapply" "sweep"  "tapply" "vapply"
```

```
#View the function Map to see whether its calls contain the string "match.fun".
> Map
function (f, …)
{
    f <- match.fun(f)
    mapply(FUN = f, …, SIMPLIFY = FALSE)
}
<bytecode: 0x21688e0>
<environment: namespace:base>
```

Search all the function arguments in the stats packages to find the function names that exactly match the string "FUN" in their arguments.

```
> find_funs("package:stats", fun_args, "^FUN$")
Using environment package:stats
[1] "addmargins"           "aggregate.data.frame" "aggregate.ts"
[4] "ave"                  "dendrapply"
```

```
#View the function avg to see whether its arguments contain the string "Fun".
> ave
function (x, …, FUN = mean)
{
    if (missing(…))
        x[] <- FUN(x)
    else {
        g <- interaction(…)
        split(x, g) <- lapply(split(x, g), FUN)
    }
    x
}
<bytecode: 0x2acba70>
<environment: namespace:stats>
```

3.1.3.12 Querying the Environments: where(), rls(), parenv()

We can use the function where() to locate the object in R environment, a bit of like the command whereis under Linux.

```
# Define a variable x.
> x <- 1
# Query the environment of the variable x.
> where("x")
<environment: R_GlobalEnv>

# Query the location of the function x.
> where("t.test")
<environment: package:stats>
attr(,"name")
[1] "package:stats"
attr(,"path")
[1] "/usr/lib/R/library/stats"

> t.test
function (x, …)
UseMethod("t.test")
<bytecode: 0x1ae9bc8>
<environment: namespace:stats>

# Query the location of the function mean.
> where("mean")
<environment: base>

# Query the location of the function where.
> where("where")
<environment: package:pryr>
attr(,"name")
[1] "package:pryr"
attr(,"path")
[1] "/home/conan/R/x86_64-pc-linux-gnu-library/3.0/pryr"
```

We can use the function rls() to show all the variables in all environments, including the variables in the current environment, those in the global environment, those in the empty environment, and those in the namespace environments.

```
# Print the variables in current environment.
> ls()
 [1] "a"               "Account"          "alice"     "a_to_b"
 [5] "b"               "compact1"         "compact2"  "examples"
 [9] "f"               "f1"               "f2"        "g"
[13] "myGeneric"       "my_long_variable" "plot2"     "union"
[17] "x"               "y"

# Print the variables in all environments.
> rls()
[[1]]
 [1] "a"                            "Account"
 [3] "alice"                        "a_to_b"
```

```
 [5] "b"                           ".__C__Account"
 [7] "compact1"                    "compact2"
 [9] ".__C__Person"                "examples"
[11] "f"                           "f1"
[13] "f2"                          "g"
[15] ".__global__"                 "myGeneric"
[17] "my_long_variable"            "plot2"
[19] ".Random.seed"               ".requireCachedGenerics"
[21] ".__T__myGeneric:.GlobalEnv"  ".__T__union:base"
[23] "union"                       "x"
[25] "y"
```

The function parenv() can be used to fine the parent environment of a function and therefore to trace up to the root of the function.

```
# Define an embedded function with 3 layers.
> adder <- function(x) function(y) function(z) x + y + z

# Call the first layer function.
> add2 <- adder(2)

# View the function.
> add2
function(y) function(z) x + y + z
<environment: 0x323c000>

# Call the second layer function.
> add3<-add2(3)
> add3
function(z) x + y + z
<environment: 0x3203558>

# View the parent environment of the internal function.
>  parenv(add3)
<environment: 0x323c000>
>  parenv(add2)
<environment: R_GlobalEnv>
```

3.1.3.13 Printing the Calling Relationship: call_tree(), ast()

We can use the function call_tree() to print the calling relationship of expressions.

```
# The calling relationship of the embedded function statement.
>  call_tree(quote(f(x, 1, g(), h(i()))))
\- ()
  \- `f
  \- `x
  \- 1
  \- ()
    \- `g
  \- ()
    \- `h
    \- ()
      \- `i
```

```
# The calling relationship of the condition statement.
> call_tree(quote(if (TRUE) 3 else 4))
\- ()
  \- `if
  \-  TRUE
  \-  3
  \-  4


# The calling relationship of the expression statement.
> call_tree(expression(1, 2, 3))
\-  1
\-  2
\-  3
```

The function ast() can print the calling relationship of statements.

```
# The embedded expression statement.
> ast(f(x, 1, g(), h(i())))
\- ()
  \- `f
  \- `x
  \-  1
  \- ()
    \- `g
  \- ()
    \- `h
    \- ()
      \- `i


# The condition statement.
> ast(if (TRUE) 3 else 4)
\- ()
  \- `if
  \-  TRUE
  \-  3
  \-  4


# The embedded expression statement.
> ast(function(a = 1, b = 2) {a + b})
\- ()
  \- `function
  \- []
    \ a = 1
    \ b = 2
  \- ()
    \- `{
    \- ()
      \- `+
      \- `a
      \- `b
  \-


# The function call.
> ast(f()()())
```

```
\- ()
  \- ()
    \- ()
      \- `f
```

3.1.3.14 Utilities for Promise Objects: uneval(), is_promise()

The promise objects are a part of the lazy loading mechanism in R which includes three parts: value, expression, and environment. When a function is being called, its arguments are matched and then every formal argument is bound to a promise object. The expression has formal arguments and the function pointer stored to a promise.

Briefly speaking, the process of lazy loading means that store the function pointer to a promise object first but don't call it immediately; when actual calling happens, the system finds the function pointer from the promise object and calls the function.

```
# Define a variable and assign it.
> x <- 10

# Check whether the variable is in promise mode.
> is_promise(x)
[1] FALSE

# Anonymous function call to check whether the variable is in promise mode.
> (function(x) is_promise(x))(x = 10)
[1] TRUE
```

The function uneval() can print the function call but don't perform the function call.

```
# Define a function.
> f <- function(x) {
+     uneval(x)
+ }

#Print the function call.
> f(a + b)
a + b

> class(f(a+b))
[1] "call"

#Print the function call.
> f(1 + 4)
1 + 4

# Lazy assignment.
> delayedAssign("x", 1 + 4)

# Don't perform the function call and only print.
> uneval(x)
1 + 4

# Perform the function call and assign.
> x
[1] 5
```

```
# Another example for lazy assignment.
> delayedAssign("x", {
+       for(i in 1:3)
+           cat("yippee!\n")
+       10
+ })

# Perform the function call and assign.
> x
yippee!
yippee!
yippee!
[1] 10
```

3.1.3.15 Data Binding: %<a-%, %<c-%, %<d-%, rebind,<<-

We can create customized operators using special function and bind data and function call through operators.

Direct binding:

```
> x %<a-% runif(1)
> x
[1] 0.06793592
> x
[1] 0.8217227
```

Constant binding:

```
> y %<c-% 4 + 2
[1] 6
> y
[1] 4
```

Lazy binding:

```
> z %<d-% (a + b)
> a <- 10
> b <- 100
> z
[1] 110
```

Re-binding:

```
# Reassign the known variable a.
> a <- 1
> rebind("a", 2)

# Reassign the unknown variable ccc and get error.
> rebind("ccc", 2)
Error: Can't find ccc

#Reassign the known variable a using <<-.
> a<<-2
> a
```

```
[1] 2
# Remove the variable ccc.
> rm(ccc)
> ccc
Error: object 'ccc' not found
#Reassign the unknown variable a using <<-.
> ccc<<-2
> ccc
[1] 2
```

Through the above complete introduction to pryr, we can see how powerful the package is. It is very helpful for us to understand the execution mechanism and data structure of R. This is a mandatory lesson before we deep into R's kernel.

3.2 Uncovering the Mystery of R Environments

Question

How do we use the environments in R?

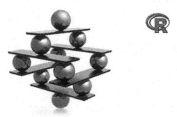

Uncovering the mystery of R environments
http://blog.fens.me/r-environments/

Introduction

The environments are unfamiliar to most R users. The lack of understanding of its execution mechanism does not affect the use of R. The environment is part of the low-level design of R in terms of computer technologies. It is mainly used for the environment loader of R. The execution that encapsulates the loader enables the user to load any 3rd part R programming packages without knowing the low-level details.

This section will uncover the mystery of R environments.

3.2.1 The Environments of R

All of the variables, objects, and functions exist in the environments in R. When a program is running, its variables and functions have their own scopes. An environment is a data structure defined by the kernel of R and comprised of a series of hierarchical frames. Each environment corresponds to one frame to distinguish runtime scopes.

The environments have some characteristics. For example, every environment has a unique name. The environments are of reference type instead of value type. Every environment has a parent except for the empty environment, which is the top-most environment without a parent. The child can inherit the variables in a parent. And so on.

The system environment used in this section:

■ Linux: Ubuntu Server 12.04.2 LTS 64bit
■ R: 3.1.1 x86_64-pc-linux-gnu (64-bit)

Let's import pryr as the assistant tool to check the hierarchy of environments. Please refer Section 3.1 for details about the pryr package.

```
# Load the pryr package.
> library(pryr)
```

3.2.1.1 Creating an Environment

Create a new environment using the new.env() function. Let's view the definition of it.

```
new.env(hash = TRUE, parent = parent.frame(), size = 29L)
```

Argument list:

■ hash: TRUE by default and use the structure of hash table.
■ parent: the parent environment of the environment being created.
■ size: the initial capability of the environment.

Run new.env() to create a new environment.

```
# Create an environment e1.
> e1 <- new.env()
# Output e1.
> e1
<environment: 0x3d7eef0>
# View the type of e1.
> class(e1)
[1] "environment"
# View the type of e1 using otype(). It is of primitive type.
> otype(e1)
[1] "primitive"
```

In the next, let's create a variable in the environment e1.

```
# Define a variable a.
> e1$a <- 10
# Output the variable a.
> e1$a
[1] 10
# List the variables in current environment.
> ls()
[1] "e1"
# List the variables in environment e1.
> ls(e1)
[1] "a"
```

Here, we met with two environments: the current environment and the environment e1. e1 was defined in current environment as a variable and the variable a was defined in e1.

3.2.1.2 The Hierarchy of Environments

The environments of R are of a hierarchical structure, where each environment has its parent environment up to the empty environment on the top. There are five types of environments in R: the global, the internal, the parent, the empty, and the package environment.

- The current environment means the user environment, where the user program is running.
- The internal environment is the one that is created temporarily. It refers to either the environment explicitly created by new.env(), or the anonymous environment.
- The parent environment means the one level higher environment.
- The empty environment is the topmost environment, which has no parent.
- The package environment is the one that is encapsulated by package.

```
# The current environment.
> environment()
<environment: R_GlobalEnv>

# The internal environment.
> e1 <- new.env()
> e1
<environment: 0x3e28948>

# The parent environment.
> parent.env(e1)
<environment: R_GlobalEnv>

# The empty environment.
> emptyenv()
<environment: R_EmptyEnv>

# The package environment.
> baseenv()
<environment: base>
```

We can use search() to view the loaded packages in the current environment.

```
# View the environments.
> search()
 [1] ".GlobalEnv"        "package:pryr"      "package:stats"
 [4] "package:graphics"  "package:grDevices" "package:utils"
 [7] "package:datasets"  "package:methods"   "Autoloads"
[10] "package:base"

# The current environment.
> .GlobalEnv
<environment: R_GlobalEnv>
> parent.frame()
<environment: R_GlobalEnv>
# View the parent environment.
# The parent environment of the environment e1.
> parent.env(e1)
<environment: R_GlobalEnv>
```

```
# The parent environment of the current environment.
> parent.env(environment())
<environment: package:pryr>
attr(,"name")
[1] "package:pryr"
attr(,"path")
[1] "/home/conan/R/x86_64-pc-linux-gnu-library/3.0/pryr"


# The parent environment of the base package environment.
> parent.env(baseenv())
<environment: R_EmptyEnv>


# The parent environment of the empty environment. Since it has no
parent, there is error thrown.
> parent.env(emptyenv())
Error in parent.env(emptyenv()) : the empty environment has no parent
```

In that the environments have hierarchical relationship, we can print the hierarchy from the customized environment e1 to the empty environment on the top.

```
# Print the parent environments recursively.
> parent.call<-function(e){
+    print(e)
+    if(is.environment(e) & !identical(emptyenv(),e)){
+      parent.call(parent.env(e))
+    }
+ }


#Run the function.
> parent.call(e1)
<environment: 0x366bf18>
<environment: R_GlobalEnv>
<environment: package:pryr>
attr(,"name")
[1] "package:pryr"
attr(,"path")
[1] "/home/conan/R/x86_64-pc-linux-gnu-library/3.0/pryr"
<environment: package:stats>
attr(,"name")
[1] "package:stats"
attr(,"path")
[1] "/usr/lib/R/library/stats"
<environment: package:graphics>
attr(,"name")
[1] "package:graphics"
attr(,"path")
[1] "/usr/lib/R/library/graphics"
<environment: package:grDevices>
attr(,"name")
[1] "package:grDevices"
attr(,"path")
[1] "/usr/lib/R/library/grDevices"
<environment: package:utils>
attr(,"name")
```

```
[1] "package:utils"
attr(,"path")
[1] "/usr/lib/R/library/utils"
<environment: package:datasets>
attr(,"name")
[1] "package:datasets"
attr(,"path")
[1] "/usr/lib/R/library/datasets"
<environment: package:methods>
attr(,"name")
[1] "package:methods"
attr(,"path")
[1] "/usr/lib/R/library/methods"
<environment: 0x20cb5d0>
attr(,"name")
[1] "Autoloads"
<environment: base>
<environment: R_EmptyEnv>
```

By recursively look up the parent environments, we see the hierarchy of the entire environment space, as shown in Figure 3.1.

From the environment hierarchy in Figure 3.1 we can also find the loading sequence of R packages. The base package was loaded first and then the six basic package were loaded through base::Autoloads(). The pryr package on the higher layer was loaded by us manually. Finally, the environment R_GlobalEnv was used as the current environment and the internal environment is the lower layer environment of R_GlobalEnv.

3.2.2 The Characteristics of Environments

As mentioned above, the environments have some characteristics. Let's introduce them one by one.

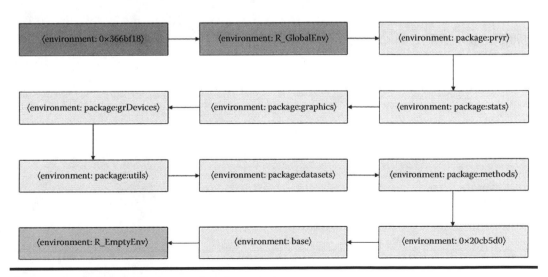

Figure 3.1   The hierarchy space of environments.

3.2.2.1 The Object Names in Each Environment Must Be Unique

Define the variable x and do some operations to x.

```
# Define a variable x.
> x<-10;x
[1] 10
# View the address of x.
> address(x)
[1] "0x2874068"
# Change the assignment to x.
> x<-11;x
[1] 11
# View the address of x.
> address(x)
[1] "0x28744c8"
```

We can see that every time the variable x was assigned, its address changes, but the name is still x. Define another variable x in a different environment.

```
# Create an environment e1.
> e1<-new.env()
# Define a variable x in e1.
> e1$x<-20
# Output x.
> x;e1$x
[1] 12
[1] 20
```

The variables with same name can exist in different environments and they cannot exist in same environment.

3.2.2.2 The Assignment of Environment Variables

What would happen if we assign the environment variable e1 to another variable f and then modify the internal variable of the environment?

```
# Assign e1 to f.
> f <- e1
# Modify the value of variable a in e1.
> e1$a <- 1111
# View the value of a in the environment f.
> f$a
[1] 1111
# Compare f and e1. They are identical.
> identical(f,e1)
[1] TRUE

# View the addresses of e1 and f. They are same.
> e1
<environment: 0x3e28948>
> f
<environment: 0x3e28948>
```

From the above results we can see that the assignment between environments is a reference passing and do not create a new environment variable.

3.2.2.3 Defining the Environment with Higher Level

The empty environment is the top most environment and the next is the environment of the base package. We can try to create an environment that is closer to the top, making its parent environment as the environment of the base package.

```
# Create an environment e2, with the parent environment as the
environment of base.
> e2 <- new.env(parent = baseenv());e2
<environment: 0x37cab18>

# View the parent environment list of e2.
> parent.call(e2)
<environment: 0x37cab18>
<environment: base>
<environment: R_EmptyEnv>
```

Therefore, the environment e2 is at the third level of the environment space.

3.2.2.4 The Child Environment Inherits the Variables in Its Parent Environment

Define a variable x in the current environment and then reassign x in the child environment e1.

```
# Define a variable x in the current environment.
> x<-1:5
# Create a new environment e1.
> e1 <- new.env()
# Define a variable x in e1.
> e1$x<-1

# Define a function in e1 and reassign the variable x in the parent environment.
> e1$fun<-function(y){
+    print('e1::fun')
+    x<<-y
+ }
# Run the function in e1 to assigned x with 50.
> e1$fun(50)
[1] "e1::fun"

# The value of variable x in current environment has been modified.
> x
[1] 50
# The value of the variable x in e1 has no change.
> e1$x
[1] 1
```

If we want to change the value of variables in parent environment, we need to use the assignment symbol <<-. Many R users are confused with the difference of <<- and <-. Here is the answer.

3.2.3 Accessing the Environments

R provides us with some basic functions to help us understand and use the environments.

- new.env() creates an environment.
- is.environment() checks where a variable is an environment.
- environment views() the definition of the environment of a function.
- environmentName() views the name of an environment.
- env.profile() views the properties of an environment.
- is() views the objects in an environment.
- get() gets the object in specific environment.
- rm() deletes the objects in an environment.
- assign() assigns the values in an environment.
- exists() checks there the specific object exists in an environment.

In the following let us perform the access operations for environments.

```
# New an environment.
> e1<-new.env()
# Check whether e1 is an environment.
> is.environment(e1)
[1] TRUE

# View the current environment.
> environment()
<environment: R_GlobalEnv>

# View the environment of a function.
> environment(ls)
<environment: namespace:base>

# View the name of an environment.
> environmentName(baseenv())
[1] "base"
> environmentName(environment())
[1] "R_GlobalEnv"

#View the name of the environment e1.
> environmentName(e1)
[1] ""

# Set the name of e1.
> attr(e1,"name")<-"e1"
> environmentName(e1)
[1] "e1"

# View the properties of e1.
> env.profile(e1)
$size
[1] 29
$nchains
[1] 1
```

```
$counts
 [1] 0 0 0 0 1 0 0 0 0 0 0 0 0 0 0 0 0 0 0 0 0 0 0 0 0 0 0 0 0 0 0 0 0
```

Following code shows the object operations in environment.

```
# Release all the objects defined in current environment.
> rm(list=ls())
# Define an environment and 3 variables: x, y and e1$x.
> e1<-new.env()
> x<-1:5;y<-2:10
> e1$x<-10

# View the objects in current environment.
> ls()
[1] "e1" "x"  "y"
# View the variables in e1.
> ls(e1)
[1] "x"

# Get the value of x in current environment.
> get("x")
[1] 1 2 3 4 5
# Get the value of x in e1.
> get("x",envir=e1)
[1] 10

# Get the value of y in e1. The y is inherit from current environment.
> get("y",envir=e1)
[1]  2  3  4  5  6  7  8  9 10

# If inherit is prohibited, then getting the value of y from e1 throws error.
> get("y",envir=e1,inherits=FALSE)
Error in get("y", envir = e1, inherits = FALSE) : object 'y' not found

# Reassign x.
> assign('x',77);x
[1] 77

# Reassign value to the x in e1.
> assign('x',99,envir=e1);e1$x
[1] 99

# Add a variable y to e1 without inheritance.
> assign('y',99,envir=e1,inherits=FALSE);
> y
[1]  2  3  4  5  6  7  8  9 10
> e1$y
[1] 99

# Remove the variable x in e1 and y in current environment.
> rm(x,envir=e1)
> e1$x
NULL
> x
[1] 77
```

```
# View the current environment and e1.
>  ls()
[1] "e1" "x"
> ls(e1)
[1] "y"

# Check whether the x object exists in current environment.
> exists('x')
[1] TRUE

# Check whether the x object exists in e1.
> exists('x',envir=e1)
[1] TRUE

# Check whether the x object exists in e1 without inheritance.
> exists('x',envir=e1,inherits=FALSE)
[1] FALSE
```

Besides, the where() function in pryr package can locate the environment of the specified object.

```
# View the environment where the mean() function was defined.
> where(mean)
Error: is.character(name) is not TRUE
> where("mean")
<environment: base>

# View the environment where the where function was defined.
> where("where")
<environment: package:pryr>
attr(,"name")
[1] "package:pryr"
attr(,"path")
[1] "/home/conan/R/x86_64-pc-linux-gnu-library/3.0/pryr"

# View the environment where the variable x was defined.
> where("x")
<environment: R_GlobalEnv>

# View the environment where the variable y was defined. The variable y
was defined in e1, which is a child of current environment so the call
cannot access y.
> where("y")
Error: Can't find y
> e1$y
[1] 99

# View the variable y in e1.
 > where("y",e1)
<environment: 0x2545db0>
```

This section described the definition, structure, and some simple applications of R's environments, which helps us to understand the underlying architecture of R. The usage of environment is critical for us to develop R packages. Once you are good at using environments, the packages developed by you not only have clear structure but also run efficiently.

3.3 Revealing the Function Environments in R*

Question

How do we use the function environments in R?

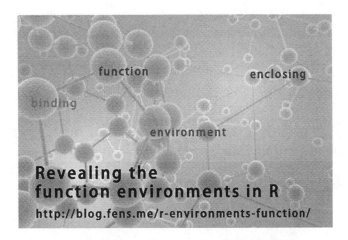

Introduction

Following the previous section, this section continues to introduce the function environments. The function environments in R are dynamic and allow for us to create more complex applications with less code.

3.3.1 The Function Environments of R

In R language, variables, objects, and functions exist in environments. In the meantime, functions have their own environments. We can define variables, object, and functions in functions recursively. These recursions form the environment system of R we are using now.

Usually we can create an environment but most times we use the function environments.

The function environments have the following four types:

- The enclosing environment. Every function has and only has one enclosing environment pointing to the environment where the function is defined.
- The binding environment. The environment to specify a name for a function by binding the function to a function variable.
- The execution environment. The environment generated dynamically in memory when executing the function. It will be destroyed automatically once the execution ends.
- The calling environment. The environment where a function is called. For example, in fun1<-function(){fun2()}, the function fun2() is called in fun1().

The system environment used in this section:

- Win7 64bit
- R: 3.1.1 x86_64-w64-mingw32/x64 (64-bit)

---

* Note: For the definitions to function environments in this section, refer *Advanced R* by Hadley Wickham.

3.3.2 *The Enclosing Environment*

The enclosing environment is easy to understand. It is a statically defined function environment. It points to where the function is located when the function is defined. We can view the enclosing environment by environment().

Let us define a function f1() in the current environment and then view the enclosing environment of function f1(). It's R_GlobalEnv.

```
> y <- 1
> f1 <- function(x) x + y
> environment(f1)
<environment: R_GlobalEnv>
```

Define another function f2() and let f2() call f1(). We can see that the enclosing environment of f2() is R_GlobalEnv.

```
> f2 <- function(x){
+    f1()+y
+ }
> environment(f2)
<environment: R_GlobalEnv>
```

So we can see that the enclosing environment of a function was configured when the function was defined. It is not related to a specific runtime environment.

3.3.3 *The Binding Environment*

The binding environment means to bind the definition of the function to its calling through a function variable. For example, if we create an environment e and define a function g in e, we bind a function to the variable g. We can call this function by finding the variable g in environment e.

```
# Create an environment.
> e <- new.env()
# Bind a function to e$g.
> e$g <- function() 1
# View the definition of function g.
> e$g
function() 1
# Run the function g.
> e$g()
[1] 1
```

Define an embedded function e$f() in environment e

```
#bind a function to e$f().
> e$f <- function() {
+      function () 1
+ }

# View the definition of function f().
> e$f
```

```
function() {
    function () 1
}
```

```
# Call function f() which returns the definition of the embedded
  anonymous function.
> e$f()
function () 1
<environment: 0x000000000dbc0a28>
```

```
# Call function f() and the embedded anonymous function to get the final
  result.
> e$f()()
[1] 1
```

View the enclosing environments of functions g() and f().

```
# The enclosing environments of functions g() and f().
> environment(e$g)
<environment: R_GlobalEnv>
> environment(e$f)
<environment: R_GlobalEnv>
```

```
# The enclosing environment of the anonymous function inside f().
> environment(e$f())
<environment: 0x000000000d90b0b0>
```

```
# The enclosing environment of the parent function of the anonymous
  function.
> parent.env(environment(e$f()))
<environment: R_GlobalEnv>
```

We can see that the enclosing environment where the functions e$g() and e$f() were defined is the current environment R_GlobalEnv and that the enclosing environment of the anonymous function e$f() is a child of the current environment, that is, the environment of function e$f().

3.3.4 The Execution Environment

An execution environment is the memory environment created when a function is called. The execution environment is temporary. When the function execution is completed, the execution environment is destroyed automatically. The variables, objects, and functions defined in the execution environment are created dynamically and are also destroyed with the release of memory.

Define a function g() with temporary variable a and parameter x.

```
# Define function g.
> g <- function(x) {
+     if (!exists("a", inherits = FALSE)) {
+         a<-1
+     }
+     a<-a+x
+     a
+ }
```

```
# Call g.
> g(10)
[1] 11
> g(10)
[1] 11
```

Both callings return the same result 11. We can see that the variable a in function g() is temporary and not persistent so a is fresh every time the function g() is called.

Let us view the execution more closely with the addition of some output messages.

```
> g <- function(x) {
+       # Add some comments.
+       message("Runtime function")
+       # print the execution environment
+       print(environment())
+       if (!exists("a", inherits = FALSE)) {
+           a<-1
+       }
+       a<-a+x
+       a
+ }
```

```
# Call function g.
> g(10)
Runtime function
<environment: 0x000000000e447380>
[1] 11
> g(10)
Runtime function
<environment: 0x000000000d2fa218>
[1] 11
```

Again, we still function g() two times and see that the two different environment addresses 0x000000000e447380 and 0x000000000d2fa218 were output by print(environment()), which indicates that the execution environment of a function is allocated temporarily in memory.

3.3.5 The Calling Environment

A calling environment is the environment where the function is called. Anonymous function is usually called in its enclosing environment where it was defined. Let us define an embedding function h() with an anonymous function inside it.

```
# The h() function.
> h <- function() {
+       x <- 5
+       # The anonymous function.
+       function() {
+           x
```

```
+       }
+ }

# Call function h() and assign the anonymous function to r1().
> r1 <- h()
# Define a variable x in current environment.
> x <- 10
# Call function r1().
> r1()
[1] 5
```

The result is 5, which shows that the variable x that r1() obtained is the x defined in the enclosing environment of the anonymous function, instead of the x defined in the current environment where r1 is located.

Let us modify the code a little. We define two variables with same name x in function h() and assign the second x using operator <<-, which means to assign value to the variable x in parent environment.

```
> h <- function() {
+       x <- 10
+       x <<- 5
+       function() {
+           x
+       }
+ }
# Call the function h().
> r1 <  h()
# Call the function r1().
> r1()
[1] 10
# The variable x in the current environment.
> x
[1] 5
```

The result is 10, showing that the variable x that r1() obtained is the x defined in the enclosing environment of the anonymous function, instead of the x that is assigned through <<- and that is from the parent environment.

3.3.6 A Complete Set of Environment Operations

In the following, let us put the operations together to make a complete demonstration, as shown in Figure 3.2.

The explanation to Figure 3.2 is as follows:

1. The small solid circles represent the function that is enclosed and they always point to the environment where functions were defined.
2. The large rectangles to the left represent the loaded package environments, including R_GlobalEnv, base, R_EmptyEnv, and etc.

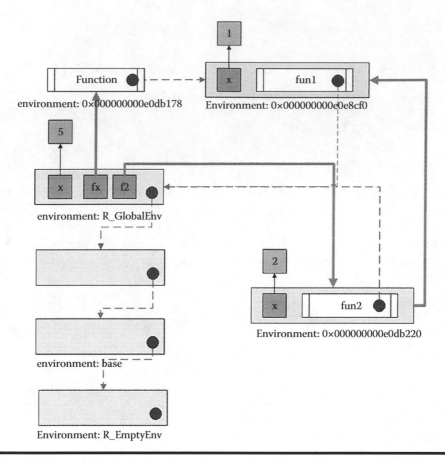

Figure 3.2 The function environments.

3. The large rectangles to the right represent the defined function environments, including environment: 0x000000000e0db220, environment: 0x000000000e0e8cf0.

4. The small rectangles in the large rectangles to the right represent the named functions, including environment: fun1 and fun2.

5. The small rectangle to the top left outside the large rectangles represents the defined anonymous function, that is, function.

6. The small squares in large rectangles represent the variables defined in the environment, including x, fx, and f2.

7. The small squares out of large rectangles represent the variable values in memory, including 5, 2, and 1.

8. The thin solid lines represent the assignments to variables.

9. The thin dashed lines represent pointing to enclosing environments.

10. The thick solid lines represent the calling processes to functions.

11. The memory addresses of function environments including 0x000000000e0db220, 0x000000000e0e8cf0 and 0x000000000e0db178 were generated when the program was running.

The following R code describes the structure in Figure 3.2.

```
# Define a variable x in current environment.
> x<-5

# Define a function f1() in current environment.
> fun1<-function(){
+    # Print the environment of fun1.
+    print("fun1")
+    print(environment())
+    # Define a variable x in the function environment of fun1().
+    x<-1
+    function() {
+       # Print the anonymous environment.
+       print("funX")
+       print(environment())
+       # Find the viable x from the direct parent environment.
+       x+15
+    }
+ }

# Define fun1 in current environment.
> fun2<-function(){
+    # Print the environment of fun2.
+    print("fun2")
+    print(environment())
+    # Define a variable x in the function environment of fun2().
+    x<-2
+    # Call fun1.
+    fun1()
+ }

# Call fun2 and bind the result to f2.
> f2<-fun2()
[1] "fun2"
<environment: 0x000000000e0db220>
[1] "fun1"
<environment: 0x000000000e0e8cf0>

# Call the anonymous function and bind it to fx().
> fx<-f2()
[1] "funX"
<environment: 0x000000000e0db178>

# Print the result of fx.
> fx
[1] 16
```

Finally we have understood the basic structure of calling relationship of R's environments through the complete scenario. Therefore, we are able to do some practical development using the features of the environments.

3.4 Managing File System with R

Question

How does R manipulate files?

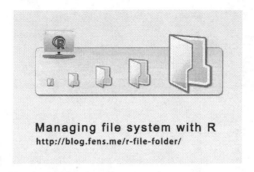

Managing file system with R
http://blog.fens.me/r-file-folder/

Introduction

As a scripting language, R has a set of functions for managing file systems as does Python. This section will introduce R's file system management in detail. Have a try and see what the difference is.

3.4.1 Introduction to File System

In computing, a file system is a mechanism that stores and organizes data. A file system makes it easy to access and retrieve data. It uses the abstract concepts such as file and directory tree to replace data blocks, the physical concept used on devices such as hard disk and CD. When storing data with the file system, users do not need to know which on data blocks the data are stored physically on hard disk (or CD), instead they just remember the directory and file name where the file is located. Before writing new data, users do not have to care about which block address on the disk is used or not. The storage management of the disk is controlled automatically by the file system. It is only necessary for users to remember which file the data is written to.

Like other programming languages, R has functionalities to manipulate the file system including manipulations to files and directories. The function APIs are defined in the base package. In the following, let us experience how to manage a file system with R.

3.4.2 Directory Manipulations

The system environment used is this section:

- Linux: Ubuntu Server 12.04.2 LTS 64bit
- R: 3.1.1 x86_64-pc-linux-gnu (64-bit)

3.4.2.1 Viewing a Directory

Enter any directory and launch the R application. View the sub-directories in the current directory.

```
# Launch the R application.
~ R
# View the current directory.
```

```
> getwd()
[1] "/home/conan/R/fs"
# View the sub-directories in the current directory.
> list.dirs()
[1] "."       "./tmp"
```

View the sub-directories and files in current directory.

```
# View the sub-directories and files in the current directory.
> dir()
[1] "readme.txt" "tmp"
```

```
# View the sub-directories and files in the specified directory.
> dir(path="/home/conan/R")
 [1] "A.txt"                           "caTools"
 [3] "chinaWeather"                    "DemoRJava"
 [5] "env"                             "FastRWeb"
 [7] "font"                            "fs"
 [9] "github"                          "lineprof"
[11] "pryr"                            "readme.txt"
[13] "RMySQL"                          "RServe"
[15] "rstudio-server-0.97.551-amd64.deb" "websockets"
[17] "x86_64-pc-linux-gnu-library"
```

```
# List only the sub-directories and files that begins with letter R.
> dir(path="/home/conan/R",pattern='^R')
[1] "RMySQL" "RServe"
```

```
# List all the sub-directories and files including the hidden ones.
> dir(path="/home/conan/R",all.files=TRUE)
 [1] "."                               ".."
 [3] ".A.txt"                          "A.txt"
 [5] "caTools"                         "chinaWeather"
 [7] "DemoRJava"                       "env"
 [9] "FastRWeb"                        "font"
[11] "fs"                              "github"
[13] "lineprof"                        "pryr"
[15] "readme.txt"                      "RMySQL"
[17] "RServe"                          "rstudio-server-0.97.551-amd64.deb"
[19] "websockets"                      "x86_64-pc-linux-gnu-library"
```

Here is another method to view the sub-directories and files in current directory, with the same result as dir().

```
> list.files()
[1] "readme.txt" "tmp"
```

```
> list.files(".",all.files=TRUE)
[1] "."          ".."          "readme.txt" "tmp"
```

View the complete directory information.

```
# View the permissions of current directory.
> file.info(".")
```

```
     size isdir mode       mtime        ctime        atime  uid  gid uname grname
. 4096   TRUE   775 2013-11-14 08:40:46 2013-11-14 08:40:46 2013-11-14
   08:41:57 1000 1000 conan   conan

# View the permissions of specified directory.
> file.info("./tmp")
       size isdir mode     mtime        ctime          atime  uid  gid uname grname
./tmp 4096   TRUE   775 2013-11-14 14:35:56 2013-11-14 14:35:56 2013-11-14
14:35:56 1000 1000 conan   conan
```

3.4.2.2 Creating a Directory

We can use the following code to create a new directory in current directory.

```
# New a directory in current directory.
> dir.create("create")
# View the sub-directories in current directory.
> list.dirs()
[1] "."         "./create" "./tmp"
```

Create a sub-directory with three levels.

```
# Error happens if we create it directly.
> dir.create(path="a1/b2/c3")
Warning message:
In dir.create(path = "a1/b2/c3"):
  cannot create dir 'a1/b2/c3', reason 'No such file or directory'

# Creating recursively gets success.
> dir.create(path="a1/b2/c3",recursive = TRUE)
> list.dirs()
[1] "."         "./a1"       "./a1/b2"     "./a1/b2/c3" "./create"  "./tmp"
```

In R, we can directly call the system commands to view the directory tree.

```
# View the directory tree through system commands.
> system("tree")

├── a1
│   └── b2
│       └── c3
├── create
├── readme.txt
└── tmp
```

3.4.2.3 Checking if a Directory Exists

We can use the function file.exists() to check whether a directory exists.

```
# The directory exists.
> file.exists(".")
[1] TRUE
> file.exists("./a1/b2")
[1] TRUE
```

```
# The directory does not exist.
> file.exists("./aa")
[1] FALSE
```

3.4.2.4 Checking the Permission of a Directory

In the Linux system, each directory or file has permission definition. We can check the permission.

```
> df<-dir(full.names = TRUE)
# Check whether files or directories exist, mode=0.
> file.access(df, 0) == 0
./a1      ./create ./readme.txt       ./tmp
TRUE         TRUE         TRUE         TRUE

# Check whether files or directories are executable, mode=1.
> file.access(df, 1) == 0
./a1      ./create ./readme.txt       ./tmp
TRUE         TRUE        FALSE         TRUE

# Check whether files or directories are writable, mode=2.
> file.access(df, 2) == 0
./a1      ./create ./readme.txt       ./tmp
TRUE         TRUE         TRUE         TRUE

# Check whether files or directories are readable, mode=4.
> file.access(df, 4) == 0
./a1      ./create ./readme.txt       ./tmp
TRUE         TRUE         TRUE         TRUE
```

We can also modify the permission of a directory.

```
# Modify the directory's permission to be read-only for all users.
> Sys.chmod("./create", mode = "0555", use_umask = TRUE)

# View the complete information of the directory, mode=555.
>  file.info("./create")
     size isdir mode      mtime      ctime      atime  uid  gid uname grname
./create 4096  TRUE   555 2013-11-14 08:36:28 2013-11-14 09:07:05 2013-11-
14 08:36:39 1000 1000 conan   conan

# Make the create directory not writable.
> file.access(df, 2) == 0
./a1      ./create ./readme.txt       ./tmp
TRUE        FALSE         TRUE         TRUE
```

3.4.2.5 Renaming a Directory

We can rename a directory using the function file.rename().

```
# Rename the directory tmp.
> file.rename("tmp", "tmp2")
[1] TRUE

# View  the directory.
> dir()
[1] "a1"          "create"        "readme.txt" "tmp2"
```

3.4.2.6 Deleting a Directory

We can delete a directory using the function unlink(). If the directory contains sub-directories or files, we need to delete them recursively.

```
# Delete the directory tmp2.
> unlink("tmp2", recursive = TRUE)
# View  the directory.
> dir()
[1] "a1"          "create"       "readme.txt"
```

3.4.2.7 Other Functions

In addition to the above basic directory manipulations, R also provides some assistant functions, such as concatenating the directory string, getting the sub-directory name at the lowest level, converting the file's extension path, normalizing the path separators and short paths of Windows or Linux, and so on.

Concatenate directory string.

```
# Concatenate directory string.
> file.path("p1","p2","p3")
[1] "p1/p2/p3"
> dir(file.path("a1","b2"))
[1] "c3"
```

Get the sub-directory name at the lowest level.

```
# Get the current directory.
> getwd()
[1] "/home/conan/R/fs"
# Get the complete directory path.
> dirname("/home/conan/R/fs/readme.txt")
[1] "/home/conan/R/fs"
# Get the name of directory at the lowest level.
> basename(getwd())
[1] "fs"
# Get the name of file at the lowest level.
>  basename("/home/conan/R/fs/readme.txt")
[1] "readme.txt"
```

Convert the file extension path.

```
# Convert ~ as user directory.
> path.expand("~/foo")
[1] "/home/conan/foo"
```

Normalize the path, used for converting the path separators in Windows or Linux.

```
# For Linux system.
> normalizePath(c(R.home(), tempdir()))
[1] "/usr/lib/R"       "/tmp/RtmpqNyjPD"
```

```
# For Windows system.
> normalizePath(c(R.home(), tempdir()))
[1]  "C:\\Program Files\\R\\R-3.0.1"
[2]  "C:\\Users\\Administrator\\AppData\\Local\\Temp\\RtmpMtSnci"
```

The short path means to shorten the length of the displayed path, only run in Windows.

```
# For Windows system.
> shortPathName(c(R.home(), tempdir()))
[1]  "C:\\PROGRA~1\\R\\R-30~1.1"
[2]  "C:\\Users\\ADMINI~1\\AppData\\Local\\Temp\\RTMPMT~1"
```

3.4.3 File Manipulations

R can manipulate not only directories but also files with rich API supports.

3.4.3.1 Viewing a File

We can view the files in current directory using the function dir().

```
# View the files in current directory.
> dir()
[1]  "create"  "readme.txt"
# Check whether a file exists.
> file.exists("readme.txt")
[1] TRUE
# The file doesn't exist.
> file.exists("readme.txt.222")
[1] FALSE
# View the file's complete information.
> file.info("readme.txt")
      size isdir mode     mtime       ctime       atime  uid  gid uname grname
readme.txt    7 FALSE  664 2013-11-14 08:24:50 2013-11-14 08:24:50 2013-
11-14 08:24:50 1000 1000 conan   conan

# View the file's access permission: exist.
>  file.access("readme.txt",0)
readme.txt
        0

# The file permission: not executable.
>  file.access("readme.txt",1)
readme.txt
       -1

# The file permission: writable.
>  file.access("readme.txt",2)
readme.txt
        0

# The file permission: readable.
>  file.access("readme.txt",4)
readme.txt
        0
```

```
# View the access permission of a file that doesn't exist: not exist.
> file.access("readme.txt222")
readme.txt222
            -1
```

Check whether an entry is file or directory.

```
# Check whether an entry is a directory.
> file_test("-d", "readme.txt")
[1] FALSE
> file_test("-d", "create")
[1] TRUE

# Check whether an entry is a file.
> file_test("-f", "readme.txt")
[1] TRUE
> file_test("-f", "create")
[1] FALSE
```

3.4.3.2 Creating a File

We can create a file using the function file.create().

```
# Create an empty file A.txt.
> file.create("A.txt")
[1] TRUE

# Create a file B.txt with content filled.
> cat("file B\n", file = "B.txt")
> dir()
[1] "A.txt"      "B.txt"      "create"      "readme.txt"

# Print A.txt.
> readLines("A.txt")
character(0)
# Print B.txt.
> readLines("B.txt")
[1] "file B"
```

Merge the content from B.txt into A.txt.

```
# Merge files.
> file.append("A.txt", rep("B.txt", 10))
 [1]  TRUE TRUE TRUE TRUE TRUE TRUE TRUE TRUE TRUE TRUE

# View the content of the file.
> readLines("A.txt")
 [1] "file B" "file B" "file B" "file B" "file B" "file B" "file B" "file
B" "file B" "file B"
```

Copy the file A.txt to C.txt.

```
# Copy file.
> file.copy("A.txt", "C.txt")
[1] TRUE
```

```
# View the content of the file.
> readLines("C.txt")
 [1] "file B" "file B" "file B" "file B" "file B" "file B" "file B" "file
B" "file B" "file B"
```

3.4.3.3 Modifying the Permission of a File

We can modify the permission of a file using the function Sys.chmod().

```
# Modify the file permission, readable, writable and executable for
owners, no permission for others.
 > Sys.chmod("A.txt", mode = "0700", use_umask = TRUE)

# View the file's fnformation.
> file.info("A.txt")
     size isdir mode    mtime        ctime      atime  uid  gid uname grname
A.txt    70 FALSE   700 2013-11-14 12:55:18 2013-11-14 12:57:39 2013-11-14
12:55:26 1000 1000 conan   conan
```

3.4.3.4 Renaming a File

We can rename a file using the function file.rename().

```
# Rename the file A.txt as AA.txt.
> file.rename("A.txt","AA.txt")
[1] TRUE
> dir()
[1] "AA.txt"     "B.txt"     "create"     "C.txt"     "readme.txt"
```

3.4.3.5 Hard Link and Soft Link

- The hard link means to link through the indexing node (inode). In the Linux file system, the file stored in disk sectors is assigned an ID, called inode index, regardless of the type of file. In Linux, multiply file names can refer to a same inode, which, generally speaking, is called the hard link. The function of the hard link is to allow a file to have multiple valid path names, so that user can create hard links to important files to prevent "mistaken deletion." As mentioned above the reason is that the inode corresponding to the specific directory has more than one hard link and only deleting one hard link does not affect the inode or other links. Only if the last one is deleted, the data blocks of the file and its directory link is released. In another words, to radically delete a file, you must delete all the related hard link files.
- The soft link, also known as the symbolic link, is similar to the shortcut in Windows. Actually it is a special text file that contains the location information of another file.

Note: The two concepts are used only in the Linux system.

```
# Hard link.
> file.link("readme.txt", "hard_link.txt")
[1] TRUE

# Soft link.
> file.symlink("readme.txt", "soft_link.txt")
[1] TRUE
```

```
# View the file directory.
> system("ls -l")
-rwx------ 1 conan conan   70 Nov 14 12:55 AA.txt
-rw-rw-r-- 1 conan conan    7 Nov 14 12:51 B.txt
dr-xr-xr-x 2 conan conan 4096 Nov 14 08:36 create
-rw-rw-r-- 1 conan conan   70 Nov 14 12:56 C.txt
-rw-rw-r-- 2 conan conan    7 Nov 14 08:24 hard_link.txt
-rw-rw-r-- 2 conan conan    7 Nov 14 08:24 readme.txt
lrwxrwxrwx 1 conan conan   10 Nov 14 13:11 soft_link.txt -> readme.txt
```

The file hard_link.txt is the hard link file of readme.txt while soft_link.txt is the soft link of it.

3.4.3.6 Deleting a File

Tow functions file.remove and unlink are available, where the function unlink is same as the operation to delete a directory.

```
# Delete files.
> file.remove("A.txt", "B.txt", "C.txt")
[1] FALSE   TRUE   TRUE

# Delete a file.
> unlink("readme.txt")

> system("ls -l")
total 12
-rwx------ 1 conan conan   70 Nov 14 12:55 AA.txt
dr-xr-xr-x 2 conan conan 4096 Nov 14 08:36 create
-rw-rw-r-- 1 conan conan    7 Nov 14 08:24 hard_link.txt
lrwxrwxrwx 1 conan conan   10 Nov 14 13:11 soft_link.txt -> readme.txt

# View the file directory.
readLines("hard_link.txt")
[1] "file A"

# Print the soft link file. Since the original file was deleted, an error
  occurred.
> readLines("soft_link.txt")
Error in file(con, "r") : cannot open the connection
In addition: Warning message:
In file(con, "r"):
  cannot open file 'soft_link.txt': No such file or directory
```

3.4.4 Some Special Directories

- R.home() is used to view the directories related to R.
- .Library is used to view the directories of R's core packages.
- .Library.site is used to view the directories of R's core packages and the installation directory of root user.
- .libPaths() is used to view the storing directories of all R packages.
- system.file() is used to view the directory where the specific package is stored.

3.4.4.1 R.home() to View the Directories Related to R

```
# Print the installation directory of the R software.
> R.home()
[1] "/usr/lib/R"

# Print the bin directory of the R software.
> R.home(component="bin")
[1] "/usr/lib/R/bin"

# Print the doc directory of the R software.
>  R.home(component="doc")
[1] "/usr/share/R/doc"
```

Locate the R files through system commands.

```
# Find the locations of R files in system.
~ whereis R
R: /usr/bin/R /etc/R /usr/lib/R /usr/bin/X11/R /usr/local/lib/R /usr/
share/R /usr/share/man/man1/R.1.gz
#Print the environment variable R_HOME.
~ echo $R_HOME
/usr/lib/R
```

As we have seen, it is easy for us to locate the directories of R using R.home().

3.4.4.2 The Package Directories of R

```
# Print the directory of the core packages.
> .Library
[1] "/usr/lib/R/library"

# Print the directories of the core packages and the installation package
directories of root user.
> .Library.site
[1] "/usr/local/lib/R/site-library" "/usr/lib/R/site-library"
[3] "/usr/lib/R/library"

# Print all the directories where the packages are stored.
> .libPaths()
[1] "/home/conan/R/x86_64-pc-linux-gnu-library/3.0"
[2] "/usr/local/lib/R/site-library"
[3] "/usr/lib/R/site-library"
[4] "/usr/lib/R/library"
```

3.4.4.3 Viewing the Directory Where the Specific Package Is Stored

```
# The directory where the base package is stored.
> system.file()
[1] "/usr/lib/R/library/base"

# The directory where the pryr package is stored.
>  system.file(package = "pryr")
[1] "/home/conan/R/x86_64-pc-linux-gnu-library/3.0/pryr"
```

Frankly speaking, it is convenient for us to manipulate the file system using R. But the function names are not normative, which costs us a lot of time to remember.

3.5 New Features in R Version 3.1

Question
How do we master the new features of R?

New features in R version 3.1
http://blog.fens.me/r-version-3-1/

Introduction
The R language keeps developing and improving. Starting from R 3.0.0, the R language has been growing with milestones. With the continuous updates for R kernel, more and more pure computing indicators are added in, which make R much closer to the enterprise commercial language family. I believe that R will be strong and keep growing stronger.

This section will introduce the new features and their usages of the version 3.1.x of R.

3.5.1 Introduction to the Versions of R 3.1

R version 3.1.0 was released on April 10, 2014, with 64 new features exposed and 16 bugs fixed. It was a large-scale release. When this book was written, the latest version 3.1.1 of R was released on July 10, 2014, with 17 new features exposed and 35 bugs fixed.

Usually, a version id of software contains three parts: X.Y.Z, separated with dots from left to right, such as R version 3.1.1.

- X represents the big version number, indicating that there are important updates or milestone features released. The releases with the big version number changed are recommended for upgrade. We'd better use the latest software. For instance, from Java5, Java6, Java7 to Java8, each upgrade has important enhancements and improvements to Java's performance and features. Of course, the upgrades with big versions may be incompatible backwards, resulting in failures of many applications developed before. For example, Python 2.7.x and Python 3.x are not compatible in Syntax. We have to use both in cases.
- Y represents the middle version number, related to some functionality. Every update with the Y version brings lots of small functionalities. Usually those small functionalities do

not abstract people's attention and do not affect the use of R, so we can upgrade when we remember to do so. It does not matter being out-of-date for just one or two Y versions.

■ Z represents the small version number, mainly related to bug fix. Each update with X or Y version may bring bugs that need to be fixed in time. So the Z version will be released often with a frequency of one to two months or three to five days. We can consider a Z version as a system patch to a specific bug. It's not necessary to patch until we encounter the bug. If you don't encounter the bug, we can choose not to upgrade with the Z version and wait for the next Y version.

Actually the rapid updates of R reflect the urgent voice of the market. If R can support concurrency, if R can support asynchronization, if R can build a stand-alone application server, if R can handle protocols like socket or http easily, and if R can handle big data…numerous functional requirements have been submitted to R's core team. I also hope the R technical team can soon overcome the technical problems , making R outstanding in enterprise applications.

The fact that R launches a new release about every four months shows the hard work of the R team. We can download the latest version of R package and check the details about version updates from the official mirror site (http://cran.rstudio.com/) as shown in Figure 3.3.

The system environment used in this section:

■ Win7 64bit
■ R: 3.1.1 x86_64-w64-mingw32/x64 (64-bit)

In the following, I will completely explain the new features of versions 3.1.0 and 3.1.1 and depict them with code.

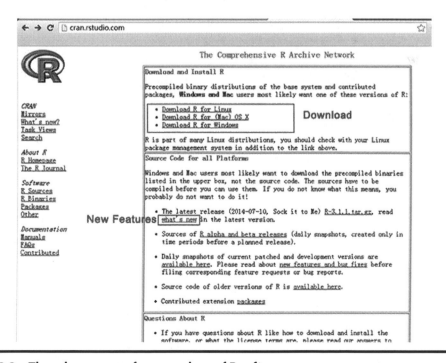

Figure 3.3 The release page of new versions of R software.

3.5.2 *The New Features of R 3.1.0 and Code Depict*

The alias of R 3.1.0 is Spring Dance. After downloading R 3.1.0, you can view the version information through version.

```
# Launch the R application.
~ R
# View the version information.
> version
platform       x86_64-w64-mingw32
arch           x86_64
os             mingw32
system         x86_64, mingw32
status
major          3
minor          1.0
year           2014
month          04
day            10
svn rev        65387
language       R
version.string R version 3.1.0 (2014-04-10)
nickname       Spring Dance
```

The new features

1. type.convert() (and hence by default read.table()) returns a character vector or factor when representing a numeric input as a double would lose accuracy. Similarly for complex inputs.

   ```
   # This function returns factor type.
   > type.convert(c('abc','bcd'))
   [1] abc bcd
   Levels: abc bcd

   # The double type loses precision.
   > type.convert(c(as.double(1.12121221111),'1.121'))
   [1] 1.121212 1.121000
   ```

2. If a file contains numeric data with unrepresentable numbers of decimal places that are intended to be read as numeric, specify colClasses in read.table() to be "numeric".
   New a file num.csv containing decimals.

   ```
   1,2,1.11
   2.1,3,4.5
   ```

   Read the file using read.table() and view the types of columns.

   ```
   # Read the file.
   > num<-read.table(file="num.csv",sep=",")
   > num
     V1 V2   V3
   1 1.0  2 1.11
   2 2.1  3 4.50
   ```

```
> class(num)
[1] "data.frame"
> class(num$V1)
[1] "numeric"
```

3. tools::Rdiff(useDiff = FALSE) is closer to the POSIX definition of diff –b (as distinct from the description in the man pages of most systems) in function Rdiff() in tools package, when the arguments useDiff = FALSE, the effect is similar as diff –b in POSIX system.
New a file num2.csv.

```
3,2,1.11
2.1,3,4.5
```

Compare the two files num.csv and num2.csv.

```
> Rdiff('num.csv','num2.csv',useDiff = FALSE)
1c1
< 1,2,1.11
---
> 3,2,1.11
[1] 1
```

4. The new function anyNA() returns same result as any(is.na()) and has better performance.

```
> is.na(c(1, NA))
[1] FALSE   TRUE
> any(is.na(c(1, NA)))
[1] TRUE
> anyNA(c(1, NA))
[1] TRUE
```

5. arrayInd() and which() have new argument useName for column name matching. I couldn't understand the meaning of this argument when I did testing.

```
> which
function (x, arr.ind = FALSE, useNames = TRUE)
```

6. is.unsorted() supports handling vectors of raw data.

```
# A sorted vector.
> is.unsorted(1:10)
[1] FALSE

# An unsorted vector.
> is.unsorted(sample(1:10))
[1] TRUE
```

7. The functions as.data.frame and as.data.frame.table for handling table supports pass arguments to provide Dimnames(sep,base). I couldn't understand what the update is when I did testing.

8. uniroot() gets new optional argument, extendInt, allows for automatically expending the search interval. It also has a new return object argument init.it.

```
# The function f1.
> f1 <- function(x) (121 - x ^2)/(x ^2+1)
# The function f2.
> f2 <- function(x) exp(-x)*(x - 12)

# Solve the roots of function f1 in interval (0,10).
> try(uniroot(f1, c(0,10)))
Error in uniroot(f1, c(0, 10)):
  f() values at end points not of opposite sign
# Solve the roots of function f2 in interval (0,2).
> try(uniroot(f2, c(0, 2)))
Error in uniroot(f2, c(0, 2)):
  f() values at end points not of opposite sign

# Extend the search interval through argument extendInt.
> str(uniroot(f1, c(0,10),extendInt="yes"))
List of 5
 $ root      : num 11
 $ f.root    : num -3.63e-06
 $ iter      : int 12
 $ init.it   : int 4
 $ estim.prec: num 6.1e-05

# Extend the search interval through argument extendInt.
> str(uniroot(f2, c(0,2), extendInt="yes"))
List of 5
 $ root      : num 12
 $ f.root    : num 4.18e-11
 $ iter      : int 23
 $ init.it   : int 9
 $ estim.prec: num 6.1e-05
```

9. When the argument f is of a factor, switch(f,) would give warning to convert to character argument.

```
> ff<-gl(3,1, labels=LETTERS[3:1])
# Warning happens.
> switch(ff[1], A = "I am A", B="Bb..", C=" is C")# -> "A"
[1] "I am A"
Warning message:
In switch(ff[1], A = "I am A", B = "Bb..", C = " is C"):
  EXPR is a "factor", treated as integer.
 Consider using 'switch(as.character( * ), …)' instead.

# Convert to character string for handling.
> switch(as.character(ff[1]), A = "I am A", B="Bb..", C=" is C")
[1] " is C"
```

10. The parser has been updated to use less memory.
11. The unary operators (+ - !) expect for names, dims and dimnames will make a copy computing when processing attributes.

```
# Create a variable x.
> x<-1:12;x
 [1]  1  2  3  4  5  6  7  8  9 10 11 12
# Do copy computing, which doesn't affect the value of the original
variable x.>
 x+5
 [1]  6  7  8  9 10 11 12 13 14 15 16 17
> !x
 [1] FALSE FALSE FALSE FALSE FALSE FALSE FALSE FALSE FALSE FALSE
FALSE FALSE

# Directly modify the value of the original variable x.> dim(x)
<- c(3,4);x
     [,1] [,2] [,3] [,4]
[1,]    1    4    7   10
[2,]    2    5    8   11
[3,]    3    6    9   12
```

12. colorRamp() and colorRampPalette() support transparent colors through setting the alpha argument to TRUE.

```
# The handle of the function that gets colors.
> cols<-colorRampPalette(c(rgb(0,0,1,1), rgb(1,0,1,0)), alpha = TRUE)
# Draw a picture with a volcano with transparent colors as shown in
  Fig. 3.4.
> filled.contour(volcano,color.palette =cols,asp = 1)
```

13. Both grid.show.layout() and grid.show.viewports() get an optional argument vp.ex for layout zooming.

```
# Load the grid package.
> library(grid)
# View the definition of the function.
> grid.show.layout
function (l, newpage = TRUE, vp.ex = 0.8, bg = "light grey",
    cell.border = "blue", cell.fill = "light blue", cell.label = TRUE,
    label.col = "blue", unit.col = "red", vp = NULL)
```

14. A new function find_gs_cmd() can be found in tools package for locating a GhostScript executable.

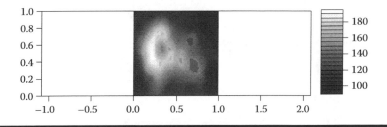

Figure 3.4   A volcano picture with transparent colors.

15. object.size() gets a method called format() for formatting the display.

```
> letters
 [1] "a" "b" "c" "d" "e" "f" "g" "h" "i" "j" "k" "l" "m" "n" "o" "p"
     "q" "r" "s" "t"
[21] "u" "v" "w" "x" "y" "z"
# View the size of letters object.
> object.size(letters)
1496 bytes
# Format the display.
> format(object.size(letters), units = "auto")
[1] "1.5 Kb"
```

16. A new font ArialMT is added for the device outputs through pdf() and postscript().
17. The files NEWS and NEWS.2 and the new text and PDF files are moved to doc directory.

```
# View the doc directory including files NEWS and NEWS.2.
> dir("C:/Program Files/R/R-3.1.0/doc")
 [1] "AUTHORS"          "CHANGES"          "CHANGES.rds"     "COPYING"
 [5] "COPYRIGHTS"       "CRAN_mirrors.csv" "FAQ"             "html"
 [9] "KEYWORDS"         "KEYWORDS.db"      "manual"          "NEWS"
[13] "NEWS.0"           "NEWS.1"           "NEWS.2"          "NEWS.pdf"
[17] "NEWS.rds"         "README.packages"  "README.Rterm"    "RESOURCES"
[21] "rw-FAQ"           "THANKS"
```

18. combn(x) supports the argument x as factor type and returns factor type.

```
# The character type.
> combn(letters[1:4], 2)
     [,1] [,2] [,3] [,4] [,5] [,6]
[1,] "a"  "a"  "a"  "b"  "b"  "c"
[2,] "b"  "c"  "d"  "c"  "d"  "d"

# The factor type.
> combn(factor(letters[1:4]), 2)
     [,1] [,2] [,3] [,4] [,5] [,6]
[1,] a    a    a    b    b    c
[2,] b    c    d    c    d    d
Levels: a b c d
```

19. The new function fileSnapshot() and changedFiles() are added to the utils package for generating file snapshot of directories and comparing the snapshot.

```
# Generate a snapshot.
> snapshot<-fileSnapshot()
> snapshot
File snapshot:
 path = D:\workspace\R\basic\r311
 timestamp =    .
 file.info = TRUE
 md5sum = FALSE
 digest = NULL
 full.names = FALSE
 args = list()
 9 files recorded.
```

```
# Add a file a.txt in current directory.
> writeBin(3L:4L,"a.txt")

# Compare the directory snapshot.
> changedFiles(snapshot)
Files added:
  a.txt
```

20. make.names() is able to handle ineligible variable names. When setting the argument unique as TRUE, the new variable names keep unique.

```
# Process the invalid variable names.
> make.names(c("a b","a.b", "a-b"))
[1] "a.b" "a.b" "a.b"

# Generate 3 unique variable names.
> make.names(c("a b","a.b", "a-b"), unique = TRUE)
[1] "a.b.1" "a.b"   "a.b.2"
```

21. New functions cospi(x), sinpi(x), and tanpi(x) are added for computing cos(pi*x) more accurately, called by lgamma(), bessel1(), and other functions.

```
> x<-1222222222222221.323232
# When x is too large, this function is used to solve the problem of
error in cos(pi*x).> cospi(x)
[1] -0.7071068
> cos(pi*x)
[1]  0.7175645
```

22. print.table(x) supports x as decimal-valued.

```
> t1 <- round(abs(rt(10, df = 1.8)),1)
> t2 <- round(abs(rt(10, df = 1.4)),1)
> print.table(table(t1,t2),zero.print = ".")
     t2
t1    0 0.1 0.4 0.5 1 1.2 1.3 1.7 8.2
  0.1 .   .   .   . .   .   .   .   1
  0.2 .   .   1   . .   .   1   .   .
  0.4 .   .   .   . 1   .   .   .   .
  0.7 .   1   .   . .   .   .   .   .
  0.8 1   .   .   . .   .   .   .   .
  1.2 .   .   .   1 .   1   .   .   .
  1.4 .   .   1   . .   .   .   .   .
  2.2 .   .   .   . .   .   1   .   .
```

23. More time-zone names are supported. One can view the time-zone list through OlsonNames() and view the time-zone bound to current system environment through Sys.timezone().

```
# Time-zone list.
> head(OlsonNames())
[1] "Africa/Abidjan"     "Africa/Accra"
[3] "Africa/Addis_Ababa" "Africa/Algiers"
[5] "Africa/Asmara"      "Africa/Asmera"
```

```
# The time-zone bound to my system.
> Sys.timezone()
[1] "Asia/Taipei"
```

24. 64-bit time_t type is supported so the system can handle date-times outside 32-bit range (before 1902 or after 2037). OS X platform is not supported in this version.

    Currently the time_t type is used on some of Unix-like 64-bit systems and Windows 64-bit.

25. A new environment configuration is added. save.defaults options can include the configuration for compression_level.

26. colSums() supports arrays and data frames columns with 2^31 or more elements.

27. The performance of as.factor() is optimized. The conversion speed of non-integer valued is accelerated.

```
> a<-rnorm(1000000,-100,100000)
> head(a)
[1]     9856.935   154567.963  -200041.134     43363.338    -74436.650
[6] -178322.313
> system.time(as.factor(a))
User System Elapse
9.50 0.03 9.64
```

28. fft() supports computation with larger amount of data, from 12 million up to 2 billion.

```
# Do FFT against 100 million numbers.
> x <- 1:100000000
> system.time(fft(x))
User System Elapse
32.96  0.22 33.32
```

29. svd() is implemented using LAPACK subroutine ZGESDD, which is the complex analogue of the routine used for the real case.

30. If you want the .tex file output by Sweave to be UTF-8 encoded, you need to add a configuration of %\SweaveUTF8 in LaTex file.

31. file.copy() has a new argument copy.date, which makes the copied file has the same modified time as the original file.

```
# Create a file b.txt by copy.
> file.copy("a.txt","b.txt",copy.date=TRUE)
[1] TRUE
# Create a file c.txt by copy.
> file.copy("a.txt","c.txt",copy.date=FALSE)
[1] TRUE
```

a.txt and b.txt have some modified time while c.txt was created newly, as shown in Figure 3.5.

32. The date-times can be printed using time-zone abbreviations through an optional argument zone in POSIXlt class. For example, the abbreviation of Paris pres-1940 may be LMT (Local Mean Time), PMT (Paris Mean Time), WET (Western European Time), or WEST (Western European Summer Time).

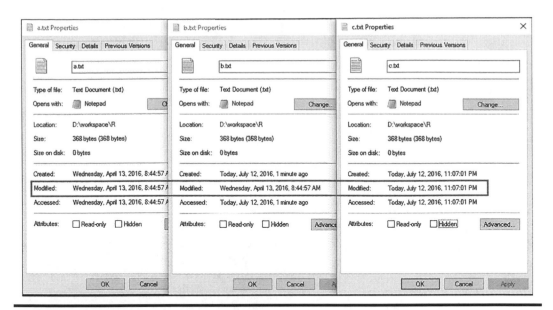

Figure 3.5   Compare the properties of three files.

33. The component gmtoff can be used to record the GMT offset on supported platforms.
34. The C function strftime() is updated to be conformed to POSIX 2008 standard, used on Windows and OS X.
35. dnorm() is more accurate for |x|>5, but slower somewhat.

```
# Execute dnorm().
> dnorm(rnorm(10,0,100))
 [1] 1.151100e-271  0.000000e+00 2.403071e-198  0.000000e+00
 [5]  0.000000e+00 3.358647e-208  0.000000e+00  0.000000e+00
 [9]  0.000000e+00 1.136764e-154
```

36. tiff() adds compression options.

```
> tiff
function (filename = "Rplot%03d.tif", width = 480, height = 480,
units = "px", pointsize = 12, compression = c("none", "rle","lzw",
"jpeg", "zip", "lzw+p", "zip+p"), bg = "white", res = NA, family =
"sans", restoreConsole = TRUE, type = c("windows","cairo"), antialias
= c("default", "none", "cleartype", "grey", "subpixel"))
```

37. read.table(), readLines(), and scan() have new argument skipNul for skipping nuls when reading data.

```
> readLines
function (con = stdin(), n = -1L, ok = TRUE, warn = TRUE, encoding =
"unknown", skipNul = FALSE)
```

38. Duplicating the right hand side values is avoided in complex assignment. This reduces the copying of replacement values.
In the meantime, some other changes reduce copying of objects.

39. The argument fast in KalmanLike(), KalmanRun(), and KalmanForecast() is replaced by argument update, returning the updated model.

```
> fit3 <- arima(presidents, c(3, 0, 0))
# The model.
> mod <- fit3$model
# Execute the Kalman prediction.
> pr <- KalmanForecast(4, mod, TRUE)
# The updated model is stored in the mod attribute.
> mod <- attr(pr, "mod")
```

40. A new argument SSinit is added in arima() and makeARIMA(), allowing the choice of likelihood functions for state space initialization.

```
> arima
function (x, order = c(0L, 0L, 0L), seasonal = list(order = c(0L, 0L,
0L), period = NA), xreg = NULL, include.mean = TRUE, transform.pars =
TRUE, fixed = NULL, init = NULL, method = c("CSS-ML", "ML", "CSS"),
n.cond, SSinit = c("Gardner1980", "Rossignol2011"), optim.method =
"BFGS", optim.control = list(), kappa = 1e+06)
```

41. A new argument noBreaks is added to warning() for simplifying the processing of output.

```
> warning
function (…, call. = TRUE, immediate. = FALSE, noBreaks. = FALSE,
domain = NULL)
```

42. A new argument encoding is added to pushback(), supporting for functions like scan(), and read.table() reading files with UTF-8 encoding.

```
# View the definition of pushback().
> pushBack
function (data, connection, newLine = TRUE, encoding = c("", "bytes",
"UTF-8"))

# Create a text pipeline.
> zz <- textConnection(LETTERS)
> readLines(zz, 2)
[1] "A" "B"
> pushBack(c("aa", "bb"), zz)
> pushBackLength(zz)
[1] 2
> readLines(zz, 1)
[1] "aa"
> pushBackLength(zz)
[1] 1
> readLines(zz, 1)
[1] "bb"
> readLines(zz, 1)
[1] "C"
> close(zz)
```

43. A new argument use.names is added to all.equal.list() to check the equality of objects by names rather than index.

```
> a<-list(a=1,b=2)
> b<-list()
> b[['a']]<-1
> b[['b']]<-2
# Equal.
> all.equal.list(a,b)
[1] TRUE


> b[['a']]<-4
# Unequal.
> all.equal.list(a,b, use.names=TRUE)
[1] "Component "a": Mean relative difference: 3"
```

44. A new argument check.attributes is added to all.equal and attr.all.equal() for comparing columns.

    A side effect is that some previously undetected errors of passing empty arguments (no object between commas) to all.equal() are detected and reported.

    There are explicit checks that check.attributes is logical, tolerance is numeric, and scale is NULL or numeric. This catches some unintended positional matching.

45. When the compared objects have different lengths, all.equal.numeric will show a "scaled difference" message but not error.

```
> all.equal.numeric(1:10,1:5)
[1] "Numeric: lengths (10, 5) differ"
```

46. all.equal() now has a POSIXt method replacing the POSIXct method.

47. seq() can use "by=quarter" to generate Date or POSIXt typed sequence.

```
# By quarter.
> seq(today,today+365,by="quarter")
[1] "2014-09-26" "2014-12-26" "2015-03-26" "2015-06-26" "2015-09-26"

# By 2 months.
> seq(today,today+365,by="2 months")
[1] "2014-09-26" "2014-11-26" "2015-01-26" "2015-03-26" "2015-05-26"
[6] "2015-07-26" "2015-09-26"
```

48. file.path() is used to adapt path separator. Unfortunately, this function has no work when I did testing.

```
> paste(getwd(),'/',sep='')
[1] "D:/workspace/R/"

> file.path(paste(getwd(),'/',sep=''))
[1] "D:/workspace/R/"
```

49. A new function agrepl() is added for fuzzy mathing.

```
> agrepl("laysy", c("1 lazy", "1", "1 LAZY"), max = 2)
[1]   TRUE FALSE FALSE
```

50. fifo is now supported on Windows.

```
# Check whether the system supports fifo.
> capabilities("fifo")
fifo
TRUE
```

51. sort.list(method="radix") supports radix sorting in case the data amount is large but the unique values number is small, which is much faster than ordinary sorting.

```
> x<-sample(1:650,1e7,replace=TRUE)
# The Ordinary sorting.
> system.time(o1<-sort.list(x))
User System Elapse
7.13 0.02 7.14
# The radix sorting.
> system.time(o2<-sort.list(x,method="radix"))
User System Elapse
0.08 0.00 0.07
# The results are exactly equal.
> all.equal(o1,o2)
[1] TRUE
```

52. Some functionality of print.ts() is available in .preformat.ts().

```
> sunsp.1 <- window(sunspot.month, end=c(1752, 12))
> m <- .preformat.ts(sunsp.1)
> print(m)
      Jan     Feb     Mar     Apr     May     Jun     Jul     Aug
1749 " 58.0" " 62.6" " 70.0" " 55.7" " 85.0" " 83.5" " 94.8" " 66.3"
1750 " 73.3" " 75.9" " 89.2" " 88.3" " 90.0" "100.0" " 85.4" "103.0"
1751 " 70.0" " 43.5" " 45.3" " 56.4" " 60.7" " 50.7" " 66.3" " 59.8"
1752 " 35.0" " 50.0" " 71.0" " 59.3" " 59.7" " 39.6" " 78.4" " 29.3"
      Sep     Oct     Nov     Dec
1749 " 75.9" " 75.5" "158.6" " 85.2"
1750 " 91.2" " 65.7" " 63.3" " 75.4"
1751 " 23.5" " 23.2" " 28.5" " 44.0"
1752 " 27.1" " 46.6" " 37.6" " 40.0"
```

53. A new argument detach is added to mcparallel(). When setting it as TRUE, the execution of code can be independent of the current session. The function calls mcfork() with estranged=TRUE to launch a child process independent of the parent process.

54. pdf() omits the circles and text at extremely small size in its output file, since some viewers failed on such files.

55. The right most break for the "months," "quarters," and "years" cases of hist.POSIXlt() has been increased by a day.

56. data.frame type supports index = 0. No error happened in such case.

```
> df<-data.frame(1:5)
> df[0,]
integer(0)
> df[1,]
[1] 1
```

57. hclust() gains a new method Ward.D2 which implements Ward's method correctly. The previous "ward" method is renamed "Ward.D."

```
> m<-matrix(sample(100,100,replace=TRUE),10);m
       [,1] [,2] [,3] [,4] [,5] [,6] [,7] [,8] [,9] [,10]
 [1,]    8   21   93   10    4   40   61   26   55    22
 [2,]  100   85   24   98   94    1   75   70   34    85
 [3,]   12   59   12   13   28   72   76   81   94    12
 [4,]   97   34   21   50   98   26   14   37   72    62
 [5,]   27   98   14   22   16   90   57   36   60    29
 [6,]   24   87   22    6   61   68   17   13   27    93
 [7,]   94   87   64    5   89   81   27    1   18    61
 [8,]   98   89    5    5   40   26   45    4   36    53
 [9,]   34   10   32   31  100   31   64   20   25    85
[10,]   19   38   71   72   35   40   75   13   54    94

> hc1<-hclust(dist(m), method="ward.D");hc1
Call:
hclust(d = dist(m), method = "ward.D")
Cluster method   : ward.D
Distance         : euclidean
Number of objects: 10

> hc2<-hclust(dist(m), method="ward.D2");hc2
Call:
hclust(d = dist(m), method = "ward.D2")
Cluster method   : ward.D2
Distance         : euclidean
Number of objects: 10

# Draw graph, as shown in Fig. 3.6.
> plot(hc1)
> plot(hc2)
```

58. The sunspot.month dataset is updated whereas susport.year remains unchanged.

```
> head(sunspot.month)
[1] 58.0 62.6 70.0 55.7 85.0 83.5
> head(sunspot.year)
[1]    5 11 16 23 36 58
```

59. The summary() method for "lm" fits warns if the fit is essentially perfect, as most of the summary may be computed inaccurately (and with platform-dependent values). I didn't receive any warn when testing on Windows.

```
> ctl <- c(4.17,5.58,5.18,6.11,4.50,4.61,5.17,4.53,5.33,5.14)
> trt <- c(4.81,4.17,4.41,3.59,5.87,3.83,6.03,4.89,4.32,4.69)
> group <- gl(2, 10, 20, labels = c("Ctl","Trt"))
> weight <- c(ctl, trt)
> lm.D9 <- lm(weight ~ group)

> summary(lm.D9)
```

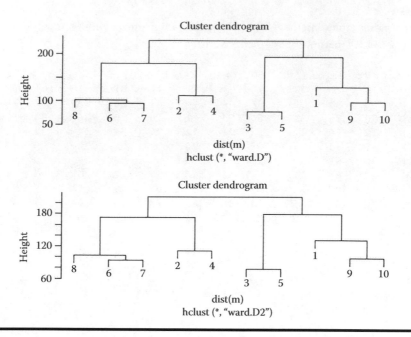

Figure 3.6 Compare the differences between two clustering algorithms.

```
Call:
lm(formula = weight ~ group)

Residuals:
    Min      1Q  Median      3Q     Max
-1.0710 -0.4938  0.0685  0.2462  1.3690

Coefficients:
            Estimate Std. Error t value Pr(>|t|)
(Intercept)   5.0320     0.2202  22.850 9.55e-15 ***
groupTrt     -0.3710     0.3114  -1.191    0.249
---
Signif. codes:  0 '***' 0.001 '**' 0.01 '*' 0.05 '.' 0.1 ' ' 1

Residual standard error: 0.6964 on 18 degrees of freedom
Multiple R-squared:  0.07308,   Adjusted R-squared:  0.02158
F-statistic: 1.419 on 1 and 18 DF,  p-value: 0.249
```

When extracting the results of the summary() function, the best way is to wrap the calls to summary() in suppressWarnings().

```
> suppressWarnings(summary(lm.D9)$cov.unscaled)
            (Intercept) groupTrt
(Intercept)         0.1     -0.1
groupTrt           -0.1      0.2
```

Hereby we have finished reviewing the 61 updates. Most of them are related to performance and some others are syntax improvements, from which we can see the priorities of R improvements.

3.5.3 Introduction and Code Depict of the New Features of R 3.1.1

The alias of R 3.1.1 is Sock it to Me, which is interesting. After downloading R 3.1.1, you can view the version information through version.

```
# Launch the R application.
~ R
# View the version information.
> version
platform        x86_64-w64-mingw32
arch            x86_64
os              mingw32
system          x86_64, mingw32
status
major           3
minor           1.1
year            2014
month           07
day             10
svn rev         66115
language        R
version.string  R version 3.1.1 (2014-07-10)
nickname        Sock it to Me
```

The new features:

1. The reports of attach() for conflicts are similar with that of library, both using the message() function.

```
# Assign women to w.
> w<-women
> w
   height weight
1      58    115
2      59    117
3      60    120
4      61    123
5      62    126
6      63    129
7      64    132
8      65    135
9      66    139
10     67    142
11     68    146
12     69    150
13     70    154
14     71    159
15     72    164
# Load the w dataset.
> attach(w)
# Use the height column in dataset directly.
> height
 [1] 58 59 60 61 62 63 64 65 66 67 68 69 70 71 72
# Load the women dataset, with column name conflicts happened.
```

```
> attach(women)
The following objects are masked from w:
    height, weight

# The conflict shows when loading package using library.
> library(xts)
Loading required program package: zoon
Loading program package: 'zoo'
The following objects are masked from 'package:base':
    as.Date, as.Date.numeric
```

2. R CMD Sweave no longer cleans any files by default. It adds new options --clean, --clean=default, and --clean=keepOuts.

```
~ C:\Users\Administrator>R CMD Sweave
Usage: R CMD Sweave [options] file

A front-end for Sweave and other vignette engines, via
buildVignette()

Options:
  -h, --help       print this help message and exit
  -v, --version    print version info and exit
  --driver=name    use named Sweave driver
  --engine=pkg::engine  use named vignette engine
  --encoding=enc   default encoding 'enc' for file
  --clean          corresponds to --clean=default
  --clean=         remove some of the created files:
                   "default" removes those the same initial name;
                   "keepOuts" keeps e.g. *.tex even when PDF is
                   produced
  --options=       comma-separated list of Sweave/engine options
  --pdf            convert to PDF document
  --compact=       try to compact PDF document:
                   "no" (default), "qpdf", "gs", "gs+qpdf", "both"
  --compact        same as --compact=qpdf

Report bugs at bugs.r-project.org
```

3. buildVignette() and buildVignettes() in tools package no longer remove any created file with clean=FALSE.

```
> library(tools)
> buildVignette
function (file, dir = ".", weave = TRUE, latex = TRUE, tangle = TRUE,
    quiet = TRUE, clean = TRUE, keep = character(), engine = NULL,
    buildPkg = NULL, ...)
```

4. The Bioconductor "version" used by setRepositories() can now be set by environment variable R_BIOC_VERSION at runtime, not just when R is installed previously. (The version of Bioconductor will be upgraded from 2.14 to 3.0.)

5. Error messages from bugs in embedded Sexpr code in Sweave documents now show the source location.

6. type.convert(), read.table(), and similar read.*() functions get a new numerals argument, specifying how numeric input is converted when its conversion to double precision loses accuracy.

```
> type.convert
function (x, na.strings = "NA", as.is = FALSE, dec = ".", numerals =
c("allow.loss", "warn.loss", "no.loss"))
.External2(C_typeconvert, x, na.strings, as.is, dec, match.
arg(numerals))
<bytecode: 0x0000000008e1b948>
<environment: namespace:utils>
```

7. The robustness of R internal code is improved, which fixes the issue that integer addition may overflow without a warning for some compilers.

```
> as.integer(2000000000)+as.integer(2000000000)
[1] NA
Warning message:
In as.integer(2e+09) + as.integer(2e+09) : NAs produced by integer
overflow
```

8. The default value of the nknots argument of smooth.spline() is changed to nknots.smspl.

```
> smooth.spline
function (x, y = NULL, w = NULL, df, spar = NULL, cv = FALSE,
    all.knots = FALSE, nknots = .nknots.smspl, keep.data = TRUE,
    df.offset = 0, penalty = 1, control.spar = list(), tol = 1e-06 *
IQR(x))
```

9. The Beta distribution-related functions dbeta(, a,b), pbeta(), qbeta() and rbeta() set the default values of a and b to 0 replacing the NaN previously. a and b are correspondent to arguments shape1 and shape2, respectively.

```
> dbeta
function (x, shape1, shape2, ncp = 0, log = FALSE)
{
    if (missing(ncp))
        .External(C_dbeta, x, shape1, shape2, log)
    else .External(C_dnbeta, x, shape1, shape2, ncp, log)
}
<bytecode: 0x000000000d54c070>
<environment: namespace:stats>
```

10. RStudio graphics device does not work correctly with the dev.new() function. So a new argument noRStudioGD is added to replace the previous default selection, which allows the graphics device development in RStudio.

```
# Warning.
> dev.new()
NULL
Warning message:
In (function ()  : Only one RStudio graphics device is permitted
```

```
# A new windows is popped up and no error happens.
> dev.new(noRStudioGD=TRUE)
```

11. Add a new function readRDS() for reading data files with rds format.

```
> d<-data.frame(a=1:10,b=10:1)
# Save the d dataset as rds format.
> saveRDS(d, "d.rds")
# Read rds data.
> d2<-readRDS("d.rds")
> d2
    a  b
1   1 10
2   2  9
3   3  8
4   4  7
5   5  6
6   6  5
7   7  4
8   8  3
9   9  2
10 10  1
```

12. Modifying internal logical scalar constants now results in an error instead of a warning.

```
# The constant pi.
> pi
[1] 3.141593
# Attempt to modify the constant pi gets an error.
> pi<<-3.2
Error: cannot change value of locked binding for 'pi'
```

13. install.packages(repos=NULL) now supports package download and installation with protocols of http or ftp.

```
> install.packages("plyr", repos = "http://cran.rstudio.com/")
trying URL 'http://cran.rstudio.com/bin/windows/contrib/3.1/
plyr_1.8.1.zip'
Content type 'application/zip' length 1151983 bytes (1.1 Mb)
opened URL
downloaded 1.1 Mb

package 'plyr' successfully unpacked and MD5 sums checked

The downloaded binary packages are in
    C:\Users\Administrator\AppData\Local\Temp\RtmpWEXnOT\
downloaded_packages
```

14. Only when options("warnPartialMatchDollar") is TRUE, the partial matching using $ operator on data frames gives warning.

```
# The default value.
> options("warnPartialMatchDollar")
```

```
$warnPartialMatchDollar
NULL
# Set it as TRUE.
> options(warnPartialMatchDollar = TRUE
> options("warnPartialMatchDollar")
$warnPartialMatchDollar
[1] TRUE

> df <- data.frame(ab=1:4,cd=1:4)
> rownames(df) <- paste0(letters[1:4],"a")
# Warning.
> df$a
[1] 1 2 3 4
Warning message:
In `$.data.frame`(df, a): Partial match of 'a' to 'ab' in data frame
# Another calling approach.
> df["a",]
```

15. package?foo is used to retrieve package information, whether loaded or not.

```
# The packaged loaded in the current environment, not including
  plyr.
> search()
 [1] ".GlobalEnv"        "tools:rstudio"     "package:stats"
 [4] "package:graphics"  "package:grDevices" "package:utils"
 [7] "package:datasets"  "package:methods"   "Autoloads"
[10] "package:base"

# This statement opens the help of the plyr package.
> package?plyr

> package?pryr
Error in `?`(package, pryr):
No files with category 'package' and topic 'pryr' (or there is error
happened when processing help documents)
```

16. General help requests now default to trying all loaded packages, not just those on the search path.

```
# View the information of package that is not loaded.
> help(package="zoo")
# View all the name matching functions in packages that are not
  loaded.
> ??zoo
```

17. Add a new function promptImport(), to generate a help page for a function that was imported from another package.

```
# Import the help document of the cat function.
> promptImport(cat)
```

Generate a file named "cat.rd." Modify the file and place it to an appropriate directory. View the generated cat.Rd file.

```
\name{cat}
\alias{cat}
\docType{import}
\title{Import from package \pkg{base}}
\description{
The \code{cat} object is imported from package \pkg{base}.
Help is available here:  \code{\link[base:cat]{base::cat}}.
}
```

Among the updates in R 3.1.1, some are new features and some others are, in fact, to fix bugs in R 3.1.0.

For users who want to learn R in depth, it is necessary to check this list each time a new upgrade is released. Sometimes the list can help solving practical issues. For example, The update 10 in R 3.1.1 for RStudio graphics device helps me to fix the problem where the game interface cannot be launched due to an error related to the graphics device.

Additionally, since for each update there is only one short message in the official document and I have limited knowledge, I cannot clearly understand all updates . The misleading words may exist in this section. Please tell me if you find them.

Chapter 4

Object-Oriented Programming

The chapter introduces the object-oriented programming (OOP) ideas in R in detail. The implementations of four types of OOP architectures are also demonstrated. The chapter helps readers to understand OOP in R in depth and to be able to establish complex business applications.

4.1 Object-Oriented Programming in R

Question

How do we understand the object oriented in R?

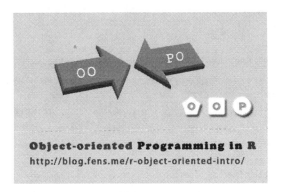

Object-oriented Programming in R
http://blog.fens.me/r-object-oriented-intro/

Introduction

The object-oriented idea is a method to understand and abstract the real world. When the code gets more and more complex for maintenance, the object-oriented idea becomes more and more important. I have experienced the process for Java and JavaScript of transforming from procedure-oriented idea to object-oriented and feel that the transformation will happen in R. Guided by industrial developments, R will develop to large-scaled enterprise applications, so the OOP will be an important direction for R's development.

4.1.1 What is the Object-Oriented Idea

The object-oriented idea is a program design specification and a methodology of programming. The object in object oriented means the collection of classes. The idea deems object as the basic programming unit, encapsulating program and data into object to increase the reusability, flexibility, and extensibility of software. The object-oriented idea is a method to understand and abstract the real world and a result of the specific stage of computer programming technologies. The computer programming was procedure oriented in the early stages. For example, to implement the arithmetic operation $2 + 3 + 4 = 9$, by designing an operation, the result is yielded.

With the development of computer technologies, the computer is used to solve more and more complex problems. Everything is an object. The object-oriented (OO) methodology abstracts things in the real world to objects and relationships to that between objects, for example, inheritance, to help people to abstract and digitally model the real world. The OO methodology allows the analysis, design, and programming to complex systems using the way that can be easily understood by people. Meanwhile, the OO methodology can improve the programming performance effectively. By the abstracts like encapsulation, inheritance, and polymorphism, we can rapidly establish a completely new system like piling blocks.

4.1.1.1 The Three Characteristics of OO: Encapsulation, Inheritance, and Polymorphism

Encapsulation: encapsulate the objective things into the abstractive classes. The classes can expose its data and methods only to trusted classes or objects for operation while hiding them against untrusted ones. An encapsulation example is shown in Figure 4.1.

Let us define two objects: teacher and student and their behaviors according to the OO methodology.

- Teacher's behaviors: lecture, assign homework, and correct homework
- Student's behaviors: attend course, do homework, and take examination

By encapsulation, we abstracted the two objective things and specified the behaviors of objects.

Inheritance: a mechanism where the child class shares the data structure and methods of the parent class. It is a relationship between classes. We can define and implement a class based on an existing class and reuse all the functionalities of the existing class and extend them without rewriting the original class. The new class created through inheritance is called "child class" or "derived class"; the inherited class is called "base class," "parent class," or "super class." And inheritance example is shown in Figure 4.2.

Usually for each course, a leader is elected from the students to assist teachers in communication with other classmates. The leader has more privilege than normal students. Through inheritance, we

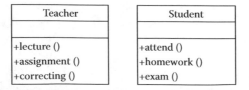

Figure 4.1 The encapsulation.

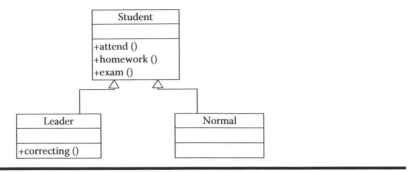

Figure 4.2   The inheritance.

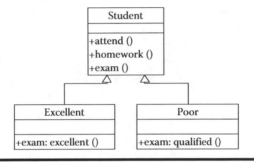

Figure 4.3   The polymorphism.

separate the normal students and the leader in two child classes. The leader has not only the behavior that the normal students have, but also the behavior to assist the teacher to correct homework.

Polymorphism means that the related but different classes through inheritance have different responses to the same event. A polymorphism example is shown in Figure 4.3.

When the final exam approaches, there are always students with good scores and those with poor scores. So for the excellent students, the exam result is excellence; for the poor students, the exam result is not good. For the related but different objects, a same behavior has different results. Therefore, we can abstract things in real world through the OO idea.

4.1.1.2 Is a and Has a

The classes in real world have certain structural relationships. Usually, there are two main relationships: is a and has a.

- is a: means inheritance. For example, diamond, circle, and square are a shape.
- has a: means comprising or clustering. For example, a computer is comprised of display, CPU, hard disk, and so on.

4.1.2 Why Do We Need to Do Object-Oriented Programming in R?

R is mainly used for statistical computing and the code size is not large, with just decades or hundreds of lines, which makes use of the procedural programming technique to fulfill tasks well. However, since R gets hot, and as more developers with engineering backgrounds join in, it begins to develop in different fields. The procedural programming technique which is used to small-sized code gets hard to deal with projects with massive amount of code, which will be definitely replaced by a new programming method, OOP.

The OOP has been widely used in the era of C++/Java. Almost more than 90% Java framework was designed according to OOP. Ten years ago, the various procedural programming codes in JavaScript brought much difficulty to frontend developers. People did not realize that JavaScript can take the OOP way until the appearance of Google's Gmail Web. Afterward, the implementations of class libraries such as jQuery and Angular.js make JavaScript perfectly support frontend programming. The birth of Node.js extends the application fields of JavaScript.

When R language is favored by people, we begin to think about how to make R become an enterprise programming language, how to build non-statistical projects, and how to effectively write code with more than 100k lines using R.

I believe the answer is to program with OO idea. Today's R is just like JavaScript 10 years ago, which needs the power from big companies and great masters to promote. According to my observations, R leading persons like Hadley Wickham have begun to develop R packages which completely involve OOP. For information on the R package memoise developed with OOP, please refer Section 3.1 in R for Programmers: Mastering the Tools.

4.1.3 The OOP in R

There are three kinds of implementations with OOP, that is, S3, S4, and RC types. S3 and S4 are based on generic functions, while RC is complete OOP implementation. All of these three OO types will be introduced in this chapter. This section takes S3 type as breakthrough points. The OOP implementations in this section are all based on S3 type.

The OOP of S3 type in R is implemented based on the generic functions, instead of class hierarchy. In the following, let us implement the three characteristics of OOP using R, taking examples of the three figures related to teacher and students in the above text.

4.1.3.1 Implementing Encapsulation Using R

Define the teacher object and behaviors encapsulated in the generic function teacher(). Define the student object and behaviors, encapsulated in the generic function student().

```
# Define the teacher object and behaviors.
> teacher <- function(x, …) UseMethod("teacher")
> teacher.lecture <- function(x) print("Lecture")
> teacher.assignment <- function(x) print("Assign homework")
> teacher.assignment <- function(x) print("Correct homework")
> teacher.default<-function(x) print("You are not teacher")

# Define the student object and behaviors.
> student <- function(x, …) UseMethod("student")
> student.attend <- function(x) print("Attend course")
> student.homework <- function(x) print("Do homework")
> student.exam <- function(x) print("Take examination")
> student.default<-function(x) print("You are not student")

# Define two variables, teacher a and student b.
> a<-'teacher'
> b<-'student'
```

```
# Set a behavior for the teacher variable.
> attr(a,'class') <- 'lecture'
# Execute the behavior of teacher.
> teacher(a)
[1] "Lecture"

# Set a behavior for the student variable.
> attr(b,'class') <- 'attend'
# Execute the behavior of student.
> student(b)
[1] "Attend course"

# Reset a behavior of the teacher variable.
> attr(a,'class') <- 'assignment'
# Execute the behavior of teacher.
> teacher(a)
[1] "Assign homework"

# Reset a behavior of the student variable.
> attr(b,'class') <- 'homework'
# Execute the behavior of student.
> student(b)
[1] "Do homework"

> attr(a,'class') <- 'correcting'
> teacher(a)
[1] "Correct homework"

> attr(b,'class') <- 'exam'
> student(b)
[1] "take examination"
```

What would happen if we assign the behaviors to a same variable?

```
# Defined a variable with classes of both teacher and student.
> ab<-'student_teacher'
# Set the behaviors of different objects.
> attr(ab,'class') <- c('lecture','homework')
#Execute the behaviors of teacher.
> teacher(ab)
[1] "Lecture"
> student(ab)
[1] "Do homework"
```

From the result, we can see that the variable ab, which has behaviors of both teacher and student, yields teacher's behavior when called by the generic function teacher(), and yields student's behavior when called by student().

4.1.3.2 Implementing Inheritance Using R

Let's add a student as the leader and add the behaviors of assisting teacher to correct homework. We can build two instances through inheritance separating the leader from the normal student.

```
# Add new behavior to the student object.
> student.correcting <- function(x) print("Assist teacher to correct
homework")
# An auxiliary variable for setting initial  value.
 > char0 = character(0)

# Implement the inheritance.
> create <- function(classes=char0, parents=char0) {
+     mro <- c(classes)
+     for (name in parents) {
+         mro <- c(mro, name)
+         ancestors <- attr(get(name),'type')
+         mro <- c(mro, ancestors[ancestors != name])
+     }
+     return(mro)
+ }

# Define a constructor to create object.
> NewInstance <- function(value=0, classes=char0, parents=char0) {
+     obj <- value
+     attr(obj,'type') <- create(classes, parents)
+     attr(obj,'class') <- c('attend','homework','exam')
+     return(obj)
+ }

# Create the parent object instance.
> StudentObj <- NewInstance()
# Create the child object instance.
> s1 <- NewInstance('Normal',classes='normal', parents='StudentObj')
> s2 <- NewInstance('Leader',classes='leader', parents='StudentObj')

# Add the behavior of correcting homework to the leader.
> attr(s2,'class') <- c(attr(s2,'class'),'correcting')

# View the object instance of normal student.
> s1
[1] "Normal"
attr(,"type")
[1] "normal"      "StudentObj"
attr(,"class")
[1] "attend"      "homework"    "exam"
# View the object instance of leader student.
> s2
[1] "Leader"
attr(,"type")
[1] "leader"      "StudentObj"
attr(,"class")
[1] "attend"      "homework"    "exam"         "correcting"
```

4.1.3.3 Implementing Polymorphism Using R

When approaching the final examination, some students get good scores, while others get poor scores. We create two instances called excellent and poor through inheritance. The two instances have the same behavior but different examination results, which demonstrates the polymorphism.

```
#Create two instances:  excellent student and poor student.
> e1 <- NewInstance('Excellent student',classes='excellent',
parents='StudentObj')
> e2 <- NewInstance('Poor student',classes='poor', parents='StudentObj')

# Modify the behavior of exam, Excellent if score > 85 and Qualified if
score <70.
> student.exam <- function(x,score) {
+      p<-"Examination Result: "
+      if(score>85) print(paste(p,"Excellent",sep=""))
+      if(score<70) print(paste(p,"Qualified",sep=""))
+ }

# Execute the exam behavior of excellent student with score of 90.
> attr(e1,'class') <- 'exam'
> student(e1,90)
[1] " Examination Result: Excellent"

# Execute the exam behavior of poor student with score of 90.
> attr(e2,'class') <- 'exam'
> student(e2,66)
[1] " Examination Result: Qualified"
```

Hereby, we have implemented the OOP through the generic function of S3 type in R.

4.1.3.4 The Procedure-Oriented Programming in R

In the following, let us do a comparison. We implement the above logics using POP code.

1. Define the two objects and their behaviors.

```
# An auxiliary variable for setting initial values.
> char0 = character(1)
# Define the teacher object and behaviors.
> teacher_fun<-function(x=char0){
+      if(x=='lecture'){
+          print("Lecture")
+      }else if(x=='assignment'){
+          print("Assign homework")
+      }else if(x=='correcting'){
+          print("Correct homework")
+      }else{
+          print("You are not teacher")
+      }
+ }

# Define the student object and behaviors.
> student_fun<-function(x=char0){
+      if(x=='attend'){
+          print("Attend course")
```

```
+        }else if(x=='homework'){
+            print("Do homework")
+        }else if(x=='exam'){
+            print("Take examination")
+        }else{
+            print("You are not student")
+        }
+ }

# Execute a behavior of teacher.
> teacher_fun('lecture')
[1] "Lecture"

# Execute a behavior of student.
> student_fun('attend')
[1] "Attend course"
```

2. Differentiate the behaviors of normal and leader students.

```
# Redefine the function of student.
> student_fun<-function(x=char0,role=0){
+        if(x=='attend'){
+            print("Attend course")
+        }else if(x=='homework'){
+            print("Do homework")
+        }else if(x=='exam'){
+            print("Take examination")
+        }else if(x=='correcting'){
+            # Leader.
+            if(role==1){
+                print("Assist teacher to correct homework")
+            }else{
+                print("You are not leader")
+            }
+        }else{
+            print("You are not student")
+        }
+ }

# Execute the behavior of leader with the role of normal student.
> student_fun('correcting')
[1] "You are not leader"

# Execute the behavior of leader with the role of leader student.
> student_fun('correcting',1)
[1] " Assist teacher to correct homework"
```

I have increased the complexity of the original function when I modified the function student_fun.

3. Take examination and differentiate the excellent and poor students with scores.

```
# Modify the function definition of student, adding the score
argument.
> student_fun<-function(x=char0,role=0,score){
```

```
+       if(x=='attend'){
+           print("Attend course")
+       }else if(x=='homework'){
+           print("Do homework")
+       }else if(x=='exam'){
+           p<-"Examination Result: "
+           if(score>85) print(paste(p,"Excellent",sep=""))
+           if(score<70) print(paste(p,"Qualified",sep=""))
+       }else if(x=='correcting'){
+           # Leader.
+           if(role==1){
+               print("Assist teacher to correct homework")
+           }else{
+               print("You are not leader")
+           }
+       }else{
+           print("You are not student")
+       }
+ }
```

```
Execute the exam function. student with score > 85 is excellent.
> student_fun('exam',score=90)
[1] "Examination Result: Excellent"
Execute the exam function. student with score < 70 is poor.
> student_fun('exam',score=66)
[1] " Examination Result: Qualified "
```

I implemented the entire logic using POP code. When programming using POP, each requirement change needs to modify the original code. In addition to increasing the code complexity, it is not helpful for long-term maintenance. You can think more about this.

This section gives you the first look at the OOP in R. Some of the code may not be strict enough. The purpose is just to give an understating about the logic. The more detailed instances of OOP will be demonstrated in the next section.

4.2 The S3-Based OOP in R

Question

How do we do OOP based on S3 object system?

The S3 Based OOP in R

http://blog.fens.me/r-class-s3/

Introduction

On OOP, R is different from other programming languages. R provides three low-level object systems: S3, S4, and RC. The S3 objects are simple and dynamic, but not obvious about structuration. The S4 objects are structured and powerful. The RC objects are the new type that is introduced in version 2.12, used to solve the object design issue that is hard to implement using S3 and S4. Starting from the S3 object, this section introduces the details of OOP in R.

4.2.1 Introduction to the S3 Objects

In R language, the OOP based on S3 objects is an implementation based on the generic function. The generic function is a special function that calls method depending on the type of input object. Different from OOP in other languages, the OOP based on S3 objects is a simulating implementation with dynamic function calling. The S3 objects were widely applied to the development packages in the early development of R. For the basic knowledge of OOP, please refer Section 4.1.

4.2.2 Creating S3 Objects

The system environment used in this section:

- Linux: Ubuntu Server 12.04.2 LTS 64bit
- R: 3.1.1 x86_64-pc-linux-gnu (64-bit)

In order for us to check the object's type conveniently, we import the pryr package as the auxiliary tool. For information about the pryr package, please refer Section 3.1.

```
# Launch the R application.
~ R
# Load the pryr package.
> library(pryr)
```

4.2.2.1 Create S3 Objects through Variables

The simplest way to create an S3 object is to add a class attribute to a variable.

```
# Define a variable.
> x<-1
# Define it as an S3 typed object.
> attr(x,'class')<-'foo'

# View the variable.
> x
[1] 1
attr(,"class")
[1] "foo"

# View the type of the variable.
> class(x)
[1] "foo"
```

```
# Check the type of x using the otype function in the pryr package.
> otype(x)
[1] "S3"
```

The second way is to create an S3 object using the structure function().

```
# Create an S3 typed object.
> y <- structure(2, class = "foo")
# View the variable.
> y
[1] 2
attr(,"class")
[1] "foo"
# View the type of the variable.
> class(y)
[1] "foo"
# Check the type of y using the otype function in the pryr package.
> otype(y)
[1] "S3"
```

4.2.2.2 Create an S3 Object with Multiple Types

S3 objects do not have a clear structural relation. One object can have more than one type. The type of an S3 object is defined through the class attribute of the variable. The class attribute can be a vector and therefore allows multiple types.

```
# Define a variable.
> x<-1
# Set more than one S3 type.
> attr(x,'class')<- c("foo", "bar")
# View the types of variable x.
> class(x)
[1] "foo" "bar"
# View the types of variable x.
> otype(x)
[1] "S3"
```

4.2.3 The Generic Function and the Method Calling

For the usage of S3 objects, usually we define a name for the generic function using UseMethod() and then determine the method to be called through the class attribute of the input argument.

In the following, let us create a generic function of S3 type for method calling of S3 objects. Define a generic function named teacher.

- Define the generic function teacher using UseMethod().
- Define the behaviors of the teacher object with the syntax of teacher.xxx, where teacher. default is the default behavior.

```
# Define the generic function teacher using UseMethod().
> teacher <- function(x, ...) UseMethod("teacher")
# Check the types of teacher using the ftype() function in the pryr
  package.
> ftype(teacher)
[1] "s3"      "generic"
# Define the internal functions of teacher.
> teacher.lecture <- function(x) print("Lecture")
> teacher.assignment <- function(x) print("Assign homework")
> teacher.correcting <- function(x) print("Correct homework")
# Define the default behavior of teacher.
> teacher.default<-function(x) print("You are not teacher")
```

When calling method, we determine which method is to be called through the class attribute of the input argument.

- Define a variable named a and set the class attribute of a as lecture.
- Pass the variable a into the generic function teacher.
- The behavior function teacher.lecture() is called.

```
# Define a variable a.
> a<-'teacher'
# Set a behavior for the variable a.
> attr(a,'class') <- 'lecture'
# Execute the behavior of teacher.
> teacher(a)
[1] "Lecture"
```

Of course we can directly call the behavior functions defined in teacher. However, doing this loses the meaning of OO encapsulation.

```
> teacher.lecture()
[1] "Lecture"
> teacher.lecture(a)
[1] "Lecture"
# Execute the default behavior function if the class attribute  of
argument is not what has been defined.
> teacher()
[1] "You are not teacher"
```

Compared with other languages, the generic function in R is method interface, and teacher. xxx is the method implementation of the interface.

4.2.4 View the Functions of S3 Objects

After doing the OO encapsulation using an S3 object, we can view the defined internal behavior functions of the S3 object.

```
# View the teacher object.
> teacher
function(x, ...) UseMethod("teacher")
```

```
# View the internal functions of the teacher object.
>  methods(teacher)
[1] teacher.assignment teacher.correcting teacher.default    teacher.lecture
```

Match the names of generic functions through the generic.function argument of methods().

```
# Find the S3 typed functions starting with "predict" in environment.
> methods(generic.function=predict)
 [1] predict.ar*            predict.Arima*          predict.arima0*
 [4] predict.glm            predict.HoltWinters*    predict.lm
 [7] predict.loess*         predict.mlm             predict.nls*
[10] predict.poly           predict.ppr*            predict.prcomp*
[13] predict.princomp*      predict.smooth.spline*  predict.smooth.
                                                    spline.fit*

[16] predict.StructTS*
```

```
Non-visible functions are asterisked
```

Match names of classes through the class argument of methods().

```
# Find the S3 typed objects with lm as the class attribute in
environment.
> methods(class=lm)
[1]  add1.lm          *alias.lm*            anova.lm           case.names.lm*
[5]  confint.lm*      cooks.distance.lm*    deviance.lm*       dfbeta.lm*
[9]  dfbetas.lm*      drop1.lm*             dummy.coef.lm*     effects.lm*
[13] extractAIC.lm*   family.lm*            formula.lm*        hatvalues.lm
[17] influence.lm*    kappa.lm              labels.lm*         logLik.lm*
[21] model.frame.lm   model.matrix.lm       nobs.lm*           plot.lm
[25] predict.lm       print.lm              proj.lm*           qr.lm*
[29] residuals.lm     rstandard.lm          rstudent.lm        simulate.lm*
[33] summary.lm       variable.names.lm*    vcov.lm*
```

```
Non-visible functions are asterisked
```

View all functions through the getAnywhere() function.

```
# View the teacher.Lecture() function.
> getAnywhere(teacher.lecture)
A single object matching 'teacher.lecture' was found
It was found in the following places
  .GlobalEnv
  registered S3 method for teacher
with value
function(x) print("Lecture")
```

```
# View the invisible function predict.ppr.
> predict.ppr
Error: object 'predict.ppr' not found
> exists("predict.ppr")
[1] FALSE
```

```
# Find the predict.ppr function through the getAnywhere() function.
```

```
> getAnywhere("predict.ppr")
A single object matching 'predict.ppr' was found
It was found in the following places
  registered S3 method for predict from namespace stats
  namespace:stats
with value
function (object, newdata, …)
{
    if (missing(newdata))
        return(fitted(object))
    if (!is.null(object$terms)) {
        newdata <- as.data.frame(newdata)
        rn <- row.names(newdata)
        Terms <- delete.response(object$terms)
        m <- model.frame(Terms, newdata, na.action = na.omit,
            xlev = object$xlevels)
        if (!is.null(cl <- attr(Terms, "dataClasses")))
            .checkMFClasses(cl, m)
        keep <- match(row.names(m), rn)
        x <- model.matrix(Terms, m, contrasts.arg = object$contrasts)
    }
    else {
        x <- as.matrix(newdata)
        keep <- seq_len(nrow(x))
        rn <- dimnames(x)[[1L]]
    }
    if (ncol(x) != object$p)
        stop("wrong number of columns in 'x'")
    res <- matrix(NA, length(keep), object$q, dimnames = list(rn,
        object$ynames))
    res[keep, ] <- matrix(.Fortran(C_pppred, as.integer(nrow(x)),
        as.double(x), as.double(object$smod), y = double(nrow(x) *
            object$q), double(2 * object$smod[4L]))$y, ncol = object$q)
    drop(res)
}
<bytecode: 0x000000000df6c2d0>
<environment: namespace:stats>
```

We can also use the getS3method() function to view the invisible function.

```
# Find the predict.ppr function through the getS3method() function.
> getS3method("predict", "ppr")
function (object, newdata, …)
{
    if (missing(newdata))
        return(fitted(object))
    if (!is.null(object$terms)) {
        newdata <- as.data.frame(newdata)
        rn <- row.names(newdata)
        Terms <- delete.response(object$terms)
        m <- model.frame(Terms, newdata, na.action = na.omit,
            xlev = object$xlevels)
```

```
            if (!is.null(cl <- attr(Terms, "dataClasses")))
                .checkMFClasses(cl, m)
            keep <- match(row.names(m), rn)
            x <- model.matrix(Terms, m, contrasts.arg = object$contrasts)
        }
        else {
            x <- as.matrix(newdata)
            keep <- seq_len(nrow(x))
            rn <- dimnames(x)[[1L]]
        }
        if (ncol(x) != object$p)
            stop("wrong number of columns in 'x'")
        res <- matrix(NA, length(keep), object$q, dimnames = list(rn,
            object$ynames))
        res[keep, ] <- matrix(.Fortran(C_pppred, as.integer(nrow(x)),
            as.double(x), as.double(object$smod), y = double(nrow(x) *
                object$q), double(2 * object$smod[4L]))$y, ncol = object$q)
        drop(res)
}
<bytecode: 0x000000000df6c2d0>
<environment: namespace:stats>
```

4.2.5 *The Inheriting Calling of S3 Objects*

There is a very simple inheriting calling for S3 objects: the NextMethod() function.

Define a generic function, node.

```
> node <- function(x) UseMethod("node", x)
> node.default <- function(x) "Default node"

# The father function.
> node.father <- function(x) c("father")
# The son function, referring to the father function through the Next
  method() function.
> node.son <- function(x) c("son", NextMethod())

# Define n1.
> n1 <- structure(1, class = c("father"))
# Pass n1 into the node function to execute the node.father() function.
> node(n1)
[1] "father"

# Define n2 and set the class attribute with two classes.
> n2 <- structure(1, class = c("son", "father"))
> node(n2)
[1] "son"    "father"
```

By passing n2 into the node() function, node.son() is called first and then node.father() is executed through NextMethod(). This actually simulates the process of calling the parent method and implements the inheritance in OOP.

4.2.6 The Disadvantages of S3 Objects

From the above introduction to S3 objects, we can conclude that they are not an implementation through completely OOP, instead the implementation is a simulation to OOP through function calling.

- Although it is easy to use S3 objects, the meaning of S3 objects is not clear when the object relationships are somewhat complex in the practice of OOP.
- The internal functions encapsulated by S3 objects can be called directly, which bypasses the check to generic functions.
- The class attribute of S3 argument can be set in any way and there is no preprocess checking.
- The S3 argument is called only by the class attribute, while other attributes cannot be executed through the class() function.
- When the S3 argument's class attribute has multiple values, only the first eligible function can be called according to the sequence of assignment.

As a result, S3 is just a simple implementation of OOP in R.

4.2.7 Use of S3 Objects

The S3 object system is widely in the early development of R. There are many S3 object in the base package.

View the S3 objects in the base package.

```
# The mean() function.
> mean
function (x, …)
UseMethod("mean")
<bytecode: 0x3f8e0f0>
<environment: namespace:base>
> ftype(mean)
[1] "s3"       "generic"

# The t() function.
> ftype(t)
[1] "s3"       "generic"

# The plot() function.
> ftype(plot)
[1] "s3"       "generic"
```

The customized S3 objects.

```
# Define a numeric type variable a.
> a <- 1
# The class of a is numeric.
> class(a)
[1] "numeric"
```

```
# Define a generic function f1.
> f1 <- function(x) {
+    a <- 2
+    UseMethod("f1")
+ }

# Define the internal function of f1.
> f1.numeric <- function(x) a
> f1(a)
[1] 2
> f1(99)
[1] 2

# Define the internal function of f1.
> f1.character <- function(x) paste("char", x)
# Pass letter a into f1.
> f1("a")
[1] "char a"
```

Through the introduction in this section, we have obtained a complete picture of the S3 object system and started the road to OOP in R. The next section will introduce the S4 object system in R.

4.3 The S4-Based OOP in R

Question

How do we do OOP based on the S4 object system?

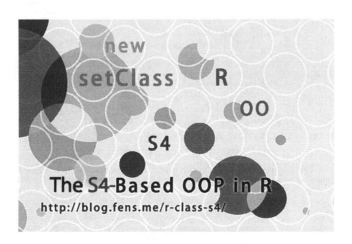

Introduction

This section will introduce the S4-based OOP in R. The S4 object system has clear structural characteristics and is more adapted to OOP. The bioconductor community makes the S4 object system as its infrastructure and only accepts the R packages complying with S4 specification.

4.3.1 Creating S4 Objects

The S4 object system is an implementation of OOP in R. The S4 objects have clear OOP characteristics including class definition, argument definition, argument checking, inheritance, instantiation, interface functions, implementation functions, and so on.

The system environment used in this section:

- Linux: Ubuntu Server 12.04.2 LTS 64bit
- R: 3.1.1 x86_64-pc-linux-gnu (64-bit)

In order for us to check the object's type conveniently, we import the pryr package as the auxiliary tool. For information about the pryr package, please refer Section 3.1.

```
# Launch the R application.
~ R
# Load the pryr package.
> library(pryr)
```

4.3.1.1 How to Create S4 Objects

The S4 object system has a special class definition function setClass() and class instantiation function new(). Let us see how to use the functions.

View the definition of setClass().

```
setClass(Class, representation, prototype, contains=character(),
         validity, access, where, version, sealed, package,
         S3methods = FALSE, slots)
```

The following is the explanation to the arguments:

- Class: define the class name
- slots: define the slots and slot types
- prototype: define the default values of slots
- contains = character(): define the parent class for inheritance
- validity: define the type checking function for slots
- where: specify the environment to store the definition
- sealed: if set to TRUE, another class with same name cannot be defined
- package: define the package name for the class
- S3methods: no longer recommended to use since R 3.0.0
- representation: no longer recommended to use since R 3.0.0
- rccess: no longer recommended to use since R 3.0.0
- version: no longer recommended to use since R 3.0.0

We can create an S4-typed object structure through the setClass() function.

```
# Define an S4 object.
> setClass("Person",slots=list(name="character",age="numeric"))
```

4.3.1.2 Creating S4 Object Instance

After defining the class through the setClass() function, we instantiate the object of the class through the new() function.

```
# Instantiate a Person object.
> father<-new("Person",name="F",age=44)
# View the father object. It has two slots: name and age.
> father
An object of class "Person"
Slot "name":
[1] "F"
Slot "age":
[1] 44

# View the class name of father. It is Person.
> class(father)
[1] "Person"
attr(,"package")
[1] ".GlobalEnv"

# View the object type of father, its type is S4.
> otype(father)
[1] "S4"
```

4.3.1.3 Creating S4 Object with Inheritance

To create an S4 object with inheritance, we can set the parent class with the contains argument of the setClass() function.

```
# Create an S4 object, person.
> setClass("Person",slots=list(name="character",age="numeric"))
# Create a child class of Person.
> setClass("Son",slots=list(father="Person",mother="Person"),contains="Pe
  rson")

# Instantiate the Person objects.
> father<-new("Person",name="F",age=44)
> mother<-new("Person",name="M",age=39)

# Instantiate a Son object.
> son<-new("Son",name="S",age=16,father=father,mother=mother

# View the name slot of the son object.
> son@name
[1] "S"

# View the name slot of the age object.
> son@age
[1] 16

# View the father slot of the age object.
> son@father
An object of class "Person"
```

```
Slot "name":
[1] "F"
Slot "age":
[1] 44

# View the mother slot of the age object.
> slot(son,"mother")
An object of class "Person"
Slot "name":
[1] "M"
Slot "age":
[1] 39

# View the type of son.
> otype(son)
[1] "S4"
# View the type of the son@name slot.
> otype(son@name)
[1] "primitive"
# View the type of the son@mother slot.
> otype(son@mother)
[1] "S4"

# Check the type of the S4 object using isS4().
> isS4(son)
[1] TRUE
> isS4(son@name)
[1] FALSE
> isS4(son@mother)
[1] TRUE
```

4.3.1.4 The Default Values of S4 Objects

The class definition usually contains slots. We can set default values to the arguments defined in slots through the prototype argument in setClass().

```
> setClass("Person",slots=list(name="character",age="numeric"))
# The age slot is NULL.
> a<-new("Person",name="a")
> a
An object of class "Person"
Slot "name":
[1] "a"
Slot "age":
numeric(0)

# Set 20 as the default value of slot age.
> setClass("Person",slots=list(name="character",age="numeric"),prototype
  = list(age = 20))
# The age slot is NUL.
> b<-new("Person",name="b")
# The default value of the age slot is 20.
> b
An object of class "Person"
```

```
Slot "name":
[1] "b"
Slot "age":
[1] 20
```

4.3.1.5 The Type Checking of S4 Objects

Set type checking function to the arguments defined in slots through the setValidity() function.

```
> setClass("Person",slots=list(name="character",age="numeric"))

# Pass invalid age type.
> bad<-new("Person",name="bad",age="abc")
Error in validObject(.Object):
  invalid class "Person" object: invalid object for slot "age" in class
"Person": got class "character", should be or extend class "numeric"

# Set the non-negative checking function for age.
> setValidity("Person",function(object) {
+     if (object@age <= 0) stop("Age is negative.")
+ })
Class "Person" [in ".GlobalEnv"]
Slots:
Name:          name          age
Class: character    numeric

# Pass age less than 0.
> bad2<-new("Person",name="bad",age=-1)
Error in validityMethod(object): Age is negative.
```

4.3.1.6 Creating Object from an Instantiated Object

The S4 objects support creating an object from an instantiated object, allowing value overriding.

```
> setClass("Person",slots=list(name="character",age="numeric"))
# Create an object instance n1.
> n1<-new("Person",name="n1",age=19);n1
An object of class "Person"
Slot "name":
[1] "n1"
Slot "age":
[1] 19

# Create instance n2 from instance n1 and modify the value of the name
slot.
> n2<-initialize(n1,name="n2");n2
An object of class "Person"
Slot "name":
[1] "n2"
Slot "age":
[1] 19
```

4.3.2 Accessing the Slots of S4 Objects

Generally we use $ to access the slots of an S3 object, while we have to use @ to access the slots of an S4 object.

```
> setClass("Person",slots=list(name="character",age="numeric"))
> a<-new("Person",name="a")

# Access a slot of an S4 object.
> a@name
[1] "a"
> slot(a, "name")
[1] "a"

# The invalid slot accessing.
> a$name
Error in a$name : $ operator not defined for this S4 class
> a[1]
Error in a[1] : object of type 'S4' is not subsettable
> a[[1]]
Error in a[[1]] : this S4 class is not subsettable
```

4.3.3 The S4 Generic Functions

Different from the S3 implementation, the S4 implementation separates the definition and implementation, as so-called separating interface from implementation in other languages. The S4 object system defines interface through setGeneric() and defines the implementing class through setMethod(), which makes it closer to the OO characteristics.

The definition and calling of the ordinary functions.

```
> work<-function(x) cat(x, "is working")
> work('Conan')
Conan is working
```

Let us see how R separates interface from implementation.
```
# Defind the Person object.
> setClass("Person",slots=list(name="character",age="numeric"))
# Defind the generic function work, i.e. the interface.
> setGeneric("work",function(object) standardGeneric("work"))

# Define the implementation function of work and set the Person object as
the argument type.
> setMethod("work", signature(object = "Person"), function(object)
cat(object@name , "is working") )
[1] "work"

# Create a Person object.
> a<-new("Person",name="Conan",age=16)
# Pass the object a into the function work.
> work(a)
Conan is working
```

The S4 object system changes the original process of function definition and calling from two steps to 4.

- Define the data object type.
- Define the interface function.
- Define the implementation function.
- Pass the data object as argument into the interface function and execute the implementation function.

So, the S4 object system is a structured OO implementation.

4.3.4 Functions for Viewing S4 Objects

After the OO encapsulation to S4 objects, we need to view the definitions of S4 objects and functions. The following code continues to use the data objects defined in previous cases of person and work.

```
# View the type of work.
> ftype(work)
[1] "s4"        "generic"

# View the definition of the work definition directly.
> work
standardGeneric for "work" defined from package ".GlobalEnv"
function (object)
standardGeneric("work")
<environment: 0x2aa6b18>
Methods may be defined for arguments: object
Use  showMethods("work")  for currently available ones.

# View the definition of the work function.
> showMethods(work)
Function: work (package .GlobalEnv)
object="Person"

# View the implementation of the work function of the Person object.
> getMethod("work", "Person")
Method Definition:
function (object)
cat(object@name, "is working")
Signatures:
        object
target  "Person"
defined "Person"

> selectMethod("work", "Person")
Method Definition:
function (object)
cat(object@name, "is working")
Signatures:
        object
target  "Person"
defined "Person"
```

```
# Check whether the Person object has the work function.
> existsMethod("work", "Person")
[1] TRUE
> hasMethod("work", "Person")
[1] TRUE
```

4.3.5 Working with the S4 Objects

In the following, let us take an example of S4 objects, defining a library of graphics functions.

Task 1. Define the data structures and computation function of the graphics library.
Suppose that the lowest level of the library sets shape as the base class, and include circle and ellipse with area and circum to be calculated.

- Define the data structure of the graphics library.
- Define the data structure of circle and calculate the area and circum.
- Define the data structure of ellipse and calculate the area and circumference.

The structure is shown in Figure 4.4.

First of all, define the base class shape and the circle class.

```
# Define the base class Shape.
> setClass("Shape",slots=list(name="character"))
# Define the Circle class, inheriting Shape. The default value of the
radius slot is 1.
> setClass("Circle",
+    contains="Shape",
+    slots=list(radius="numeric"),
+    prototype=list(radius = 1))
```

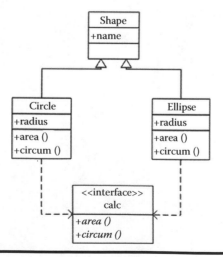

Figure 4.4   Task 1: the data structure of graphics library.

```
# Verify that the value of radius is larger than 0. Exit if the value is
less than or equal to 0.
> setValidity("Circle",function(object) {
+      if (object@radius <= 0) stop("Radius is negative.")
+ })
Class "Circle" [in ".GlobalEnv"]
Slots:
Name:      radius       name
Class:     numeric character
Extends: "Shape"

# Create two instances of Circle.
> c1<-new("Circle",name="c1")
> c2<-new("Circle",name="c2",radius=5)
```

Step 2, define the interface function and implementation function for calculating area.

```
# The generic function interface for calculating area.
> setGeneric("area",function(obj,…) standardGeneric("area"))
[1] "area"

# The function implementation for calculating area.
> setMethod("area","Circle",function(obj,…){
+      print("Area Circle Method")
+      pi*obj@radius^2
+ })
[1] "area"

# Calculate the areas of the two circles c1 and c2.
> area(c1)
[1] "Area Circle Method"
[1] 3.141593
> area(c2)
[1] "Area Circle Method"
[1] 78.53982
```

Step 3, define the interface and implementation for calculating circum.

```
# The generic function interface for calculating circum.
> setGeneric("circum",function(obj,…) standardGeneric("circum"))
[1] "circum"

# The function implementation for calculating circum.
> setMethod("circum","Circle",function(obj,…){
+      2*pi*obj@radius
+ })

[1] "circum"
# Calculate the circums of the two circles c1 and c2.
> circum(c1)
[1] 6.283185
> circum(c2)
[1] 31.41593
```

We have implemented the definition of the circle class using the above code. In the following, we implement for the ellipse class.

```
# Define the Ellipse class, inheriting Shape. The default value of radius
is c(1,1), representing the long and short radius respectively.
> setClass("Ellipse",
+    contains="Shape",
+    slots=list(radius="numeric"),
+    prototype=list(radius=c(1,1)))

# Validate the radius argument.
> setValidity("Ellipse",function(object) {
+        if (length(object@radius) != 2 ) stop("It's not Ellipse.")
+        if (length(which(object@radius<=0))>0) stop("Radius is negative.")
+ })
Class "Ellipse" [in ".GlobalEnv"]
Slots:
Name:        radius        name
Class:    numeric character
Extends: "Shape"

# Create two instances of Ellipse, e1 and e2.
> e1<-new("Ellipse",name="e1")
> e2<-new("Ellipse",name="e2",radius=c(5,1))

# The function implementation for calculating the ellipse's area.
> setMethod("area", "Ellipse",function(obj,…){
+        print("Area Ellipse Method")
+        pi * prod(obj@radius)
+ })
[1] "area"

# Calculate the areas of ellipses e1 and e2.
> area(e1)
[1] "Area Ellipse Method"
[1] 3.141593
> area(e2)
[1] "Area Ellipse Method"
[1] 15.70796
```

Following is the function implementation for calculating the circum of ellipse. This is just an approximate calculation. Please do not struggle with the calculation formula.

```
# The function implementation for calculating the circum of ellipse.
 > setMethod("circum","Ellipse",function(obj,…){
+        cat("Ellipse Circum :\n")
+        2*pi*sqrt((obj@radius[1]^2+obj@radius[2]^2)/2)
+ })
[1] "circum"

# Calculate the circums of e1 and e2.
> circum(e1)
Ellipse Circum:
[1] 6.283185
```

```
> circum(e2)
Ellipse Circum:
[1] 22.65435
```

We have finished the data structure definition of circle and ellipse and the implementations for calculating area and circum. Have you found that circle is a special case of ellipse?

Task 2. Refactor the design of circle and ellipse.
When the long radius of an ellipse is equal to its short radius, that is, the two values of radius are equal, the ellipse is a circle. By this fact, we can redesign the relationship of circle and ellipse. Ellipse is the parent class of circle, while circle is the child class of ellipse.

The structure is shown as in Figure 4.5.

```
# The base class Shape.
> setClass("Shape",
+   slots=list(name="character"))

# Ellipse inherits Shape.
> setClass("Ellipse",
+   contains="Shape",
+   slots=list(radius="numeric"),
+   prototype=list(radius=c(1,1)))

# Circle inherits Ellipse.
> setClass("Circle",
+   contains="Ellipse",
+   slots=list(radius="numeric"),
+   prototype=list(radius = 1))
```

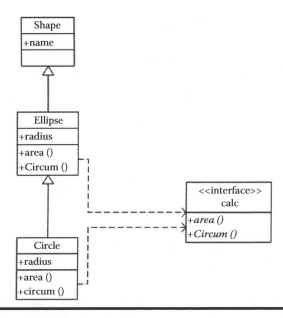

Figure 4.5   Task 2: refactor the data structure of ellipse.

```
# Define the area interface.
> setGeneric("area",function(obj,...) standardGeneric("area"))
[1] "area"

# Define the Ellipse implementation of area.
> setMethod("area","Ellipse",function(obj,...){
+     cat("Ellipse Area :\n")
+     pi * prod(obj@radius)
+ })
[1] "area"

# Define the Circle implementation of area.
> setMethod("area","Circle",function(obj,...){
+     cat("Circle Area :\n")
+     pi*obj@radius^2
+ })
[1] "area"

# Define the circum interface.
> setGeneric("circum",function(obj,...) standardGeneric("circum"))
[1] "circum"

# Define the Ellipse implementation of circum.
> setMethod("circum","Ellipse",function(obj,...){
+     cat("Ellipse Circum :\n")
+     2*pi*sqrt((obj@radius[1]^2+obj@radius[2]^2)/2)
+ })
[1] "circum"

> setMethod("circum","Circle",function(obj,...){
+     cat("Circle Circum :\n")
+     2*pi*obj@radius
+ })
[1] "circum"

# Create the instance.
> e1<-new("Ellipse",name="e1",radius=c(2,5))
> c1<-new("Circle",name="c1",radius=2)

# Calculate the area and circum of ellipse.
> area(e1)
Ellipse Area:
[1] 31.41593
> circum(e1)
Ellipse Circum:
[1] 23.92566

# Calculate the area and circum of circle.
> area(c1)
Circle Area:
[1] 12.56637
> circum(c1)
Circle Circum:
[1] 12.56637
```

That is more reasonable after refactoring, Isn't it?

Task 3. Add the graphics handling for rectangle.
Let us extend the graphics library to add rectangle and square.

- Define the data structure of rectangle and calculate its area and circum.
- Define the data structure square and calculate its area and circum.
- Square is a special case of rectangle. Define rectangle as the parent class of square and square as the child class of rectangle.

The data structure is shown as in Figure 4.6.

```
# Define the Rectangle class inheriting Shape.
> setClass("Rectangle",
+    contains="Shape",
+    slots=list(edges="numeric"),
+    prototype=list(edges=c(1,1)))

# Define the Square class inheriting Rectangle.
> setClass("Square",
+    contains="Rectangle",
+    slots=list(edges="numeric"),
+    prototype=list(edges=1))
```

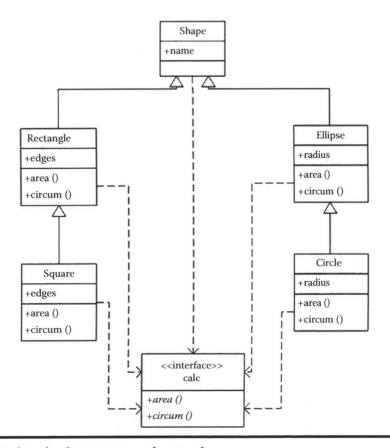

Figure 4.6   Task 3: the data structure of rectangle.

```
# Define the Rectangle implementation of area.
> setMethod("area","Rectangle",function(obj,…){
+     cat("Rectangle Area :\n")
+     prod(obj@edges)
+ })
[1] "area"

# Define the Square implementation of area.
> setMethod("area","Square",function(obj,…){
+     cat("Square Area :\n")
+     obj@edges^2
+ })
[1] "area"

# Define the Rectangle implementation of circum.
> setMethod("circum","Rectangle",function(obj,…){
+     cat("Rectangle Circum :\n")
+     2*sum(obj@edges)
+ })
[1] "circum"

# Define the Square implementation of circum.
> setMethod("circum","Square",function(obj,…){
+     cat("Square Circum :\n")
+     4*obj@edges
+ })
[1] "circum"

# Create instances.
> r1<-new("Rectangle",name="r1",edges=c(2,5))
> s1<-new("Square",name="s1",edges=2)

# Calculate the area and circum of rectangle.
> area(r1)
Rectangle Area:
[1] 10
> area(s1)
Square Area:
[1] 4

# Calculate the area and circum of square.
> circum(r1)
Rectangle Circum:
[1] 14
> circum(s1)
Square Circum:
[1] 8
```

Our graphics library now supports four shapes! Is the structured idea very clear when designing using OO structure?

Task 4. Add the shape slot and the getShape method in the base class shape.

Next, let us add a variable shape of shape type and a function getShape to check whether the instance is of shape type.

The data structure is shown as in Figure 4.7.

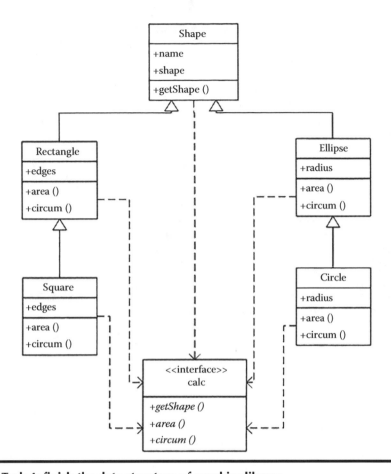

Figure 4.7   Task 4: finish the data structure of graphics library.

Without the OO structure, you need to add an argument and a checking function in all the shape definitions to meet the requirement. The modification would be very complex if there were 100 shapes. Instead, in OOP design, the requirement is very easy to meet. We only need to modify the code of the base class.

```
# Re-define the base class Shape to add a shape shot.
> setClass("Shape",
+    slots=list(name="character",shape="character"))


# Define the interface getShape.
> setGeneric("getShape",function(obj,…) standardGeneric("getShape"))
[1] "getShape"


# Define the implementation of getShape.
> setMethod("getShape","Shape",function(obj,…){
+       cat(obj@shape,"\n")
+ })
[1] "getShape"
```

Actually, the above modification is enough for us. Next, we just need to re-instantiate each graphics object. If we want to do one more step, we can modify the definitions of the objects to add the default values for the shape slot.

```
> setClass("Ellipse",contains="Shape",slots=list(radius="numeric"),protot
ype=list(radius=c(1,1),shape="Ellipse"))
> setClass("Circle",contains="Ellipse",slots=list(radius="numeric"),proto
type=list(radius = 1,shape="Circle"))
> setClass("Rectangle",contains="Shape",slots=list(edges="numeric"),proto
type=list(edges=c(1,1),shape="Rectangle"))
> setClass("Square",contains="Rectangle",slots=list(edges="numeric"),prot
otype=list(edges=1,shape="Square"))
# Instantiate a Square object and assign value to the shape slot.
> s1<-new("Square",name="s1",edges=2, shape="Square")
# Call the getShape() function of the base class.
> getShape(s1)
Square
```

Is it so easy? Once we modify the corresponding segment of code, all the shapes get the corresponding slots and methods. The following is the complete implementation:

```
setClass("Shape",slots=list(name="character",shape="character"))
setClass("Ellipse",contains="Shape",slots=list(radius="numeric"),prototyp
e=list(radius=c(1,1),shape="Ellipse"))
setClass("Circle",contains="Ellipse",slots=list(radius="numeric"),prototy
pe=list(radius = 1,shape="Circle"))
setClass("Rectangle",contains="Shape",slots=list(edges="numeric"),prototy
pe=list(edges=c(1,1),shape="Rectangle"))
setClass("Square",contains="Rectangle",slots=list(edges="numeric"),protot
ype=list(edges=1,shape="Square"))
setGeneric("getShape",function(obj,…) standardGeneric("getShape"))
setMethod("getShape","Shape",function(obj,…){
cat(obj@shape,"\n")
})

setGeneric("area",function(obj,…) standardGeneric("area"))
setMethod("area","Ellipse",function(obj,…){
  cat("Ellipse Area :\n")
  pi * prod(obj@radius)
})

setMethod("area","Circle",function(obj,…){
  cat("Circle Area :\n")
  pi*obj@radius^2
})

setMethod("area","Rectangle",function(obj,…){
  cat("Rectangle Area :\n")
  prod(obj@edges)
})
setMethod("area","Square",function(obj,…){
  cat("Square Area :\n")
  obj@edges^2
})
```

```
setGeneric("circum",function(obj,…) standardGeneric("circum"))
setMethod("circum","Ellipse",function(obj,…){
  cat("Ellipse Circum :\n")
  2*pi*sqrt((obj@radius[1]^2+obj@radius[2]^2)/2)
})
setMethod("circum","Circle",function(obj,…){
  cat("Circle Circum :\n")
  2*pi*obj@radius
})

setMethod("circum","Rectangle",function(obj,…){
  cat("Rectangle Circum :\n")
  2*sum(obj@edges)
})
setMethod("circum","Square",function(obj,…){
  cat("Square Circum :\n")
  4*obj@edges
})

e1<-new("Ellipse",name="e1",radius=c(2,5))
c1<-new("Circle",name="c1",radius=2)

r1<-new("Rectangle",name="r1",edges=c(2,5))
s1<-new("Square",name="s1",edges=2)

area(e1)
area(c1)
circum(e1)
circum(c1)

area(r1)
area(s1)
circum(r1)
circum(s1)
```

Through the above example, we have completely understood the OOP based on S4 object system in R. Have you found the universal feeling as in other languages?

In the world of the programmer, everything can be abstracted as an object.

4.4 The RC-Based OOP in R

Question

How do we program based on the RC OO system?

Introduction

This section will introduce the RC-based OOP in R. The RC object system changed the existing S3 and S4 object system from the underlying layer. It gets the generic functions removed and implements the OO characteristics completely based on classes.

4.4.1 Introduction to the RC Object System

RC is the short name for Reference classes, aka R5. It was introduced in R 2.12 and is the latest OO system of R. Differing from the legacy S3 and S4 object systems, the methods of RC are defined in classes instead of generic functions. The behaviors of RC objects are similar to that in other programming languages. The syntax of object instantiation was changed also. However, the RC object system still relies on the S4 object system, so we can simply consider RC as a further OO encapsulation of S4.

Let us redefine the following concepts from the perspective of OO:

- Class: the basic type of OO system. Class is a static structure.
- Object: the structure that is generated in memory as a result of class instantiation.
- Method: the function definition in class, without implementation through generic function.

4.4.2 Creating RC Classes and Objects

The system environment used in this section:

- Linux: Ubuntu Server 12.04.2 LTS 64bit
- R: 3.1.1 x86_64-pc-linux-gnu (64-bit)

In order for us to check the object's type conveniently, we import the pryr package as the auxiliary tool. For information about the pryr package, please refer Section 3.1.

```
# Launch the R application.
~ R
# Load the pryr package.
> library(pryr)
```

4.4.2.1 How to Create RC Classes

The RC object system takes class as its basic type. It has a dedicated function setRefClass() for defining classes and dedicated methods for generating the instantiated objects. Let us introduce how to create a class using the RC object system.

Let us view the definition of the RC class creation function setRefClass().

```
setRefClass(Class, fields = , contains = , methods =, where =, ...)
```

The following is the explanation to the arguments:

- Class: the class name
- fields: the properties and property types
- contains: the parent class, the inheritance
- methods: the methods in the class
- where: the environment in which to store the class definition

We can see that setRefClass() has less arguments than the S4 function setClass().

4.4.2.2 Creating RC Classes and Instances

```
# Define an RC class.
> User<-setRefClass("User",fields=list(name="character"))

# View the definition of User.
> User
Generator for class "User":

Class fields:
Name:          name
Class: character

Class Methods:
"callSuper", "copy", "export", "field", "getClass",
"getRefClass", "import", "initFields", "show", "trace",
"untrace", "usingMethods"

Reference Superclasses:
"envRefClass"
```

In the following, let us instantiate an RC object.

```
# Instantiate a User object, u1.
> u1<-User$new(name="u1")
# View the object u1.
> u1
Reference class object of class "User"
Field "name":
[1] "u1"

# Check the type of class User.
> class(User)
[1] "refObjectGenerator"
attr(,"package")
[1] "methods"
> is.object(User)
[1] TRUE
> otype(User)
[1] "RC"

# Check the type of u1.
> class(u1)
[1] "User"
attr(,"package")
[1] ".GlobalEnv"
> is.object(u1)
[1] TRUE
> otype(u1)
[1] "RC"
```

4.4.2.3 Creating an RC with Inheritance

```
# Create an RC class, User.
> User<-setRefClass("User",
+   fields=list(name="character"))
```

```
# Create User's child class, Member.
> Member<-setRefClass("Member",
+    contains="User",
+    fields=list(manager="User"))

# Instantiate User.
> manager<-User$new(name="manager")
# Instanciate Member.
> member<-Member$new(name="member",manager=manager)

# View the member object.
> member
Reference class object of class "Member"
Field "name":
[1] "member"
Field "manager":
Reference class object of class "User"
Field "name":
[1] "manager"

# View the name property of the member object.
> member$name
[1] "member

# View the manager property of the member object.
> member$manager
Reference class object of class "User"
Field "name":
[1] "manager"

# Check the types of the member's properties.
> otype(member$name)
[1] "primitive"
> otype(member$manager)
[1] "RC"
```

4.4.2.4 The Default Values of RC Objects

In RC class, there is a specific constructor $initialize() which runs automatically when instantiating object. We can set the default values for properties through it.

```
# Define an RC class.
> User<-setRefClass("User",
+      # Define 2 properties.
+      fields=list(name="character",level='numeric'),
+      methods=list(
+          # The constructor.
+          initialize = function(name,level){
+              print("User::initialize")
+              add default values for properties
+              name <<- 'conan'
+              level <<- 1
+          }
+      )
+ )
```

```
# Instantiate an object, u1.
> u1<-User$new()
[1] "User::initialize"

# View the object v1. Its properties were assigned with default values.
> u1
Reference class object of class "User"
Field "name":
[1] "conan"
Field "level":
[1] 1
```

4.4.3 *Object Assignment*

```
# Define a User class.
> User<-setRefClass("User",
+    fields=list(name="character",age="numeric",gender="factor"))

# Define a variable with type of factor.
> genderFactor<-factor(c('F','M'))

# Instantiate u1.
> u1<-User$new(name="u1",age=44,gender=genderFactor[1])
# View the value of the age property.
> u1$age
[1] 44
```

Assign the age property of u1.

```
# Reassign.
> u1$age<-10
# The value of the age property was changed.
> u1$age
[1] 10
```

Assign u1 to u2.

```
# Assign u1 to u2.
> u2<-u1
# View the age property of u2.
> u2$age
[1] 10
# Reassign.
> u1$age<-20
# View the values of the age property of u1 and u2. Both were changed.
> u1$age
[1] 20
> u2$age
[1] 20
```

This is because when assigning u1 to u2, the reference of u1 rather than the value itself was passed. This is same as the object assignment in other languages. If we want to assign the value instead of reference, we can use the following method.

```
# Assign the value to u3 by calling u1's built-in method copy().
> u3<-u1$copy()
# View the value of u3's age property.
> u3$age
[1] 20

# Reassign.
> u1$age<-30
# View the age property of u1. It was changed.
> u1$age
[1] 30
# View the age property of u3. It remains unchanged.
> u3$age
[1] 20
```

Mastering the reference relations between objects can not only reduce the memory duplications during value passing, but also improve the efficiency of the program.

4.4.4 Defining the Object Methods

In the object systems of S3 and S4, we implement the object behaviors through generic functions. The biggest problem of this implementation is that the code of function and object definitions are separated from each other, so we have to finish the method calling by checking the object type. As opposite, in the RC object system, methods can be defined inside the class and the method calling can be finished through the instantiated object.

```
# Define an RC object with its methods.
> User<-setRefClass("User",
+        fields=list(name="character",favorite="vector"),
+        # The methods argument.
+        methods = list(
+             # Add a favorite.
+             addFavorite = function(x) {
+                  favorite <<- c(favorite,x)
+             },
+             # Delete a favorite.
+             delFavorite = function(x) {
+                  favorite <<- favorite[-which(favorite == x)]
+             },
+             # Redefine the favorite list.
+             setFavorite = function(x) {
+                  favorite <<- x
+             }
+        )
+ )

# Instantiate an object u1.
> u1<-User$new(name="u1",favorite=c('movie','football'))

# View the object u1.
> u1
Reference class object of class "User"
Field "name":
```

```
[1] "u1"
Field "favorite":
[1] "movie"     "football"
```

In the following, let's do some method operations.

```
# Delete a favorite.
> u1$delFavorite('football')
# View the favorite list.
> u1$favorite
[1] "movie"
# Add a favorite.
> u1$addFavorite('shopping')
> u1$favorite
[1] "movie"     "shopping"
# Reset the favorite list.
> u1$setFavorite('reading')
> u1$favorite
[1] "reading"
```

RC defines methods directly inside class. The methods are accessed by the instantiated objects. So, the scope of the methods was limited when defining them, which reduces the system cost of runtime checking.

4.4.5 The Built-In Methods and Properties of RC Objects

For the instantiated RC objects, there are some built-in methods besides the customized methods. The copy() method used in copy assignment to properties is one of them.

4.4.5.1 The Built-In Methods

■ initialize: the initialization method of class used to set the default values of properties, only used in the method of class definition.
■ callSuper: used to call the same named method in parent class, only used in the method of class definition.
■ copy: used to clone all properties of the instantiated object.
■ initFields: used to assign values to object properties.
■ field: used to view/assign values to properties.
■ getClass: used to view the class definition of object.
■ getRefClass: same as getClass.
■ show: used to view the current object.
■ export: used to properties values with class as scope.
■ import: used to assign the properties values of one object to another.
■ trace: used to trace the method calling in object, used for debugging.
■ untrace: used to cancel the existing trace.
■ usingMethods: used to implement the method calling, only used in the method of class definition. This method has bad effect to program's robustness and hence is not recommended.

In the following, let us try these built-in methods. First of all, define a parent class User with two properties name and level, two functional methods addLevel and add HighLevel and a constructor method initialize.

```
# The User class.
> User<-setRefClass("User",
+    fields=list(name="character",level='numeric'),
+    methods=list(
+      initialize = function(name,level){
+        print("User::initialize")
+        name <<- 'conan'
+        level <<- 1
+      },
+      addLevel = function(x) {
+        print('User::addLevel')
+        level <<- level+x
+      },
+      addHighLevel = function(){
+        print('User::addHighLevel')
+        addLevel(2)
+      }
+    )
+)
```

Define the child class Member inheriting the parent class User. It contains the same named method addLevel which overrides the parent method. The addLevel method calls the same named method of parent.

```
# The child class Member.
> Member<-setRefClass("Member",contains="User",
+    # The properties of child class.
+    fields=list(age='numeric'),
+    methods=list(
+      # Override the same named method of parent.
+      addLevel = function(x) {
+        print('Member::addLevel')
+        # Call the same named method of parent.
+        callSuper(x)
+        level <<- level+1
+      }
+    )
+)
```

Instantiate objects u1 and m1.

```
# Instantiate u1.
> u1<-User$new(name='u1',level=10)
[1] "User::initialize"

# View the object u1. $new() does not assign values.
> u1
Reference class object of class "User"
Field "name":
[1] "conan"
```

```
Field "level":
[1] 1

# Assign values to properties through $initFields().
> u1$initFields(name='u1',level=10)
Reference class object of class "User"
Field "name":
[1] "u1"
Field "level":
[1] 10

# Instantiate m1.
> m1<-Member$new()
[1] "User::initialize"

> m1$initFields(name='m1',level=100,age=12)
Reference class object of class "Member"
Field "name":
[1] "m1"
Field "level":
[1] 100
Field "age":
[1] 12
```

Call the $copy() method, copying the object properties and passing values.

```
# Copy properties to u2.
> u2<-u1$copy()
[1] "User::initialize"

# Call the addLevel method to add 1 to level. u1 has been changed.
> u1$addLevel(1);u1
[1] "User::addLevel"
Reference class object of class "User"
Field "name":
[1] "u1"
Field "level":
[1] 11

# The level properties has no reference relation and has no change.
> u2
Reference class object of class "User"
Field "name":
[1] "u1"
Field "level":
[1] 10
```

View and assign value to the level property through the field() method.

```
# View the level property.
> u1$field('level')
[1] 11
# Assign 1 to the level property.
> u1$field('level',1)
```

```
# View the value of the level property.
> u1$level
[1] 1
```

View class definition of the u1 object through the getRefClass() and getClass() methods.

```
# The definition of class reference.
> m1$getRefClass()
Generator for class "Member":

Class fields:
Name:        name        level        age
Class: character   numeric    numeric

Class Methods:
"addHighLevel", "addLevel", "addLevel#User", "callSuper", "copy",
"export", "field", "getClass", "getRefClass", "import", "initFields",.
"initialize", "show", "trace", "untrace", "usingMethods"

Reference Superclasses:
"User", "envRefClass"

# The class definition.
> m1$getClass()
Reference Class "Member":

Class fields:
Name:        name        level        age
Class: character   numeric    numeric

Class Methods:
"addHighLevel", "addLevel", "addLevel#User", "callSuper", "copy",
"export", "field", "getClass", "getRefClass", "import", "initFields",.
"initialize", "show", "trace", "untrace", "usingMethods"

Reference Superclasses:
"User", "envRefClass"

# View the difference between types through the otype() function.
> otype(m1$getRefClass())
[1] "RC"
> otype(m1$getClass())
[1] "S4"
```

View the properties values of object through the $show() method. It works same as the show() function. When outputting object directly, the system calls the object's $show() method.

```
> m1$show()
Reference class object of class "Member"
Field "name":
[1] "m1"
Field "level":
[1] 100
Field "age":
[1] 12
```

```
> show(m1)
Reference class object of class "Member"
Field "name":
[1] "m1"
Field "level":
[1] 100
Field "age":
[1] 12

> m1
Reference class object of class "Member"
Field "name":
[1] "m1"
Field "level":
[1] 100
Field "age":
[1] 12
```

Trace the method calling through $trace() and then cancel the trace binding through $untrace().

```
# Tracke the addLevel() method.
> m1$trace("addLevel")
Tracing reference method "addLevel" for object from class "Member"
[1] "addLevel"

# When calling the addLevel() method, the system prints Tracing
m1$addLevel(1), indicate that the tracing takes effects.
> m1$addLevel(1)
Tracing m1$addLevel(1) on entry
[1] "Member::addLevel"
[1] "User::addLevel"

# When calling the addHighLevel() method in parent class, the system
prints Tracing m1$addHighLevel(2), indicate that the tracing takes
effects.
> m1$addHighLevel()
[1] "User::addHighLevel"
Tracing addLevel(2) on entry
[1] "Member::addLevel"
[1] "User::addLevel"

$ Cancel the tracing to the addLevel() method.
> m1$untrace("addLevel")
Untracing reference method "addLevel" for object from class "Member"
[1] "addLevel"
```

Call the $export() method to view the property values taking the class as the scope.

```
> m1$export('Member')
Reference class object of class "Member"
Field "name":
[1] "m1"
Field "level":
[1] 105
```

```
Field "age":
[1] 12

# View the properties in the User class, not including the age property
in current scope.
> m1$export('User')
[1] "User::initialize"
Reference class object of class "User"
Field "name":
[1] "m1"
Field "level":
[1] 105
```

Assign the property values of one object to another through the $import() method.

```
# Instantiate m2.
> m2<-Member$new()
[1] "User::initialize"
> m2
Reference class object of class "Member"
Field "name":
[1] "conan"
Field "level":
[1] 1
Field "age":
numeric(0)

# Assign the values of the m1 object to m2.
> m2$import(m1)
> m2
Reference class object of class "Member"
Field "name":
[1] "m1"
Field "level":
[1] 105
Field "age":
[1] 12
```

4.4.5.2 The Built-In Properties

The instantiated RC object has two built-in properties:

- ■ .self, the instantiated object itself
- ■ .refClassDef, the definition type of the class

```
# The $.self property.
> m1$.self
Reference class object of class "Member"
Field "name":
[1] "m1"
Field "level":
[1] 105
Field "age":
[1] 12
```

```
# m1$.self and m1 are identical.
> identical(m1$.self,m1)
[1] TRUE
# View the type.
> otype(m1$.self)
[1] "RC"

# The .refClassDef property.
> m1$.refClassDef
Reference Class "Member":

Class fields:
Name:        name      level        age
Class: character    numeric    numeric

Class Methods:
"addHighLevel", "addLevel", "addLevel#User", "callSuper", "copy",
"export", "field", "getClass", "getRefClass", "import", "initFields",.
"initialize", "show", "trace", "untrace", "usingMethods"

Reference Superclasses:
"User", "envRefClass"

# The result is identical to that of $getClass.
> identical(m1$.refClassDef,m1$getClass())
[1] TRUE

# View the type.
> otype(m1$.refClassDef)
[1] "S4"
```

4.4.6 *The Auxiliary Functions of RC Classes*

After defining the RC classes, we can use some auxiliary functions to view the properties and methods of these classes. The $new() function above used to create instantiated objects is one of them. The following content introduces these auxiliary functions in detail:

- new, used for object instantiation
- help: used to inquire the method calling in class
- methods: used to list all the methods defined in class
- fields: used to list all the methods defined in class
- lock: used to add lock to property. The locked property in instantiated object can only be assigned once. The value cannot be changed one assigned
- trace: used to trace method
- accessors: used to attach get/set methods to property

In the following, let us try the auxiliary functions, still taking the User class defined previously.

```
# Define the User class.
> User<-setRefClass("User",
+     fields=list(name="character",level='numeric'),
+     methods=list(
```

```
+        initialize = function(name,level){
+          print("User::initialize")
+          name <<- 'conan'
+          level <<- 1
+        },
+        addLevel = function(x) {
+          print('User::addLevel')
+          level <<- level+x
+        },
+        addHighLevel = function(){
+          print('User::addHighLevel')
+          addLevel(2)
+        }
+    )
+)

# Instantiate an object u1.
> u1<-User$new()
# List the properties of the User class.
> User$fields()
        name          level
  "character"    "numeric"

# List the methods of the User class.
> User$methods()
  [1] "addHighLevel" "addLevel"       "callSuper"
  [4] "copy"          "export"         "field"
  [7] "getClass"      "getRefClass"    "import"
 [10] "initFields"    "initialize"     "show"
 [13] "trace"         "untrace"        "usingMethods"

# View the method callings of the User class.
> User$help("addLevel")
Call:
$addLevel(x)

> User$help("show")
Call:
$show()
```

Add get/set methods to the properties in the User class.

```
# Add get/set methods to the level property.
> User$accessors("level")
# Add get/set methods to the name property.
> User$accessors("name")

# List all the methods.
> User$methods()
  [1] "addHighLevel" "addLevel"       "callSuper"
  [4] "copy"          "export"         "field"
  [7] "getClass"      "getLevel"       "getName"
 [10] "getRefClass"   "import"         "initFields"
 [13] "initialize"    "setLevel"       "setName"
 [16] "show"          "trace"          "untrace"
 [19] "usingMethods"
```

Trace the addLevel method through the $trace() function.

```
> User$trace('addLevel')
Tracing reference method "addLevel" for class
"User"
[1] "addLevel"

# Instantiate the u3 object.
> u3<-User$new(name='u3',level=1)
[1] "User::initialize"

# Call the addLevel method, sending tracing log Tracing u3$addLevel(2).
> u3$addLevel(2)
Tracing u3$addLevel(2) on entry
[1] "User::addLevel"
```

Specify the level property as constant through the $lock() function.

```
# Lock the level property.
> User$lock("level")
# View the locked property in the User class.
> User$lock()
[1] "level"

# Instantiate the u3 object. The level property is assigned once during
the instantiation.
> u3<-User$new()
[1] "User::initialize"
> u3
Reference class object of class "User"
Field "name":
[1] "conan"
Field "level":
[1] 1

$ Error happens when trying to reassign the level property.
> u3$level=1
Error: invalid replacement: reference class field 'level' is read-only
> u3$addLevel(2)
[1] "User::addLevel"
Error: invalid replacement: reference class field 'level' is read-only
```

4.4.7 Working with the RC Object System

In the following, let us define a set of animal study models using the RC object system.

Task 1. Define the data structure animals and their voicing methods.
Suppose animal as the base class. The animals to study include cat, dog, and duck. We need to define the data structure of animals and the voicing methods bark(), respectively. The structure is shown in Figure 4.8.

Define the data structure of animals, including the structures of base class and three animals.

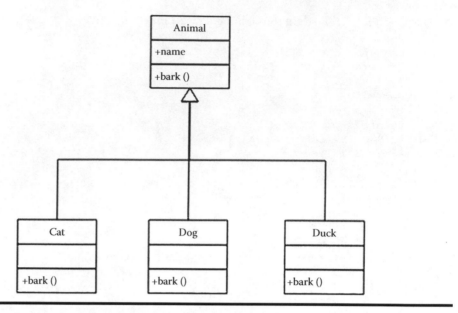

Figure 4.8 Task 1: the animal system data structure.

```
# Create the Animal class, including the name property, the constructor
inintialize() and the voicing method bark().
> Animal<-setRefClass("Animal",
+   fields=list(name="character"),
+   methods=list(
+     initialize = function(name) name <<- 'Animal',
+     bark = function() print("Animal::bark")
+   )
+)


# Create the Cat class, inheriting the Animal class and overwriting the
methods intitialize() and bark().
> Cat<-setRefClass("Cat",contains="Animal",
+   methods=list(
+     initialize = function(name) name <<- 'cat',
+     bark = function() print(paste(name,"is miao miao"))
+   )
+)


# Create the Dog class, inheriting the Animal class and overwriting the
methods intitialize() and bark().
> Dog<-setRefClass("Dog",contains="Animal",
+   methods=list(
+     initialize = function(name) name <<- 'dog',
+     bark = function() print(paste(name,"is wang wang"))
+   )
+)


# Create the Duck class, inheriting the Animal class and overwriting the
methods intitialize() and bark().
> Duck<-setRefClass("Duck",contains="Animal",
```

```
+    methods=list(
+       initialize = function(name) name <<- 'duck',
+       bark = function() print(paste(name,"is ga ga"))
+    )
+)
```

Next, let us instantiate the objects and study their voices.

```
# Create a cat instance.
> cat<-Cat$new()
> cat$name
[1] "cat"

# The voice of cat.
> cat$bark()
[1] "cat is miao miao"

# Create a dog instantance and name the dog Huang.
> dog<-Dog$new()
> dog$initFields(name='Huang')
Reference class object of class "Dog"
Field "name":
[1] "Huang"
> dog$name
[1] "Huang"

# The voice of dog.
> dog$bark()
[1] "Huang is wang wang"
# Create a duck instance.
> duck<-Duck$new()
# The voice of duck.
> duck$bark()
[1] "duck is ga ga"
```

Task 2. Define the appearance of animals.
The appearance of animals contains head, body, limbs, wings, and so on. All the three animals have limbs. Cat and dog have four limbs and duck has two limbs and two wings. The structure is shown in Figure 4.9.

We need to modify the original structure.

```
# Define the animal class, adding the limbs property with default value of 4.
> Animal<-setRefClass("Animal",
+    fields=list(name="character",limbs='numeric'),
+    methods=list(
+       initialize = function(name) {
+           name <<- 'Animal'
+           limbs<<-4
+       },
+       bark = function() print("Animal::bark")
+    )
+)
```

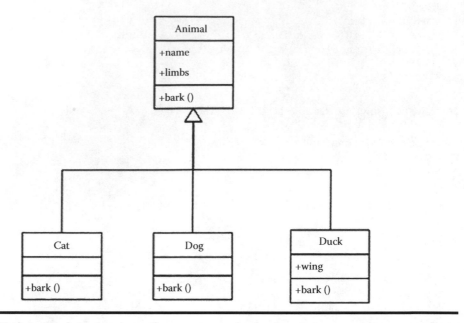

Figure 4.9   Task 2: the data structure of appearance of animal system.

```
# Call the callSuper() method to call the same named method in parent
class in the initialize() method in class Cat.
> Cat<-setRefClass("Cat",contains="Animal",
+     methods=list(
+        initialize = function(name) {
+          callSuper()
+          name <<- 'cat'
+        },
+        bark = function() print(paste(name,"is miao miao"))
+     )
+)

# Call the callSuper() method to call the same named method in parent
class in the initialize() method in class Dog.
> Dog<-setRefClass("Dog",contains="Animal",
+     methods=list(
+        initialize = function(name) {
+          callSuper()
+          name <<- 'dog'
+        },
+        bark = function() print(paste(name,"is wang wang"))
+     )
+)

# Call the callSuper() method to call the same named method in parent
class in the initialize() method in class Duck.
> Duck<-setRefClass("Duck",contains="Animal",
+     fields=list(wing='numeric'),
+     methods=list(
+        initialize = function(name) {
```

```
+          name <<- 'duck'
+          limbs<<- 2
+          wing<<- 2
+       },
+     bark = function() print(paste(name,"is ga ga"))
+   )
+)
```

Instantiate objects and view the property values of the three animals.

```
# Instantiate the cat object with limbs = 4.
> cat<-Cat$new();cat
Reference class object of class "Cat"
Field "name":
[1] "cat"
Field "limbs":
[1] 4

# Instantiate the dog object with limbs = 4.
> dog<-Dog$new()
> dog$initFields(name='Huang')
Reference class object of class "Dog"
Field "name":
[1] "Huang"
Field "limbs":
[1] 4
> dog
Reference class object of class "Dog"
Field "name":
[1] "Huang"
Field "limbs":
[1] 4

# Instantiate the duck object with limbs = 2 and wings = 2.
> duck<-Duck$new();duck
Reference class object of class "Duck"
Field "name":
[1] "duck"
Field "limbs":
[1] 2
Field "wing":
[1] 2
```

Task 3. Define the behaviors of animals
Cat, dog, and duck can walk on land but have different actions. The special actions are

- Cat: climbing tree
- Dog: swimming
- Duck: swimming and short flying

The structure is shown in Figure 4.10.
In the following, let us modify the model according to different actions of animals.

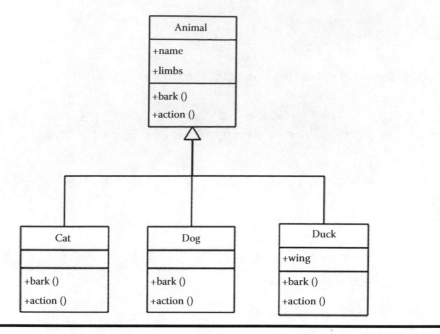

Figure 4.10   Task 3: the structure of actions of animal system.

```
# Define the Animal class, adding the action method as general action
walk on land.
> Animal<-setRefClass("Animal",
+     fields=list(name="character",limbs='numeric'),
+     methods=list(
+       initialize = function(name) {
+         name <<- 'Animal'
+         limbs<<-4
+       },
+       bark = function() print("Animal::bark"),
+       action = function() print("I can walk on the foot")
+     )
+)

# Define the Cat class, overwriting the action() method and adding the
action climbing.
> Cat<-setRefClass("Cat",contains="Animal",
+     methods=list(
+       initialize = function(name) {
+         callSuper()
+         name <<- 'cat'
+       },
+       bark = function() print(paste(name,"is miao miao")),
+       action = function() {
+         callSuper()
+         print("I can Climb a tree")
+       }
+     )
+   )
```

```
# Define the Dog class, overwriting the action() method and adding the
action swimming.
> Dog<-setRefClass("Dog",contains="Animal",
+    methods=list(
+       initialize = function(name) {
+          callSuper()
+          name <<- 'dog'
+       },
+       bark = function() print(paste(name,"is wang wang")),
+       action = function() {
+           callSuper()
+           print("I can Swim.")
+       }
+    )
+)

# Define the Duck class, overwriting the action() method and adding the
actions swimming and short flying.
> Duck<-setRefClass("Duck",contains="Animal",
+     fields=list(wing='numeric'),
+     methods=list(
+       initialize = function(name) {
+         name <<- 'duck'
+         limbs<<- 2
+         wing<<- 2
+       },
+       bark = function() print(paste(name,"is ga ga")),
+       action = function() {
+         callSuper()
+         print("I can swim.")
+         print("I also can fly a short way.")
+       }
+     )
+)
```

Instantiate the objects and call the action method.

The actions of cat.

```
# Instantiate cat.
> cat<-Cat$new()
# The actions of cat.
> cat$action()
[1] "I can walk on the foot"
[1] "I can Climb a tree"
```

The actions of dog.

```
> dog<-Dog$new()
> dog$action()
[1] "I can walk on the foot"
[1] "I can Swim."
The actions of duck.
> duck<-Duck$new()
> duck$action()
[1] "I can walk on the foot"
```

```
[1] "I can swim."
[1] "I also can fly a short way."
```

Through the above example, we should have understood the OOP based on the RC object system in R. The RC object system provides a complete OOP implementation. Chapter 6 will apply the RC-based OOP to game framework development.

4.5 The R6-Based OOP in R

Question

How do we program based on the R6 OO system?

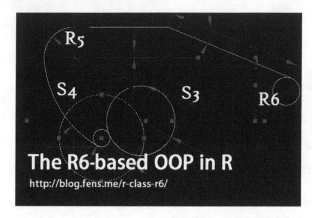

Introduction

What is R6? Is it a new type? In fact I encountered R6 by accident. R6 is an OO package of R. The R6 type is very close to but lighter than RC type. R6 classes are not created based on the S4 object system, so it is more effective to build OO system using R6.

4.5.1 A First Look at R6

R6 is a standalone R package, which is different from the native OO system types S3, S4, and RC that we are familiar with. In the OO system of R, the R6 type is similar to the RC type. But R6 is not built on the S4 object system, so we do not need the methods package when developing R packages using the R6 type. In the opposite manner, when developing R packages using the RC type, we need to set the dependencies to the methods package, which will be observed in Section 6.5.

Compared to the RC type, the R6 type is more conforming with the OO configurations in other programming languages. It supports the public and private member of classes, the active binding of functions, and the inheritance across different packages. The emergence of packages like R6 is because of the incomplete OO system design of the RC type. In the following, let us experience the OO system programming based on R6.

4.5.2 Creating R6 Classes and Instantiating Objects

The system environment used in this section:

- Win7 64bit
- R: 3.1.1 x86_64-w64-mingw32/x64 (64-bit)

Install the R6 package first. In this section, the R6 package version is 2.0 and the source code of the corresponding Github project is available at https://github.com/wch/R6/releases/tag/v2.0. In the meantime, in order for us to check the object's type conveniently, we import the pryr package as the auxiliary tool. For information about the pryr package, please refer Section 3.1.

First of all, install the R6 packages.

```
# Launch the R application.
~ R
# Install the R6 package.
> install.packages("R6")
# Load the R6 package.
> library(R6)
# Load the pryr package.
> library(pryr)
```

Note. R6 supports both Win7 and Linux.

4.5.2.1 How Do We Create R6 Classes?

The R6 object system takes class as its basic type. It has a dedicated function R6Class () for defining classes and dedicated methods for generating the instantiated objects. In the following, let us create a class using the RC object system. First, view the definition of the class creation function R6Class().

```
> R6Class
function (classname = NULL, public = list(), private = NULL,
    active = NULL, inherit = NULL, lock = TRUE, class = TRUE,
    portable = TRUE, parent_env = parent.frame())
```

Argument list:

- classname: used to define the class names.
- public: used to define the public members, including public methods and properties.
- private: used to define the private members, including public methods and properties.
- active: the list of actively binding functions.
- inherit: used to specify the parent class with inheritance.
- lock: used to lock or unlock the environment to store the class variables. If the environment is locked, it cannot be modified.
- class: used to determine whether to encapsulate the properties into classes. The default value is TRUE. If FALSE, then the class properties exist in an environment.
- portable: used to determine the class is portable. The default value is TRUE, which means that we need to call the self and private objects to access the members in class.
- parent_env: used to specify the parent environment for the object.

For the definition of the R6Class function, we can see that the arguments have more OO characteristics than that of the setRefClass() function in the RC object system.

4.5.2.2 Creating R6 Classes and Instantiate Objects

Let's create a simplest R6 class containing only one public method.

```
# Define an R6 class.
> Person <- R6Class("Person",
+   public=list(
+     # Define a public method hello.
+     hello = function(){
+       print(paste("Hello"))
+     }
+   )
+)

# View the definition of Person.
> Person
<Person> object generator
  Public:
    hello: function
  Parent env: <environment: R_GlobalEnv>
  Lock: TRUE
  Portable: TRUE

# Check the type of Person.
> class(Person)
[1] "R6ClassGenerator"
```

Next, instantiate the Person object using the $new() function.

```
# Instantiate a Person object u1.
> u1<-Person$new()
# View the u1 object.
> u1
<Person>
  Public:
    hello: function
# Check the type of u1.
> class(u1)
[1] "Person" "R6"
```

Check the types of the Person class and the u1 object through the otype function of the pryr package.

```
# View the type of Person.
> otype(Person)
[1] "S3"
#View the type of u1.
> otype(u1)
[1] "S3"
```

Totally unexpectedly, both Person and u1 are of S3 type. If R6 is built on S3 object system, then we can explain the difference between R6 and RC, and know that R6 is more efficient with functionalities of value passing and inheriting.

4.5.2.3 The Public and Private Members

The members of a class include both properties and methods. The R6 class definition can specify the public and private members separately. Let us specify the public member by modifying the definition of the Person class, adding a public property name in the public argument, and printing the value of name through the hello() method. This makes the R6 class more like JavaBean in Java. We need to use the self object to call when accessing the public member in class.

```
> Person <- R6Class("Person",
+   public=list(
+     # The public property.
+     name=NA,
+     # The constructor.
+     initialize = function(name){
+       self$name <- name
+     },
+     #The public method.
+     hello = function(){
+       print(paste("Hello",self$name))
+     }
+   )
+)


# Instantiate an object.
> conan <- Person$new('Conan')
# Call the hello() method.
> conan$hello()
[1] "Hello Conan"
```

Next, we define the private members. Add the private argument to the Person class and call the private member variables using the private object.

```
> Person <- R6Class("Person",
+   # The public members.
+   public=list(
+     name=NA,
+     initialize = function(name,gender){
+       self$name <- name
+       # Assign value to the private property.
+       private$gender<- gender
+     },
+     hello = function(){
+       print(paste("Hello",self$name))
+      # Call the private method.
+       private$myGender()
+     }
+   ),
+   # The private members.
+   private=list(
+     gender=NA,
+     myGender=function(){
+       print(paste(self$name,"is",private$gender))
```

```
+      }
+    )
+ )
# Instantiate an object.
> conan <- Person$new('Conan','Male')
# Call the hello method.
> conan$hello()
[1] "Hello Conan"
[1] "Conan is Male"
```

When adding the private members in the Person class, we define the private property gender and the private method myGender() through the private argument. It is notable that we need to use the private object to access the private member. When directly accessing the public and private properties, we can see that the public property returns the correct value where the private one returns NULL and the private method is invisible.

```
# The public property.
> conan$name
[1] "Conan"
# The private property.
> conan$gender
NULL
# The private method.
> conan$myGender()
Error: attempt to apply non-function
```

Furthermore, let us check what the self and private objects actually are. Add a public method member() in the Person class to print the self and private object.

```
> Person <- R6Class("Person",
+    public=list(
+      name=NA,
+      initialize = function(name,gender){
+        self$name <- name
+        private$gender<- gender
+      },
+      hello = function(){
+        print(paste("Hello",self$name))
+        private$myGender()
+      },
+      # The method for testing.
+      member = function(){
+        print(self)
+        print(private)
+        print(ls(envir=private))
+      }
+    ),
+    private=list(
+      gender=NA,
+      myGender=function(){
+        print(paste(self$name,"is",private$gender))
+      }
+    )
+ )
```

```
>
> conan <- Person$new('Conan','Male')
# Call the member() method.
> conan$member()
# The output of print(self).
<Person>
  Public:
    hello: function
    initialize: function
    member: function
    name: Conan

# The output of print(private).
<environment: 0x0000000008cfc918>
# The output of print(ls(envir=private)).
[1] "gender"   "myGender"
```

From the test result, we can see that the self object is the instantiated object itself and the private object is an environment that is a child of the environment where the self object is located. Therefore, the private members can only be called in the current class and they are not available to access from outside. Storing the private properties and methods in environment to prevent them from the external calls through the access control of environment is a trick technique often used in R package development.

4.5.3 The Active Binding of R6 Classes

The active binding is a special way of calling function. It transforms the function access into property access. The active binding functions belong to public members. The active binding is implemented through setting the active argument in the class definition. Let us add two active binding functions active and rand in the Person class.

```
> Person <- R6Class("Person",
+    public = list(
+      num = 100
+    ),
+    The active binding.
+    active = list(
+      active  = function(value) {
+        if (missing(value)) return(self$num +10 )
+        else self$num <- value/2
+      },
+      rand = function() rnorm(1)
+    )
+)

> conan <- Person$new()
# View the public property.
> conan$num
[1] 100
# Call the active binding function active(). The result is
num+10=100+10=110
s> conan$active
[1] 110
```

When passing the argument to the active binding function active, we should use the assignment sign instead of ().

```
# Passings argument.
> conan$active<-100
# View the public property num.
> conan$num
[1] 50
# Call the active binding function active(). The result is
num+10=50+10=60
> conan$active
[1] 60
# If we access using function calling, an error occurs prompting no such
function.
> conan$active(100)
Error: attempt to apply non-function
```

Let us call the rand function and check what happens.

```
# Call the rand function.
> conan$rand
[1] -0.4767338
> conan$rand
[1] 0.1063623
# Error happens when passing argument.
> conan$rand<-99
Error in (function ()   : unused argument (quote(99))
```

No error happens if we access the rand function directly. But error happens when passing argument into the rand() function. The operation is not allowed since there is no argument defined in the rand() function.

The active binding transforms function accessing into property accessing. It makes function operation more flexible.

4.5.4 The Inheritance of R6 Classes

Inheritance is one of the basic features of OO. The R6 OO system also supports inheritance. During the creation, a class can inherit another as its parent. Let us first create a parent class Person, containing the public and private members.

```
> Person <- R6Class("Person",
+    # The public members.
+    public=list(
+      name=NA,
+      initialize = function(name,gender){
+        self$name <- name
+        private$gender <- gender
+      },
+      hello = function(){
+        print(paste("Hello",self$name))
+        private$myGender()
+      }
+    ),
```

```
+     # The private members.
+   private=list(
+     gender=NA,
+     myGender=function(){
+       print(paste(self$name,"is",private$gender))
+     }
+   )
+ )
```

Create a child class Worker inhering the parent class Person and add a public method bye() in it.

```
> Worker <- R6Class("Worker",
+   # Inheritance, referring to the parent class.
+   inherit = Person,
+   public=list(
+     bye = function(){
+       print(paste("bye",self$name))
+     }
+   )
+ )
```

Instantiate the parent and child classes to check whether the inheritance relation takes effect.

```
# Instantiate the parent class.
> u1<-Person$new("Conan","Male")
> u1$hello()
[1] "Hello Conan"
[1] "Conan is Male"

# Instantiate the child class.
> u2<-Worker$new("Conan","Male")
> u2$hello()
[1] "Hello Conan"
[1] "Conan is Male"
> u2$bye()
[1] "bye Conan"
```

We can see that the inheritance takes effect. We did not define the hello() method in the child class. The child class instance u2 can directly use the hello() method in the parent class. In the meanwhile, the bye() method in the child class uses the name property defined in the parent class and gets the correct result.

In the following, let us define a child class method with the same name as that in the parent class and then view the function calling to see whether the feature of function overwriting in inheritance takes effect. Modify the Work class to add private property and method in the child class.

```
> Worker <- R6Class("Worker",
+   inherit = Person,
+   public=list(
+     bye = function(){
+       print(paste("bye",self$name))
+     }
+   ),
```

```
+    private=list(
+      . gender=NA,
+      myGender=function(){
+        print(paste("worker",self$name,"is",private$gender))
+      }
+    )
+ )
```

Instantiate the child class and then call the hello() method.

```
> u2<-Worker$new("Conan","Male")
# Call the hello() method.
> u2$hello()
[1] "Hello Conan"
[1] "worker Conan is Male"
```

Since the private method myGender() in the child class overwrote the private method myGender() in the parent class, the hello() method called the child class implementation of the myGender() method, ignoring that in the parent class.

To call the parent class method in child, one way is using the super object to call the method using the syntax of super$xx().

```
> Worker <- R6Class("Worker",
+    inherit = Person,
+    public=list(
+      bye = function(){
+        print(paste("bye",self$name))
+      }
+    ),
+    private=list(
+      gender=NA,
+      myGender=function(){
+        #Call the method in parent class.
+        super$myGender()
+        print(paste("worker",self$name,"is",private$gender))
+      }
+    )
+ )

> u2<-Worker$new("Conan","Male")
> u2$hello()
[1] "Hello Conan"
[1] "Conan is Male"
[1] "worker Conan is Male"
```

The child class method myGender() calls the myGender() method in parent class using the super object. From the output, we can see that the same named method in parent class was called as well.

4.5.5 The Static Properties of R6 Objects

In OOP, all variables are actually objects. We can define an instantiated object as a property of another class to form a reference relation chain of objects. Please note that when assigning a property with an object of another R6 class, the value of the property is the reference of the object instead of the object instance itself. By this rule, we can implement the static property of an object, which means to share object instance among various instances, similar to the static property in Java.

Let us depict it with code for understanding. Define two classes A and B. A has a public property x and B has a public property a that is an instantiated object of A.

```
> A <- R6Class("A",
+   public=list(
+     x = NULL
+   )
+ )
>
> B <- R6Class("B",
+   public = list(
+     a = A$new()
+   )
+ )
```

Run the program to implement the call from B's instantiated object b to A's instantiated object a and assign value to the variable x.

```
# Instantiate B's object b.
> b <- B$new()
# Assign value to the variable x.
> b$a$x <- 1
# View the value of the variable x.
> b$a$x
[1] 1

# Instantiate B's object b2.
> b2 <- B$new()
# Assign value to the variable x.
> b2$a$x <- 2
# View the value of the variable x.
> b2$a$x
[1] 2

# Change also happened in the variable x of the object a of the instance b.
> b$a$x
[1] 2
```

From the result, we can see that the share of object a among multiple B objects has been implemented. When the object b2 changed the value of x of the object a, the value of x of the object b also got changed. We should avoid assigning value using the initialization().

```
> C <- R6Class("C",
+   public = list(
+     a = NULL,
+     initialize = function() {
+       a <<- A$new()
+     }
+   )
+ )
```

```
> cc <- C$new()
> cc$a$x <- 1
> cc$a$x
[1] 1

> cc2 <- C$new()
> cc2$a$x <- 2
> cc2$a$x
[1] 2

# The value of x doesn't change.
> cc$a$x
[1] 1
```

The object a created through the initialize() method is a reference to a single environment, so this way cannot implement the share of reference object.

4.5.6 The Portable Type of R6 Classes

In the definition of R6 classes, the portable argument is used to set whether a R6 class is a portable or non-portable type. There are two obvious differences between the two types:

■ The portable type supports inheritance across R packages, while the non-portable types has bad compatibility when inheriting across R packages.
■ The members in a portable type need the self and private objects to be accessed, self$x, private$y for instances, where the members in a non-portable type can be accessed directly through the variable x,y and assigned with ≪–.

The version used in this section is the version 2.0 of R6, so the portable types are created by default. When considering whether there is cross package inheritance, we can make a choice between the portable and the non-portable types.

Let us compare the RC type, the portable type of R6, and the non-portable type of R6. Define a simple class containing a property x and two methods getx() and setx().s

```
# The definition of RC type.
> RC <- setRefClass("RC",
+   fields = list(x = 'Hello'),
+   methods = list(
+     getx = function() x,
+     setx = function(value) x <<- value
+   )
+ )
> rc <- RC$new()
> rc$setx(10)
> rc$getx()
[1] 10
```

Create a non-portable R6 type with identical behaviors.

```
# The non-portable R6 type.
> NR6 <- R6Class("NR6",
```

```
+    portable = FALSE,
+    public = list(
+      x = NA,
+      getx = function() x,
+      setx = function(value) x <<- value
+    )
+ )
> np6 <- NR6$new()
> np6$setx(10)
> np6$getx()
[1] 10
```

Create a portable R6 type with identical behaviors.

```
# The portable R6 type.
> PR6 <- R6Class("PR6",
+    portable = TRUE,
+    public = list(
+      x = NA,
+      getx = function() self$x,
+      setx = function(value) self$x <- value
+    )
+ )
> pr6 <- PR6$new()
> pr6$setx(10)
> pr6$getx()
[1] 10
```

From this case, we can see that the difference between portable and non-portable R6 types is the usage of the self object.

4.5.7 *The Dynamic Binding of R6 Classes*

In statically typed programming languages, once a class is defined, we cannot modify the properties and method, except for the special operations with high cost like reflection. However, for dynamically typed programming languages, there is no such limitation and we can arbitrarily modify the structure of a class or an instantiated object. As a dynamic language, R supports the modification to dynamic variables. The R3/R4-based types can dynamically add function definition through generic function but the RC types do not, which makes me feel that R's OO design is too weird.

This case has been considered by the R6 package. It provides a way to dynamically specify member variable through the $set() function.

```
> A <- R6Class("A",
+    public = list(
+      x = 1,
+      getx = function() x
+    )
+ )
# Add the getx2() method dynamically.
> A$set("public", "getx2", function() self$x*2)
> s <- A$new()
```

```
# View the structure of instantiated object.
> s
<A>
  Public:
    getx: function
    getx2: function
    x: 1
# Call the getx2() method.
> s$getx2()
[1] 20
```

In same way, the property can be modified dynamically. Let us change the value of the x property dynamically.

```
# Change the value of the x property dynamically.
> A$set("public", "x", 10, overwrite = TRUE)
> s <- A$new()
# View the value of the x property.
> s$x
[1] 10
# Call the getx() method. The variable x was lost in the portable type.
> s$getx()
Error in s$getx() : object 'x' not found
```

The class A is a portable type by default, so we should access the variable x through the self object. Otherwise, error would happen if we modify the member dynamically. Let us replace x with self$x in the getx method.

```
> A <- R6Class("A",
+   public = list(
+     x = 1,
+     # Replace with self$x.
+     getx = function() self$x
+   )
+ )
> A$set("public", "x", 10, overwrite = TRUE)
> s <- A$new()
> s$x
[1] 10
# Call the getx() method.
> s$getx()
[1] 10
```

It is recommended to use the self and private object to access the members for both portable and non-portable types.

4.5.8 The Printing Function of R6 Classes

R6 provides a default method print() for printing. When printing the instantiated objects, the default method print() is called, somehow like the default toString() method in Java class. We can overwrite the print() method and use customized print prompt.

```
> A <- R6Class("A",
+  public = list(
+    x = 1,
+    getx = function() self$x
+  )
+ )
> a <- A$new()
# Use the default printing method.
> print(a)
<A>
  Public:
    getx: function
    x: 1
```

Customize the printing method to overwrite the default.

```
> A <- R6Class("A",
+    public = list(
+      x = 1,
+      getx = function() self$x,
+      print = function(…) {
+        cat("Class <A> of public ", ls(self), " :", sep="")
+        cat(ls(self), sep=",")
+        invisible(self)
+      }
+    )
+ )
> a <- A$new()
> print(a)
Class <A> of public getxprintx :getx,print,x
```

The customized method can be used to overwrite the default one to output what we want.

4.5.9 The Storage of the Instantiated Objects

R6 is built on the S3 OO system and S3 is a relative loose type. This would run into an inundation of variables in the user environment. R6 privates a solution to store the properties and methods defined in a class into an S3 object by setting the argument class in R6Class(). This solution is by default. Another solution is to store the properties and methods defined in a class into a single environment. Let us try the default scenario where with class = TRUE, the instantiated object a is actually an S3 class.

```
> A <- R6Class("A",
+  class=TRUE,
+  public = list(
+    x = 1,
+    getx = function() self$x
+  )
+ )
> a <- A$new()
> class(a)
[1] "A"   "R6"
> a
```

```
<A>
  Public:
    getx: function
    x: 1
```

With class = FALSE, the instantiated object a is an environment to store the data of class variables.

```
> B <- R6Class("B",
+   class=FALSE,
+   public = list(
+     x = 1,
+     getx = function() self$x
+   )
+ )
> b <- B$new()
> class(b)
[1] "environment"
> b
<environment: 0x000000000d83c970>
> ls(envir=b)
[1] "getx" "x"
```

The storage of instantiated objects takes another consideration. Since the class variables are stored in an environment, we can manually find the environment and add or modify variables in it. This brings risks to the security of our program. To prevent the security issue, we can use the argument lock in R6Class() to lock the environment to deny the dynamic modification. The default value is TRUE to prevent the modification.

```
> A <- R6Class("A",
+   # Lock the environment.
+   lock=TRUE,
+   public = list(
+     x = 1
+   )
+ )
> s<-A$new()
# View the variables in the environment s.
> ls(s)
[1] "x"
# Attempt to add new variable results in error.
> s$aa<-11
Error in s$aa <- 11 : cannot add bindings to a locked environment
# Attempt to remove existing variable results in error.
> rm("x",envir=s)
Error in rm("x", envir = s):
  cannot remove bindings from a locked environment
```

If we do not lock the environment with lock = FALSE, the environment remains open completely. We can modify any variables.

```
> A <- R6Class("A",
+ # Don't lock the environment.
+   lock=FALSE,
```

```
+  public = list(
+     x = 1
+  )
+  )
> s<-A$new()
# View the variables in the environment s.
> ls(s)
[1] "x"
# Add a variable.
> s$aa<-11
> ls(s)
[1] "aa" "x"
# Remove a variable.
> rm("x",envir=s)
> ls(s)
[1] "aa"
```

Through the introduction to R6, we have mastered the basic knowledge of the R6 OO system. In the following, let us apply the knowledge of R6-based OOP through a simple example.

4.5.10 A Case of R6 OO System

Let us build a use case of book classification using R6-based OO system.

Task 1. Define the static structures of books.
Take Book as the parent class and R, Java, and PHP as the three child classes. There are private properties and public methods in the Book class. The inheritance relation is shown in Figure 4.11.

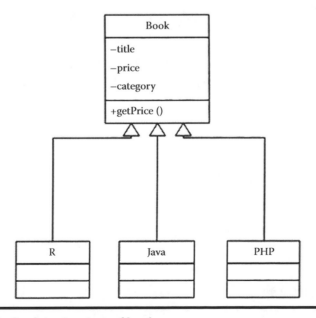

Figure 4.11   Task 1: the data structure of books.

Define the data structure of book system, including the structure of the parent class and three types of books.

```
# The parent class.
> Book <- R6Class("Book",
+    private = list(
+       title=NA,
+       price=NA,
+       category=NA
+    ),
+    public = list(
+      initialize = function(title,price,category){
+        private$title <- title
+        private$price <- price
+        private$category <- category
+      },
+      getPrice=function(){
+        private$price
+      }
+    )
+ )


# The child class, R book.
> R <- R6Class("R",
+    inherit = Book
+ )
# The child class, JAVA book.
> Java <- R6Class("JAVA",
+    inherit = Book
+ )
# The child class, PHP book.
> Php <- R6Class("PHP",
+    inherit = Book
+ )
```

Create three instantiated objects: R book *R for Programmers: Mastering the Tools,* Java book *Thinking in Java*, and PHP book *Head First PHP & MySQL*, and get the prices.

```
> r1<-R$new("R for Programmers: mastering the tools ",59,"R")
> r1$getPrice()
[1] 59

> j1<-Java$new("Thinking in Java ",108,"JAVA")
> j1$getPrice()
[1] 108

> p1<-Java$new("Head First PHP & MySQL",98,"PHP")
> p1$getPrice()
[1] 98
```

Task 2. Discounting promotion of various books during the period of 11.11.
Let us define a discounting strategy to promote the sales of books. The strategy is a pure fiction.

- All the books are discounted by 10%.
- Java books are discounted by 30%, additional discount not supported.
- To promote sales, R books are discounted by 30% and support additional discount.
- PHP books have no special discount.

According to the strategy, all books can be discounted, so discounting can be considered a behavior of the Book object. R, Java, and PHP have their own discounting strategies, so this is a scenario of polymorphism.

Modify the definition of the parent class by adding the discounting method discount(), with default as 0.9, satisfying strategy one.

```
> Book <- R6Class("Book",
+    private = list(
+        title=NA,
+        price=NA,
+        category=NA
+    ),
+    public = list(
+        initialize = function(title,price,category){
+            private$title <- title
+            private$price <- price
+            private$category <- category
+        },
+        getPrice=function(){
+            p<-private$price*self$discount()
+            print(paste("Price:",private$price,", Sell out:",p,sep=""))
+        },
+        discount=function(){
+            0.9
+        }
+    )
+ )
```

Modify the three child classes with their own discounting strategies:

- Add the discount() method to the child class Java to overwrite the discount() method of parent class, making Java books discounted by 30% to satisfy strategy 2.
- Add the discount() method to the child class R by calling the discount() method of parent class in it, making R books discounted by 30% in addition to the discount by 10% to satisfy strategy 3.
- No modification to the child class PHP, completely following strategy 1.

```
> Java <- R6Class("JAVA",
+    inherit = Book,
+    public = list(
+        discount=function(){
+            0.7
+        }
+    )
+ )
> R <- R6Class("R",
```

```
+    inherit = Book,
+    public = list(
+      discount=function(){
+         super$discount()*0.7
+      }
+    )
+ )
> Php <- R6Class("PHP",
+    inherit = Book
+ )
```

View the discounted prices of the three books:

```
> r1<-R$new("R for Programmers: Mastering the Tools",59,"R")
> r1$getPrice()
# 59 * 0.9 *0.7= 37.17
[1] "Price:59, Sell out:37.17"
>
> j1<-Java$new("Thinking in Java",108,"JAVA")
> j1$getPrice()
# 108 *0.7= 75.6
[1] "Price:108, Sell out:75.6"
>
> p1<-Php$new("Head First PHP & MySQL",98,"PHP")
> p1$getPrice()
# 98 *0.9= 88.2
[1] "Price:98, Sell out:88.2"
```

The R book is discounted most, with the privilege of repeated discounts of 30% off and 10% off, 59 * 0.9 * 0.7 = 37.17. The Java book has the privilege of discount of 30% off, 108 * 0.7 = 75.6. The PHP has the privilege of discount of 10% off, 98 * 0.9 = 88.2.

Through the above example, we have implemented the three features: encapsulation, inheritance, and polymorphism in OOP using the R6 object system and hence proved that R6 is a complete OO implementation. The R6 object system provides a compatibly OO implementation and is more close to the OO definitions in other programming languages. R6 is built on the base of S3 and is therefore more efficient than RC.

Totally, we have introduced four OO architectures in R. Choose the one that you understand best. There is always at least one solution for you.

DEVELOPING
R PACKAGE

Chapter 5

Developing R Packages

This chapter introduces ways to develop R packages. It is complex to develop the low-level R functions from scratch, while the devtools package helps in streamlining the development. The case to develop the Daily China Weather R package brings readers a comprehensive understanding of the process of R package development.

5.1 Developing an R Package from Scratch

Question

How do we develop an R package?

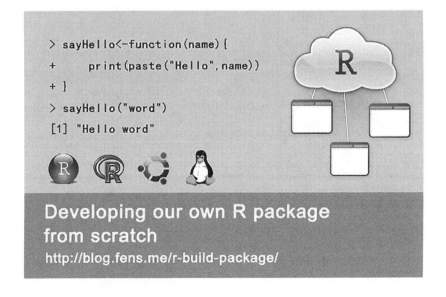

```
> sayHello<-function(name){
+       print(paste("Hello",name))
+ }
> sayHello("word")
[1] "Hello word"
```

Developing our own R package
from scratch
http://blog.fens.me/r-build-package/

Introduction

R is a product of the collaboration of worldwide developers. Up to July 2016, there were nearly 8722 packages available to download from the Internet. We are for now the users of R. Someday we will be R developers sharing our knowledge with the world with the R toolkits we have ourselves developed. In this section, we will learn how to develop an R package.

5.1.1 Developing R Packages with Linux Command Line

Let us develop an R package from scratch. All the works can be done with only a Linux command line. The development of R packages is divided into six steps: creating, building, installing, using, checking, and uninstalling. To know how hard it is to develop an R package, you must try to do it.

The system environment used in this section:

- Linux: Ubuntu Server 12.04.2 LTS 64bit
- R: 3.1.1 x86_64-pc-linux-gnu (64-bit)

Note: R package development supports both Win7 and Linux platforms. To reduce the publishing error, it is recommended to develop it on Linux. Section 6.5 of this book will introduce how to develop R packages on Win7.

5.1.1.1 Creating the Project sayHello

How do we start to develop an R package?

Step 1: Name your project as, for example, sayHello and then create project directory /home/conan/R/demo.

```
# Create a directory.
~ mkdir /home/conan/R/demo
# Enter the directory.
~ cd /home/conan/R/demo
```

Step 2: New a file sayHello.R in the project and customize a function sayHello as the first function in your package.

```
# New sayHello.R
~ vi sayHello.R

sayHello<-function(name){
    print(paste("Hello",name))
}
```

Step 3: Generate the skeleton of the project sayHello using the native function package.skeleton.

```
# Launch the R application.
~ R
# Clean up the variables in current environment.
> rm(list=ls())
# Set project directory.
> setwd("/home/conan/R/demo")
```

```
# Generate the project skeleton using the script of sayHello.
> package.skeleton(name="sayHello",code_files="/home/conan/R/demo/
  sayHello.R")
Creating directories...
Creating DESCRIPTION...
Creating NAMESPACE...
Creating Read-and-delete-me...
Copying code files...
Making help files...
Done.
Further steps are described in './sayHello/Read-and-delete-me'.
```

We can see that a directory sayHello was generated in current directory.

```
# View the current directory.
~ ls -l
drwxrwxr-x 4 conan conan 4096 Sep 28 09:57 sayHello
-rw-rw-r-- 1 conan conan   59 Sep 28 09:48 sayHello.R

# View the sayHello directory.
~ ls -l sayHello
-rw-rw-r-- 1 conan conan  281 Sep 28 09:57 DESCRIPTION
drwxrwxr-x 2 conan conan 4096 Sep 28 09:57 man
-rw-rw-r-- 1 conan conan   31 Sep 28 09:57 NAMESPACE
drwxrwxr-x 2 conan conan 4096 Sep 28 09:57 R
-rw-rw-r-- 1 conan conan  420 Sep 28 09:57 Read-and-delete-me

# View the man directory.
~ ls -l sayHello/man
-rw-rw-r-- 1 conan conan 1043 Sep 28 09:57 sayHello-package.Rd
-rw-rw-r-- 1 conan conan 1278 Sep 28 09:57 sayHello.Rd

# View the R directory.
~ ls -l sayHello/R
-rw-rw-r-- 1 conan conan 59 Sep 28 09:57 sayHello.R
```

Explanations to files and directories:

- DESCRIPTION file: the project's description file, used to specify the project's global configuration.
- NAMESPACE file: the project's namespace, used to specify the project's input and output functions.
- Read-and-delete-me file: the readme file. Can be deleted.
- Man directory: directory to store the help files for functions.
- R directory: directory to store the source files.
- Man/sayHello.Rd: the help file for the sayHello function, with syntax of LaTex, used to generate PDF documents.
- Man/sayHello-package.Rd: the help file for the sayHello package. Can be deleted.

Step 4: Edit the DESCRIPTION file to specify the project's global configuration.

```
~ vi sayHello/DESCRIPTION
```

```
Package: sayHello
Type: Package
Title: R package demo for sayHello
Version: 1.0
Date: 2014-09-28
Author: Dan Zhang
Maintainer: Dan Zhang <bsspirit@gmail.com>
Description: This package provides a package demo
License: GPL-3
```

Explanations to configuration items:

Package: project name(package name); Type: project type; Title: project title; Version: project version number. Date: project creation date; Author: project author; Maintainer: the main contributors, multiple persons allowed: Description: the description to project in details, more text recommended; license: the license the project is published under.

By editing the DESCRIPTION file, we define the profile for the entire project.

Step 5: Edit the NAMESPACE file, which is used to specify the project's input and output functions.

```
~ vi sayHello/NAMESPACE
export(sayHello)
```

There is only one function sayHello in our project and it is open to users, so we just need one line definition.

Step 6: Edit the sayHello.Rd file. This is to write the help documentation for the sayHello function. The syntax of LaTex is used here.

```
~ vi sayHello/man/sayHello.Rd

\name{sayHello}
\alias{sayHello}
\title{a sayHello function demo}
\description{
a sayHello function demo
}
\usage{
sayHello(name)
}
\arguments{
   \item{name}{a word}
}
\details{
nothing
}
\value{
no return
}
\references{
nothing
}
\author{
Dan Zhang
```

```
}
\note{
nothing
}
\seealso{
nothing
}
\examples{
sayHello("world")
}
\keyword{ sayHello }
```

Step 7: Delete the files that can be ignored. Otherwise warnings would occur during the checking process.

```
~ rm sayHello/Read-and-delete-me
~ rm sayHello/man/sayHello-package.Rd
```

By walking through the 7 steps above, we have finished the tasks to create the R package of sayHello.

5.1.1.2 Building the sayHello Project

After creating an R package project, the next is to pack the project to generate R package installation file. Install the file through command line and load it into the user environment so that we can use our own package just like using other packages.

Let us switch to the command line to execute the packing command.

```
# Execute the packing command.
~ R CMD build sayHello

* checking for file 'sayHello/DESCRIPTION'...OK
* preparing 'sayHello':
* checking DESCRIPTION meta-information...OK
* checking for LF line-endings in source and make files
* checking for empty or unneeded directories
* building 'sayHello_1.0.tar.gz'
```

The packing process got succeeded, generating the installation package sayHello_1.-.tar.gz under the current directory.

```
# View the files under the current directory.
~ ls -l
drwxrwxr-x 4 conan conan 4096 Sep 28 10:34 sayHello
-rw-r--r-- 1 conan conan  622 Sep 28 11:01 sayHello_1.0.tar.gz
-rw-rw-r-- 1 conan conan   59 Sep 28 09:48 sayHello.R
```

5.1.1.3 Installing the sayHello Package Locally

Everything goes well installing the sayHello package in the local environment.

```
# Install sayHello_1.0.tar.gz through the command line.
~ R CMD INSTALL sayHello_1.0.tar.gz
```

```
* installing to library '/home/conan/R/x86_64-pc-linux-gnu-library/3.1'
* installing *source* package 'sayHello'...
** R
** preparing package for lazy loading
** help
*** installing help indices
** building package indices
** testing if installed package can be loaded
* DONE (sayHello)
```

View the installation direction of R and find the sayHello directory.

```
~ ls /home/conan/R/x86_64-pc-linux-gnu-library/3.1
bitops devtools evaluate memoise RCurl sayHello
```

5.1.1.4 Using the Function in the sayHello Package

Launch the R application, load the sayHello package, and then execute the function in the package.

```
# Launch the R application.
~ R
# Load the sayHello package.
> library(sayHello)

# Execute the function in the package.
> sayHello("Conan")
[1] "Hello Conan"

# View the system help documentation for the sayHello function.
> ?sayHello

sayHello          package:sayHello          R Documentation
a sayHello function demo
Description:
     a sayHello function demo
Usage:
     sayHello(name)
Arguments:
     name: a word
Details:
     nothing
Value:
     no return
Note:
     nothing
Author(s):
     Dan Zhang
References:
     nothing
See Also:
     nothing
Examples:
     sayHello("world")
```

Hereby, we have finished the package building successfully and installed and used it locally.

5.1.1.5 Checking R Packages

To publish the R package to CRAN, there is a long way for us to go. Before submitting the R package, we must do a check. A package with any error or warning fails the check. The process of checking is the most complicated and error-prone step in package development. It is rare to pass the check in the first time. So we need to be patient to solve every issue.

Switch back to the command line again. A PDF file will be generated during the checking process. The generation of PDF is dependent on LaTex. So we must install the LaText dependency package.

```
#Install LaTex.
~ sudo apt-get install texlive-full
```

After finishing the installation of LaTex, execute the check command.

```
~ R CMD check sayHello_1.0.tar.gz

* using log directory '/home/conan/R/demo/sayHello.Rcheck'
* using R version 3.1.1 (2014-07-10)
* using platform: x86_64-pc-linux-gnu (64-bit)
* using session charset: UTF-8
* checking for file 'sayHello/DESCRIPTION'...OK
* checking extension type...Package
* this is package 'sayHello' version '1.0'
* checking package namespace information...OK
* checking package dependencies...OK
* checking if this is a source package...OK
* checking if there is a namespace...OK
* checking for executable files...OK
* checking for hidden files and directories...OK
* checking for portable file names...OK
* checking for sufficient/correct file permissions...OK
* checking whether package 'sayHello' can be installed...OK
* checking installed package size...OK
* checking package directory...OK
* checking DESCRIPTION meta-information...OK
* checking top-level files...OK
* checking for left-over files...OK
* checking index information...OK
* checking package subdirectories...OK
* checking R files for non-ASCII characters...OK
* checking R files for syntax errors...OK
* checking whether the package can be loaded...OK
* checking whether the package can be loaded with stated
dependencies...OK
* checking whether the package can be unloaded cleanly...OK
* checking whether the namespace can be loaded with stated
dependencies...OK
* checking whether the namespace can be unloaded cleanly...OK
* checking loading without being on the library search path...OK
* checking dependencies in R code .OK
* checking S3 generic/method consistency...OK
* checking replacement functions...OK
```

```
* checking foreign function calls...OK
* checking R code for possible problems...OK
* checking Rd files...OK
* checking Rd metadata...OK
* checking Rd cross-references...OK
* checking for missing documentation entries...OK
* checking for code/documentation mismatches...OK
* checking Rd \usage sections...OK
* checking Rd contents...OK
* checking for unstated dependencies in examples...OK
* checking examples...OK
* checking PDF version of manual...OK
```

Once the R package passes the checking process, we can submit it to CRAN. Although the example looks simple, we cannot say the checking process is easy. Various strange warnings or errors may occur. Actually I have fixed many problems and finally passed the checking process.

5.1.1.6 Uninstalling R Packages

By now we have installed the package. But sometimes we need to uninstall the packages, especially for those under development. It is very easy to uninstall R packages. Two ways are available: one is through the command line and another is through function.

Uninstall through the command line. Execute the following command.

```
~ R CMD REMOVE sayHello
Removing from library '/home/conan/R/x86_64-pc-linux-gnu-library/3.1'

# View the installation directory of R packages.
~ ls /home/conan/R/x86_64-pc-linux-gnu-library/3.1
bitops devtools evaluate memoise RCurl
```

Uninstall through function. Call the remove.packages() function in the R environment.

```
# Uninstall the R package.
> remove.packages("sayHello")
Removing package from '/home/conan/R/x86_64-pc-linux-gnu-library/3.1'
(as 'lib' is unspecified)

# Unable to load the sayHello package.
> library(sayHello)
Error in library(sayHello) : there is no package called 'sayHello'
```

5.1.2 Developing R Packages in RStudio

The above operations were finished using Linux's command line with companion of VI editor. It is somehow hard for those who are unfamiliar with Linux commands and VI. Another way is through RStudio. Section 1.5 in R for programmers: Mastering the tools has introduced how to configure RStudio Server on Linux. RStudio provides us a graphical Integrated Development Environment (IDE) for developing R packages more conveniently.

5.1.2.1 Developing R Packages in RStudio

Let us repeat the above operations in RStudio. The open-source version of RStudio Server is used in this section. The UI of RStudio Server is same as RStudio. Please note that the term RStudio in the rest of this section refers to RStudio Server. Its version is v0.98.

Create a project in RStudio, with name as sayHello2, type as Package, source files including sayHello.R, as shown in Figure 5.1.

RStudio automatically calls the project files generated by the package.skeleton() function. Of course, we can manually create them, as shown in Figure 5.2.

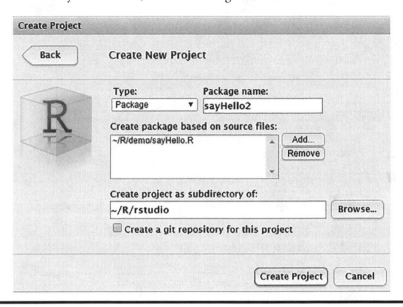

Figure 5.1   Create project through RStudio.

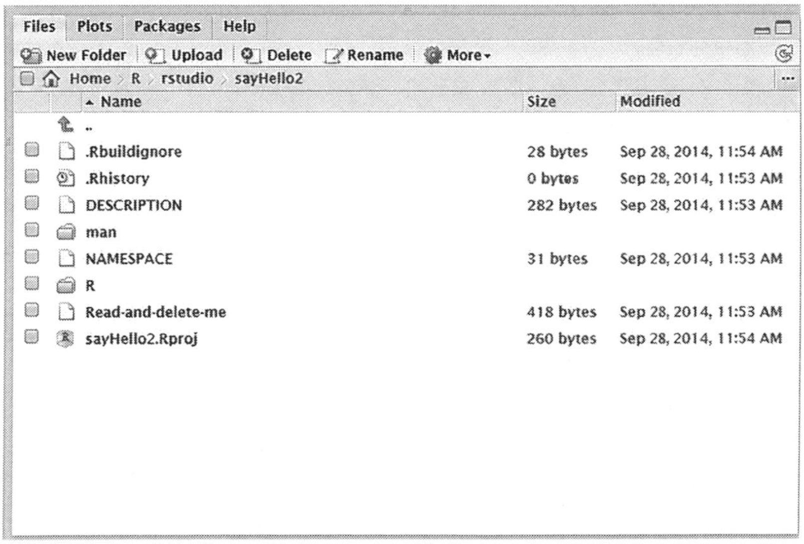

Figure 5.2   RStudio project directory.

5.1.2.2 Editing the Project Description Files DESCRIPTION and NAMESPACE

The package name is sayHello2. Other configurations are same as those above. The explanations are same as those in command line operations.

Edit the DESCRIPTION file.

```
Package: sayHello2
Type: Package
Title: R package demo for sayHello
Version: 1.0
Date: 2014-09-28
Author: Dan Zhang
Maintainer: Dan Zhang <bsspirit@gmail.com>
Description: This package provides a package demo
License: GPL-3
```

Exit the NAMESPACE file.

```
export(sayHello)
```

5.1.2.3 Editing the R Program Code R.sayHello.R

Modify the code with Hello replacing hi to be distinguished from the previous sayHello function.

```
sayHello<-function(name){
  print(paste("Hello",name))
}
```

5.1.2.4 Editing the Help Documentation man/sayHello.Rd

Same as the documentation in command line operations.

```
\name{sayHello}
\alias{sayHello}
\title{a sayHello function demo}
\description{
a sayHello function demo
}
\usage{
sayHello(name)
}
\arguments{
  \item{name}{a word}
}
\details{
nothing
}
\value{
no return
}
\references{
nothing
}
```

```
\author{
Dan Zhang
}
\note{
nothing
}
\seealso{
nothing
}
\examples{
sayHello("world")
}
\keyword{ sayHello }
```

5.1.2.5 Executing Build & Reload

Click the Build & Reload button in the Build menu in RStudio to execute the process of build and reload, as shown in Figure 5.3.

We can see that the two warnings are about the file sayHello2-package.Rd. Delete the two files man/sayHello2-package.Rd and Read-and-delete-me, which are unrelated to packing. Execute Build & Reload again and no warning appears. From the operation log, we can see that sayHello2 has been installed in the R environment, as shown in Figure 5.4.

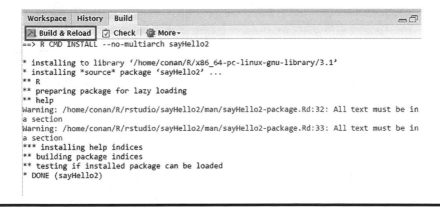

Figure 5.3   Execute Build & Reload through RStudio.

Figure 5.4   Execute Build & Reload through RStudio again.

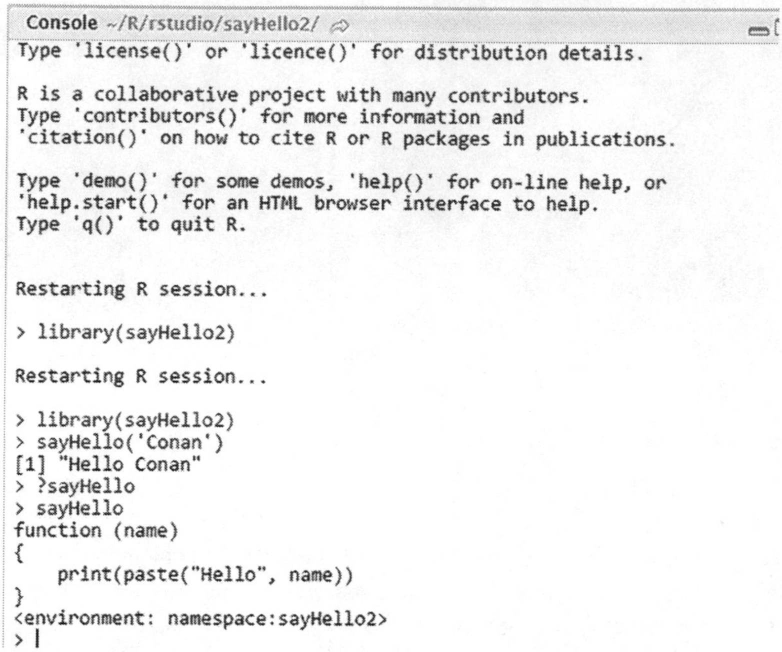

Figure 5.5   Load the sayHello package and execute the function.

The next operation is simpler. Load the sayHello package and execute the sayHello() function in the Console pane, as shown in Figure 5.5.

```
> library(sayHello2)
> sayHello('Conan')
[1] "Hello Conan"
> ?sayHello
> sayHello
function (name)
{
    print(paste("Hello", name))
}
<environment: namespace:sayHello2>
```

5.1.2.6 Executing the Check Operation

Last, let us test the check operation. In RStudio, click the check button to execute the check operation, as shown in Figure 5.6.

We have finished the R package development with green lights all the way, basically by clicks. With RStudio, we can execute the process of developing, packing, checking, etc. more conveniently. RStudio is exactly a magic tool for R development.

This section has introduced the complete process of developing R packages according to the R development specification. Although each step in the specification is clear, the operations still bring many difficulties to R package development due to its complexity and because of not enough support from official functions. The next section will introduce the development process redefined by Hadley Wickham, which helps us to finish the development of R packages more simply.

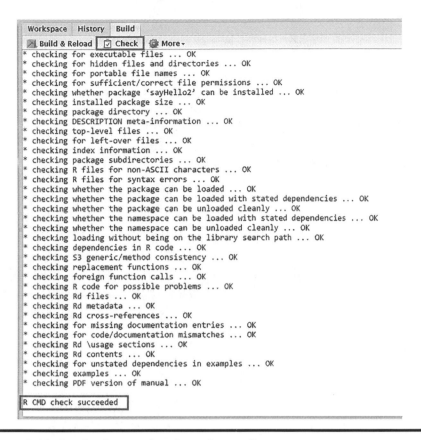

```
 Workspace  History  Build
 ⬛ Build & Reload   ☑ Check   ⚙ More▾
* checking for executable files ... OK
* checking for hidden files and directories ... OK
* checking for portable file names ... OK
* checking for sufficient/correct file permissions ... OK
* checking whether package 'sayHello2' can be installed ... OK
* checking installed package size ... OK
* checking package directory ... OK
* checking DESCRIPTION meta-information ... OK
* checking top-level files ... OK
* checking for left-over files ... OK
* checking index information ... OK
* checking package subdirectories ... OK
* checking R files for non-ASCII characters ... OK
* checking R files for syntax errors ... OK
* checking whether the package can be loaded ... OK
* checking whether the package can be loaded with stated dependencies ... OK
* checking whether the package can be unloaded cleanly ... OK
* checking whether the namespace can be loaded with stated dependencies ... OK
* checking whether the namespace can be unloaded cleanly ... OK
* checking loading without being on the library search path ... OK
* checking dependencies in R code ... OK
* checking S3 generic/method consistency ... OK
* checking replacement functions ... OK
* checking foreign function calls ... OK
* checking R code for possible problems ... OK
* checking Rd files ... OK
* checking Rd metadata ... OK
* checking Rd cross-references ... OK
* checking for missing documentation entries ... OK
* checking for code/documentation mismatches ... OK
* checking Rd \usage sections ... OK
* checking Rd contents ... OK
* checking for unstated dependencies in examples ... OK
* checking examples ... OK
* checking PDF version of manual ... OK
R CMD check succeeded
```

Figure 5.6   Finish the check operation through RStudio.

5.2 Streamlining R Package Development

Question

Is there easy way to develop R packages?

Introduction

From the previous section, we have learned how to build an R extension package from the low level according to R's specification. However the complexity in development brings frustration to R users without a programming background. It would be good if there were a simple way to streamline the development. We are lucky, since Hadley Wickham has redesigned a specification for R package development and provided lots of auxiliary functions to make R package development easier. We can see further on the shoulders of the giants.

5.2.1 Introducing the Development Flow

On the shoulders of the giants to develop R packages, we have three weapons: devtools, roxygen2, and testthat.

- Devtools: a collection of various developing tools to make development faster and very practical.
- Roxygen2: generating documentation through comments, getting rid of LaTex.
- Testthat: unit tests to make R packages stable and robust, relieving the pains during upgrading.

The re-streamlined development process contains the following five steps: (1) Coding; (2) Debugging program; (3) Performing; (4) Authoring documentation; and (5) Building project.

5.2.2 Coding

In the following, the entire development will be finished through the combination of Linux commands and vi editor. Those who need GUI to program can refer the process of developing R packages in RStudio introduced in Section 5.1.

The system environment used in this section:

- Linux: Ubuntu Server 12.04.2 LTS 64bit
- R: 3.1.1 x86_64-pc-linux-gnu (64-bit)

Note: R package development supports both Win7 and Linux platforms. To reduce the publishing issue, it is recommended to develop on Linux.

5.2.2.1 Installing the Packages: Devtool, roxygen2, and testthat

Before installing the three packages, we need to install the dependency package, for example, libcurl4-opensll-dev.

```
~ sudo apt-get install libcurl4-openssl-dev
```

Launch the R application to install the developing packages: devtools, roxygen2, and testthat.

```
# Launch the R application.
~ R

# Install the packages.
> install.packages("devtools")
```

```
> install.packages("roxygen2")
> install.packages("testthat")

# Load the packages.
> library(devtools)
> library(roxygen2)
> library(testthat)

# View the packages in current environment.
> search()
 [1] ".GlobalEnv"         "package:testthat"  "package:roxygen2"
 [4] "package:devtools"   "package:stats"     "package:graphics"
 [7] "package:grDevices"  "package:utils"     "package:datasets"
[10] "package:methods"    "Autoloads"         "package:base"
```

5.2.2.2 Creating a Project chinaWeather

Create a project named chinaWeather. Use the create() function in devtools package instead of the package.skeleton() function to create the skeleton of the project.

```
# Set the working directory.
> setwd("/home/conan/R")
# Create the project.
> create("/home/conan/R/chinaWeather")
Creating package chinaWeather in /home/conan/R
No DESCRIPTION found. Creating with values:

Package: chinaWeather
Title: What the package does (one line)
Version: 0.1
Authors@R: "First Last <first.last@example.com> [aut, cre]"
Description: What the package does (one paragraph)
Depends: R (>= 3.1.1)
License: What license is it under?
LazyData: true
Adding RStudio project file to chinaWeather
```

Check the directories and files created.

```
# Set the project directory as working directory.
> setwd("/home/conan/R/chinaWeather")
# View the files in the directory.
> dir()
[1] "DESCRIPTION" "NAMESPACE" "R"
```

5.2.2.3 Creating the Project Files

Edit the file DESCRIPTION.
```
~ vi /home/conan/R/chinaWeather/DESCRIPTION

Package: chinaWeather
Type: Package
```

```
Title:a visualized package for chinaWeather
Version: 0.1
Authors@R: "Dan Zhang <bsspirit@gmail.com> [aut, cre]"
Description: a visualized package for chinaWeather
Depends: R (>= 3.1.1)
License: GPL-2
LazyData: true
Date: 2014-09-28
```

Create an R program file chinaWeather.R. Define a function filename() in the file, no Chinese in comments.

```
~ vi /home/conan/R/chinaWeather/R/chinaWeather.R

# Define a filename from current date.
filename<-function(date=Sys.time()){
  paste(format(date, "%Y%m%d"),".csv",sep="")
}
```

Here we defined a simple function filename(), generating a file name according to the specified date.

5.2.3 Debugging Program

With the help of devtools, we can directly load the project directory to R's runtime environment, not like the previous one, when we could not load the package until building.

```
# Load the project directory.
> load_all("/home/conan/R/chinaWeather")
Loading chinaWeather

# View the filename() function.
> filename
function(date=Sys.time()){
  paste(format(date, "%Y%m%d"),".csv",sep="")
}
<environment: namespace:chinaWeather>

# Call the filename() function.
> filename()
[1] "20140928.csv"

# Pass argument and call the filename() function.
> day<-as.Date("20110701",format="%Y%m%d")
> filename(day)
[1] "20110701.csv"
```

5.2.4 Carrying Unit Tests

Usually we need to write the unit testing code to make project code robust. How do we carry unit tests against the filename() function?

Create a directory for unit testing.

```
# Create the tests directory to store the unit testing files.
~ mkdir -p /home/conan/R/chinaWeather/inst/tests
```

Create a file test.chinaWeather.R to implement the unit testing against the function in file chinaWeather.R. We need to notice the name of the unit testing file, starting with the test and appending the name of the source file.

Create file test.chinaeWeather.R and then define the unit testing code with the test_that() function.

```
~ vi /home/conan/R/chinaWeather/inst/tests/test.chinaWeather.R

library(testthat)
context("filename: current of date")

test_that("filename is current of date", {
  daystr<-paste(format(Sys.Date(), "%Y%m%d"),".csv",sep="")
  expect_that(filename(), equals(daystr))

  day<-as.Date("20110701",format="%Y%m%d")
  expect_that(filename(day), equals("20110701.csv"))
})
```

Run the unit testing program.

```
# Load the unit testing program.
> source("/home/conan/R/chinaWeather/inst/tests/test.chinaWeather.R")
# Executing the unit test.
> test_file("/home/conan/R/chinaWeather/inst/tests/test.chinaWeather.R")
filename: current of date : ..

# Executing the unit testing for all the files under the directory.
> test_dir("/home/conan/R/chinaWeather/inst/tests/",reporter = "summary")
filename: current of date : ..
```

If the result of the unit test is correct, no special information is output. Let us add an erroneous case and see what the difference is. Modify the file test.chinaWeather.R by adding the following code.

```
~ vi /home/conan/R/chinaWeather/inst/tests/test.chinaWeather.R

library(testthat)
context("filename: current of date")

test_that("filename is current of date", {
  daystr<-paste(format(Sys.Date(), "%Y%m%d"),".csv",sep="")
  expect_that(filename(), equals(daystr))

  day<-as.Date("20110701",format="%Y%m%d")
  expect_that(filename(day), equals("20110701.csv"))
})

test_that("filename is current of date, bad test", {
  day<-as.Date("20110701",format="%Y%m%d")
  expect_that(filename(day), equals("20110702.csv"))
})
```

Launch the unit testing program again. The error shows.

```
> source("/home/conan/R/chinaWeather/inst/tests/test.chinaWeather.R")
Error: Test failed: 'filename is current of date, bad test'
```

```
Not expected: filename(day) not equal to "20110702.csv"
1 string mismatches:
x[1]: "20110702.csv"
y[1]: "20110701.csv"

> test_file("/home/conan/R/chinaWeather/inst/tests/test.chinaWeather.R")
filename: current of date : ..1

1. Failure(@test.chinaWeather.R#14): filename is current of date, bad test -----
filename(day) not equal to "20110702.csv"
1 string mismatches:
x[1]: "20110702.csv"
y[1]: "20110701.csv"
```

Remove the erroneous unit testing code and continue the following operations. If there are many functions in a package, there is plenty of unit testing code. We may need to carry automatic testing against the entire package. By setting the directories of source code and testing code, we call the auto_test() function to scan all the files under the two directories.

Configure the automatic unit tests.

```
# The directory of source code.
> src<-"/home/conan/R/chinaWeather/R/"
# The directory of unit testing code.
> test<-"/home/conan/R/chinaWeather/inst/tests/"
# Execute the unit tests.
> auto_test(src,test)
filenam: current of date : ..
```

Besides, we can carry unit tests against the entire package. This is because the directories of source files and testing files are unchanged in same project.

```
> test("/home/conan/R/chinaWeather")
Testing chinaWeather
filename: current of date : ..
```

Hereby, we have finished the unit testing very smoothly. Next, let us author the help documentation.

5.2.5 Authoring Documentation

In the previous section, we need to manually author the LaTex-based help documentation, which is separated from code and located in different files. So it is very hard to author such documentation. In this section, we use the roxygen2 to author documentation, generating LaTex-based help documentation by commenting source code. This solution is closer to the programmer's idea and commenting is much easier than writing LaTex.

Edit the source code file chinaWeather.R by adding comments to the function.

```
# Edit the file.
~ vi /home/conan/R/chinaWeather/R/chinaWeather.R

#' Define a filename from current date.
#'
#' @param date input a date type
#' @return character a file name
```

```
#' @keywords filename
#' @export
#' @examples
#' filename()
#' filename(as.Date("20110701",format="%Y%m%d"))
filename<-function(date=Sys.time()){
    paste(format(date, "%Y%m%d"),".csv",sep="")
}
```

Then, we generate LaTex-based help documentation through the roxgenize() function in the roxygen2 package.

```
# Generate the LaTex document using the roxygenize() function.
> roxygenize("/home/conan/R/chinaWeather")
First time using roxygen2 4.0. Upgrading automatically...
Writing NAMESPACE
Writing filename.Rd
```

View the LaTex file generated.

```
# View the help file.
~ cat /home/conan/R/chinaWeather/man/filename.Rd

% Generated by roxygen2 (4.0.2): do not edit by hand
\name{filename}
\alias{filename}
\title{Define a filename from current date.}
\usage{
filename(date = Sys.time())
}
\arguments{
\item{date}{input a date type}
}
\value{
character a file name
}
\description{
Define a filename from current date.
}
\examples{
filename()
filename(as.Date("20110701",format="\%Y\%m\%d"))
}
\keyword{filename}
```

This solution bypasses the LaTex syntax although it is not very different from the original one. This way we can focus on programming R packages.

5.2.6 Building the Project

The above processes (loading project, carrying unit testing, and generating help document) can be simpler, just calling three functions.

```
# Load the project program.
> load_all("/home/conan/R/chinaWeather")
```

```
# Execute the unit testing.
> test("/home/conan/R/chinaWeather")
# Generate the help document.
> document("/home/conan/R/chinaWeather")
```

The process of checking the program can be done with one function calling.

```
# Execute the program checking.
> check("/home/conan/R/chinaWeather")
Updating chinaWeather documentation
Loading chinaWeather
'/usr/lib/R/bin/R' --vanilla CMD build '/home/conan/R/chinaWeather' \
  --no-manual --no-resave-data


* checking for file '/home/conan/R/chinaWeather/DESCRIPTION'...OK
* preparing 'chinaWeather':
* checking DESCRIPTION meta-information...OK
* checking for LF line-endings in source and make files
* checking for empty or unneeded directories
* building 'chinaWeather_0.1.tar.gz'


'/usr/lib/R/bin/R' --vanilla CMD check \
  '/tmp/RtmpJ37lsJ/chinaWeather_0.1.tar.gz' --timings

* using log directory '/tmp/RtmpJ37lsJ/chinaWeather.Rcheck'
* using R version 3.1.1 (2014-07-10)
* using platform: x86_64-pc-linux-gnu (64-bit)
* using session charset: UTF-8
* checking for file 'chinaWeather/DESCRIPTION'...OK
* checking extension type...Package
* this is package 'chinaWeather' version '0.1'
* checking package namespace information...OK
* checking package dependencies...OK
* checking if this is a source package...OK
* checking if there is a namespace...OK
* checking for executable files...OK
* checking for hidden files and directories...OK
* checking for portable file names...OK
* checking for sufficient/correct file permissions...OK
* checking whether package 'chinaWeather' can be installed...OK
* checking installed package size...OK
* checking package directory...OK
* checking DESCRIPTION meta-information...OK
* checking top-level files...OK
* checking for left-over files...OK
* checking index information...OK
* checking package subdirectories...OK
* checking R files for non-ASCII characters...OK
* checking R files for syntax errors...OK
* checking whether the package can be loaded...OK
* checking whether the package can be loaded with stated dependencies...OK
* checking whether the package can be unloaded cleanly...OK
* checking whether the namespace can be loaded with stated dependencies...OK
* checking whether the namespace can be unloaded cleanly...OK
```

```
* checking loading without being on the library search path...OK
* checking dependencies in R code...OK
* checking S3 generic/method consistency...OK
* checking replacement functions...OK
* checking foreign function calls...OK
* checking R code for possible problems...OK
* checking Rd files...OK
* checking Rd metadata...OK
* checking Rd line widths...OK
* checking Rd cross-references...OK
* checking for missing documentation entries...OK
* checking for code/documentation mismatches...OK
* checking Rd \usage sections...OK
* checking Rd contents...OK
* checking for unstated dependencies in examples...OK
* checking examples...OK
* checking PDF version of manual...OK
```

Everything is OK. All get passed.

5.2.7 Publishing the Program

There are several platforms (CRAN, RForge, and Github) to select to publish the R packages that have finished the development if we are willing to open the source to others.

- CRAN: familiar to most people, maintained by R Core Team, with restrict review process.
- RForge: another publishing platform for R packages.
- Github, a publishing platform maintained through the devtools package, suitable for personal publishing without review.

There are many reviews on CRAN and RForge that do not allow free publishing. Let us publish the program on Github first. After optimizing the functionalities, we will commit to CRAN orRForge. The operation to uploading the project to Github is not related to R. Install the git tool locally and create a resource pool on the Github community, and then upload the project through the git tool. I have created a new resource pool chinaWeatherDemo at https://github.com/bsspirit/chinaWeatherDemo, as shown in Figure 5.7.

Commit the local code onto github.

```
# Enter the project directory.
~ cd /home/conan/R/chinaWeather
# Initialize the git project.
~ git init
Initialized empty Git repository in /home/conan/R/chinaWeather/.git/

# Manage the local file with git.
~ git add.
# Commit files to local library with git.
~ git commit -m 'init'
[master (root-commit) 43deae4] init
6 files changed, 62 insertions(+)
create mode 100644 .gitignore
```

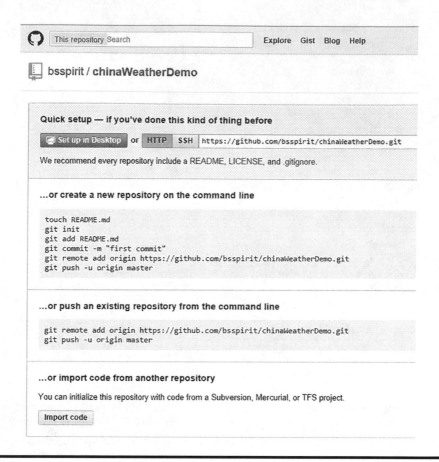

Figure 5.7   Create project on Github.

```
create mode 100644 DESCRIPTION
create mode 100644 NAMESPACE
create mode 100644 R/chinaWeather.R
create mode 100644 inst/tests/test.chinaWeather.R
create mode 100644 man/filename.Rd

# Bind the local library to the remote library on Github.
~ git remote add origin https://github.com/bsspirit/chinaWeather
# Sync the local library to the remote.
~ git push -u origin master
To https://github.com/bsspirit/chinaWeatherDemo.git
 * [new branch]        master -> master
Branch master set up to track remote branch master from origin.
```

Thus, the code is uploaded onto Github. Next we can easily download the R project from Github and install.

By now, my chinaWeather package has been published to Github. Anyone who wants to use can install it with the following commands.

```
# Load devtools.
> library(devtools)
```

```
# Download and install the chinaWeather package.
> install_github("bsspirit/chinaWeatherDemo")
Downloading github repo bsspirit/chinaWeatherDemo@master
Installing chinaWeather
'/usr/lib/R/bin/R' --vanilla CMD INSTALL \
   '/tmp/RtmpJ37lsJ/devtools646348c59273/bsspirit-chinaWeatherDemo-43deae4'
   \--library='/home/conan/R/x86_64-pc-linux-gnu-library/3.1'
--install-tests

* installing *source* package 'chinaWeather'...
** R
** inst
** preparing package for lazy loading
** help
*** installing help indices
** building package indices
** testing if installed package can be loaded
* DONE (chinaWeather)
Reloading installed chinaWeather

# Load the chinaWeather package.
> library(chinaWeather)
# Call the filename() function.
> filename()
[1] "20140928.csv"

# View the help document.
> ?filename
filename               package:chinaWeather               R Documentation
Define a filename from current date.
Description:
     Define a filename from current date.
Usage:
     filename(date = Sys.time())
Arguments:
     date: input a date type
Value:
     character a file name
Examples:
     filename()
     filename(as.Date("20110701",format="%Y%m%d"))
```

We have finished the entire process of R package development. With the three tooling packages: devtools, roxygen2, and testthat, the development becomes much more efficient than completely manual operations. Take more with less cost. I hope more friends can go forward on the shoulders of giants and create amazing works!

5.3 Weather Visualization with R

Question

How do we develop a weather visualization app with R?

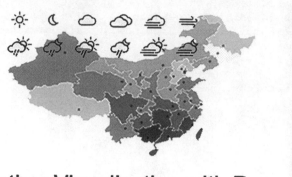

Weather Visualization with R

http://blog.fens.me/r-app-china-weather
http://apps.weibo.com/chinaweatherapp

Introduction

Lots of people think that R is just a toy without capability to develop enterprise apps. They don't know R at all, and even have no clear picture on how to develop enterprise apps. In my first time using R, I have embedded it as visualization engine into Fans-showing, one of my Weibo apps. And then I developed a contest website for data-mining algorithms with R as the algorithm engine which supports online programming and running. The third R app I developed is the Weibo app Chinaweather which will be shared in this section. This time I still take R as the visualization engine and, furthermore, use R to finish tasks such as data crawling, xml documents parsing, data processing, etc. Besides, my 4th, 5th, and 6th apps take R as the core as well, which are about quantitative investment and will be introduced in R for Programmers: Quantitative investment.

From my experience, R has already the ability to develop enterprise apps. However I don't use R in all programming tasks. Most of my projects combine multiple programming languages. The future of programming is to mix the advantages from each language.

The development of the Weibo app Chinaweather will be introduced with two sections. Section 5.3 shows functional implementations with R and Section 5.4 focuses on R package development.

5.3.1 Project Introduction

As the discipline-crossing in computer science, the mixed-language programming is the working mode that I have always advocated. Today, many programming languages have emerged. Minority languages in various market segments grow like bamboo shoots. They have obvious advantages in specific fields compared with universal programming languages. For example, coding the logistic regression in Java is very difficult, but it is easy for R. As another example, PHP or Nodejs can implement a website with little effort, where Java costs heavily with a lot of complex code. So, using a universal language is not always the best solution to a specific app. Your app would be cool and different and you could combine multiple languages.

The Weibo app Chinaweather to be introduced in this section is an implementation of mixed-language programming.

5.3.1.1 Project Introduction

The simple motivation of this project is showing the weather of each province of China to provide weather information to those who are going to travel. First of all, let us list what functionalities are to be implemented and what problems we need to solve.

- Weather data: where to find, how to download, and how to store.
- Scheduled task: the weather data needs to be updated daily and the picture needs to be generated once a day.
- Visualization of map and weather data: we need to combine the China region map and the weather data in one graph to make it easy to understand for users.
- Showing on Web: the picture generated by visualization is static and needs to be published on the Web to be shown.
- Weibo: combine with Weibo to make more users see and use this app.
- User interactivity: users can view pictures with different dates and types. They can also share the pictures through Weibo.

Although the app is small, it is fully equipped. We need an all-sided solution to implement the app.

5.3.2 System Architecture Design

Actually the functionalities described above can be implemented with a single language. We can easily establish a website with only PHP. There is SDK for us to call to connect Weibo. Data crawling and storing is not too difficult. The only thing somehow blocking us is how to implement the visualization of map and weather data. With R only, it is easy to implement data crawling and storing. We can conveniently implement the visualization of map and weather data. However, establishing the website will be the bottleneck. R is a single-threaded synchronous computing model. The high concurrency of web app would simply crush the R program. Considering the above, by combining R and PHP, we can get around the weakness of the languages and take advantage of the different features of the languages that work well. We can make a difference with the technology of mixed-language programming.

To implement the functional requirements, we need to design a set of system architecture as shown in Figure 5.8.

The system architecture above is explained as following:

- Launch the data crawler through the timer to download data from Yahoo's weather data source.
- The crawler downloads data to the local server to parse and store app related data to the CSV file.
- The visualization program reads the data of weather and map to generate a static picture as the output of visualization.
- The end users load the web app through Weibo and view the static picture.
- The end users share the app through Weibo to make more people see the app.

In the following, the app architecture is split in terms of the features of languages, with R implementing crawling, data processing, and visualization, and PHP implementing web development, Weibo API connectivity, and user interactivity, as shown in Figure 5.9.

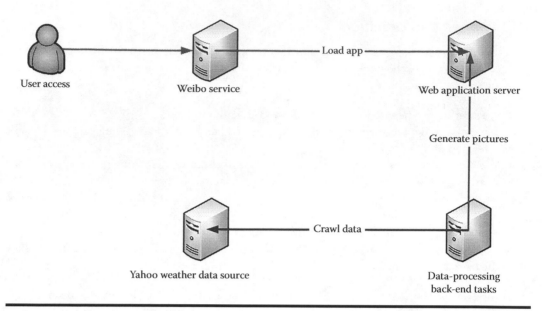

Figure 5.8   Simple system architecture.

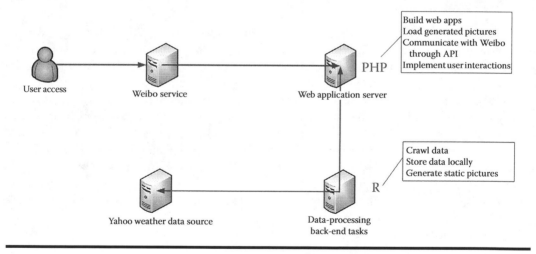

Figure 5.9   System architecture considering language advantages.

The complexity is small since the app does not require communications between R and PHP. The Weibo app Fans-showing combined three languages including R, PHP, and Java. The app used Java to implement the mid-layer to schedule the communications between R and PHP. By separating functionalities according to the features of different languages, we can make best use of the languages to finish things with the most suitable languages.

On the back-end, the timer functionality was implemented by CRON on Linux. The data crawling was finished by R program through the RCurl package. The output data, with format of XML, was then parsed with the XML package and stored locally in CSV format. Next, the R program processed the data and loaded the packages ggmap, mapdata, and maptools with companion of the plot() function to output the final picture which was stored on the local server.

Let us discuss the PHP program on the front-end. It is easy to build a website with PHP. A YII framework was used to finish the website rapidly. Weibo logon and sharing were implemented by API operations with Weibo SDK. Finally, Nginx + Spawn provided the PHP runtime environment, where Nginx corresponded to load balance and picture loading, and switched features in terms of the PHP access rules.

With reasonable architecture design and appropriate language combination, we have implemented the Weibo app Chinaweather. In fact, we can build various creative website apps with mixed-language programming, with prerequisites of mastering different languages.

I would like to speak more on this. As I know, many programmers take it easy in their own technical fields. Once mastering the core techniques of one language and getting some development experience, they will not learn a second language. They always consider themselves as the center of the world, with the capability to implement all the functionalities. Indeed they have great resolution, but they have run into a myth where they are trapped by the existing techniques. They cannot and will not see the change outside. I used to be such a programmer.

I admit that Java is an omnipotent language. However, don't you think it very costly to implement all with Java? The stronger the generality of a language is, the weaker its application is to specific fields. This is why I got out from the single-language route of Java. Actually, after mastering a language, learning another one is not so hard. The one who is addicted to mastered techniques will soon be overstepped by a new generation and new technique.

5.3.3 R Program Implementation

Next, let us introduce parts of the R program development. Before coding, we need to tidy up the development flow and have a program design, knowing which functionalities to implement with R and which third part packaged to use.

Figure 5.10 shows the calling relationship between programs, where six parts will be implemented by R: data crawling, locally storing, map loading, data visualization, and static picture generating.

Figure 5.10 attaches the R packages or functions to each step. Besides, we can define functions according to the flow. Having planned the entire program, we can easily code for corresponding steps.

5.3.3.1 Data Crawling

The task of data crawling is to periodically download the weather data of cities and/or regions, parse data to reserve only fields we are concerned with, and store data locally. There are many public weather data sources for free on the Internet. As for me, there are two very convenient data sources that can be used: Yahoo and Google. I choose Yahoo as weather data source since the Google API is frequently unavailable in mainland China.

The access address of the Yahoo weather data source is as follows.

https://query.yahooapis.com/v1/public/yql?q=select%20*%20from%20weather.forecast%20where%20woeid%20%3DWOEID

The parameter WOEID represents the city code. To get the weather data of Beijing, replace WOEID with Beijing's code 2151330 and access in the browser.

https://query.yahooapis.com/v1/public/yql?q=select%20*%20from%20weather.forecast%20where%20woeid%20%3D2151330

Opening the web address in the browser, we can see that the data is published in XML format, as shown in Figure 5.11.

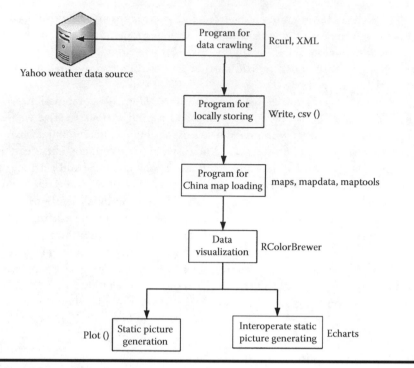

Figure 5.10   R program design.

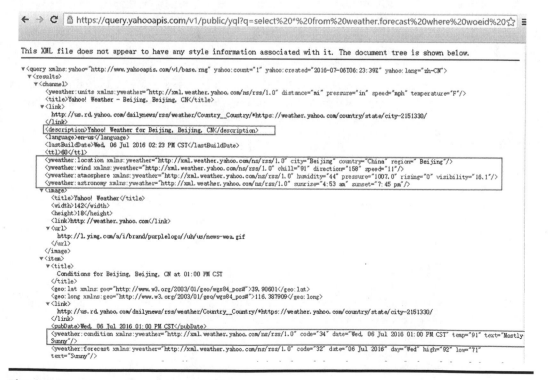

Figure 5.11   The weather data of Beijing.

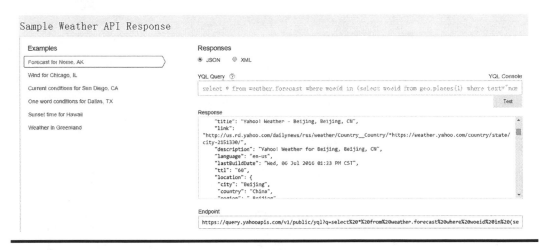

Figure 5.12   Yahoo's weather API query wizard page.

In case you don't understand the URL address that is bit complex, you can access the weather API query wizard page provided by Yahoo. The address is https://developer.yahoo.com/weather/. On the page, you can use YQL Query, an SQL-like statement to query the corresponding weather data by just inputting the city's name. In the meanwhile you can also choose JSON or XML as the format of returned value. In this section, we chose XML as the format for the following program development (Figure 5.12).

We need to parse the returned XML document to find the data we require for extraction. In R, we use the RCurl package to perform HTTP web access and retrieve the entire XML document and use the XML package to parse the DOM tree of the XML document to find what we require.

The system environment used in this section:

■ Win7 64bit
■ R: 3.1.1 x86_64-w64-mingw32/x64 (64-bit)

After getting a clear picture of the business logic and technical implementation, we start to code. Only a dozen lines of code can implement the functionality of data crawling and XML document parsing.

```
#Load the libraries.
> library(RCurl)
> library(XML)

> getWeather<-function (x){
+       # Yahoo's data source address.
+       url<-paste("https://query.yahooapis.com/v1/public/yql",
+              "?q=select%20*%20from%20weather.forecast%20where%20woeid%20
                %3D",
+              x,
+              "&format=xml",
+              "&env=store%3A%2F%2Fdatatables.org%2Falltableswithkeys",
                sep="")

+    # Load the XML data through https request.
```

```
+    xml <- getURL(url,.opts = list(ssl.verifypeer = FALSE))
+    # Parse the XML document.
+    doc <- xmlTreeParse(xml,useInternal = TRUE)

+     # Extract the data in the yweather:atmosphere through XPath
         expression.
+    ans<-getNodeSet(doc, "//yweather:atmosphere",c(yweather="http://xml.
weather.yahoo.com/ns/rss/1.0"))

+    # The humidity.
+    humidity<-as.numeric(sapply(ans, xmlGetAttr, "humidity"))
+    # The visibility.
+    visibility<-as.numeric(sapply(ans, xmlGetAttr, "visibility"))
+    # The pressure.
+    pressure<-as.numeric(sapply(ans, xmlGetAttr, "pressure"))
+    # Change of the pressure.
+    rising<-as.numeric(sapply(ans, xmlGetAttr, "rising"))
+    # Extract data in the yweather:condition label through XPath
         expression.
+    ans<-getNodeSet(doc, "//item/yweather:condition",c(yweather=" http://
xml.weather.yahoo.com/ns/rss/1.0"))
         #The weather condition code.
+    code<-sapply(ans, xmlGetAttr, "code")

+    ans<-getNodeSet(doc, "//item/yweather:forecast[1]",c(yweather="
http://xml.weather.yahoo.com/ns/rss/1.0"))
         #The lowest temperature.
+    low<-as.numeric(sapply(ans, xmlGetAttr, "low"))
         #The highest temperature.
+    high<-as.numeric(sapply(ans, xmlGetAttr, "high"))
+
+    print(paste(x,'==>',low,high,code,humidity,visibility,pressure,rising))
         #Return data with format of data.frame
+    cbind(low,high,code,humidity,visibility,pressure,rising)
+ }
```

Run the program and view the result.

```
# Execute the data crawling program.
> w<-getWeather(2151330)
[1] "2151330 ==> 9 13 21 59 4.1 1016.4 0"

# The result set returned.
> w

    low high code humidity visibility pressure rising
[1,]  "9" "13" "21" "59"     "4.1"      "1016.4" "0"
```

According to the functional requirement, we just need to store seven fields of data, with others in the XML document filtered out.

5.3.3.2 Locally Storing

After being downloaded and filtered in the data crawling step, the data has been transformed into data.frame. Next, let us use the write.csv() function to store the data into a local file as backup.

The process of locally storing will generate a CSV file, name the file, and row-bind data of multiple cities into the file to store. In the following, we define two functions. The filename() function is used to generate a name for the new file and the loadDate() function is used to load data of multiple cities and row-bind them into a file.

We have prepared the city list in advance. Hereby I choose 34 Chinese cities for which we get weather data information. To crawl data for more cities, simply supplement the list. The content of city list file WOEID.csv is as following:

```
beijing,2151330,Beijing,116.4666667,39.9
shanghai,2151849,Shanghai,121.4833333,31.23333333
tianjin,2159908,Tianjin,117.1833333,39.15
chongqing,20070171,Chongqing,106.5333333,29.53333333
harbin,2141166,Heilongjiang,126.6833333,45.75
changchun,2137321,Jilin,125.3166667,43.86666667
shenyang,2148332,Liaoning,123.4,41.83333333
hohhot,2149760,Inner-Mongolia,111.8,40.81666667
shijiazhuang,2171287,Hebei,114.4666667,38.03333333
wulumuqi,26198317,Xinjiang,87.6,43.8
lanzhou,2145605,Gansu,103.8166667,36.05
xining,2138941,Qinghai,101.75,36.63333333
xian,2157249,Shaanxi,108.9,34.26666667
yinchuan,2150551,Ningxia,106.2666667,38.33333333
zhengzhou,2172736,Henan,113.7,34.8
jinan,2168327,Shandong,117,36.63333333
taiyuan,2154547,Shanxi,112.5666667,37.86666667
hefei,2127866,Anhui,117.3,31.85
wuhan,2163866,Hubei,114.35,30.61666667
changsha,26198213,Hunan,113,28.18333333
nanjing,2137081,Jiangsu,118.8333333,32.03333333
chengdu,2158433,Sichuan,104.0833333,30.65
guiyang,2146703,Guizhou,106.7,26.58333333
kunming,2160693,Yunnan,102.6833333,25
nanning,2166473,Guangxi,108.3333333,22.8
lasa,26198235,Tibet,91.16666667,29.66666667
hangzhou,2132574,Zhejiang,120.15,30.23333333
nanchang,26198151,Jiangxi,115.8666667,28.68333333
guangzhou,2161838,Guangdong,113.25,23.13333333
fuzhou,2139963,Fujian,119.3,26.08333333
taipei,2306179,Taiwan,121.5166667,25.05
haikou,2162779,Hainan,110.3333333,20.03333333
hongkong,2165352,Hongkong,114.1666667,22.3
macau,1887901,Macau,113.5,22.2
```

The explanation to the fields is as follows:

- Column 1: city's English name
- Column 2: WOEID code
- Column 3: the English name of the province where the city is located
- Column 4: longitude (east longitude as default)
- Column 5: latitude (north latitude as default)

Following is the R implementation of functions to generate data file.

```
# Name the file according to date.
> filename<-function(date=Sys.time()){
+       paste(format(date, '%Y%m%d'),'.csv',sep='')
+ }

# Read the city list, call the data crawling function and row-bind the
data into one file.
> loadDate<-function(date){
+       print(paste('Date','=>',date))
+       city<-read.csv(file="WOEID.csv",header=FALSE,fileEncoding="utf-8",
encoding="utf-8")
+       # Load the city list.
+       names(city)<-c('en','woeid','prov','long','lat')
+       city<-city[-nrow(city),]
+
+       wdata<-do.call(rbind, lapply(city$woeid,getWeather))
+       w<-cbind(city,wdata)
+       write.csv(w,file=filename(date),row.names=FALSE,fileEncoding="utf-8")
+ }
# Choose date.
> date=Sys.time();date #date<-as.Date('2016-07-19')
[1] "2016-07-19 16:01:12"

# Crawling data.
> loadDate(date)
[1] "Date => 2016-07-19 16:01:12"
[1] "2151330 => 73 78 12 100 4.1 1003 0"
[1] "2151849 => 76 87 4 95 10.8 1004 0"
[1] "2159908 => 74 85 26 78 16.1 1005 0"
[1] "20070171 => 75 80 47 100 3.6 951 0"
[1] "2141166 => 66 86 26 60 16.1 994 0"
[1] "2137321 => 67 85 28 60 16.1 985 0"
[1] "2148332 => 70 87 26 55 16.1 1004 0"
[1] "2149760 => 61 69 11 89 14.9 882 0"
[1] "2171287 => 71 75 12 100 3.7 997 0"
[1] "26198317 => 53 68 28 57 10 1003.04 0"
[1] "2145605 => 58 81 32 40 16.1 812 0"
[1] "2138941 => 55 79 32 35 16.1 755 0"
[1] "2157249 => 72 87 30 65 16.1 950 0"
[1] "2150551 => 61 84 34 41 16.1 880 0"
[1] "2172736 => 73 77 12 100 3.6 987 0"
[1] "2168327 => 75 87 26 77 16.1 995 0"
[1] "2154547 => 65 69 12 100 3.7 898 0"
[1] "2127866 => 78 82 4 100 4.1 1000 0"
[1] "2163866 => 78 88 4 100 4.8 997 0"
[1] "26198213 => 81 90 26 83 16.1 996 0"
[1] "2137081 => 77 87 4 98 5.8 999 0"
[1] "2158433 => 73 91 34 68 16.1 941 0"
[1] "2146703 => 71 79 12 91 12.3 874 0"
[1] "2160693 => 63 70 47 93 12.1 794 0"
[1] "2166473 => 81 90 4 86 16.1 988 0"
[1] "26198235 => 45 62 11 71 3 1000 0"
[1] "2132574 => 77 88 4 95 9.9 1003 0"
[1] "26198151 => 79 92 26 80 16.1 1002 0"
[1] "2161838 => 80 88 4 95 11.3 1004 0"
```

```
[1] "2139963 ⟹ 77 89 4 85 16.1 989 0"
[1] "2306179 ⟹ 76 86 4 94 11.3 988 0"
[1] "2162779 ⟹ 82 92 4 85 16.1 1002 0"
[1] "2165352 ⟹ 81 84 4 100 4.4 1000 0"
```

After running, the program generates a file named 20141001.csv in current directory. Open the file and we see the data for generating visualization pictures.

```
"en","woeid","prov","long","lat","low","high","code","humidity","visibility",
"pressure","rising"
"beijing",2151330,"Beijing",116.4666667,39.9,"73","78","12","100","3.8",
"1003","0"
"shanghai",2151849,"Shang
hai",121.4833333,31.23333333,"76","87","4","96","8.3","1004","0"
"tianjin",2159908,"Tian
jin",117.1833333,39.15,"74","85","26","80","16.1","1006","0"
"chongqing",20070171,"Chongq
ing",106.5333333,29.53333333,"75","80","47","100","3.6","951","0"
"harbin",2141166,"Heilongji
ang",126.6833333,45.75,"66","86","26","60","16.1","994","0"
"changchun",2137321,"Ji
lin",125.3166667,43.86666667,"67","85","26","60","16.1","986","0"
"shenyang",2148332,"Liaon
ing",123.4,41.83333333,"70","87","26","58","16.1","1004","0"
"hohhot",2149760,"Inner-Mongo
lia",111.8,40.81666667,"61","69","12","88","15","883","0"
"shijiazhuang",2171287,"He
bei",114.4666667,38.03333333,"71","75","12","100","3.7","998","0"
"wulumuqi",26198317,"Xinji
ang",87.6,43.8,"53","68","30","57","10","1003.72","0"
"lanzhou",2145605,"Ga
nsu",103.8166667,36.05,"58","81","32","49","16.1","813","0"
"xining",2138941,"Qing
hai",101.75,36.63333333,"55","79","32","45","16.1","756","0"
"xian",2157249,"Shaa
nxi",108.9,34.26666667,"72","87","30","72","16.1","951","0"
"yinchuan",2150551,"Ning
xia",106.2666667,38.33333333,"61","84","34","47","16.1","881","0"
"zhengzhou",2172736,"He
nan",113.7,34.8,"73","77","47","100","3.6","988","0"
"jinan",2168327,"Shand
ong",117,36.63333333,"75","87","26","76","16.1","996","0"
"taiyuan",2154547,"Sha
nxi",112.5666667,37.86666667,"65","69","12","100","3.6","899","0"
"hefei",2127866,"Anhui",117.3,31.85,"78","82","4","100","3.8","1000","0"
"wuhan",2163866,"Hu
bei",114.35,30.61666667,"78","88","4","100","4.2","998","0"
"changsha",26198213,"Hu
nan",113,28.18333333,"81","90","23","83","16.1","996","0"
"nanjing",2137081,"Jian
gsu",118.8333333,32.03333333,"77","87","4","95","6.8","999","0"
"chengdu",2158433,"Sich
uan",104.0833333,30.65,"73","91","34","72","16.1","942","0"
"guiyang",2146703,"Guiz
hou",106.7,26.58333333,"71","79","11","90","15.9","875","0"
```

```
"kunming",2160693,"Yun
nan",102.6833333,25,"63","70","47","98","5.9","795","0"
"nanning",2166473,"Guan
gxi",108.3333333,22.8,"81","90","4","89","16.1","989","0"
"lasa",26198235,"Ti
bet",91.16666667,29.66666667,"48","63","4","66","10","1001.69","0"
"hangzhou",2132574,"Zheji
ang",120.15,30.23333333,"77","88","4","95","8.8","1003","0"
"nanchang",26198151,"Jian
gxi",115.8666667,28.68333333,"79","92","26","80","16.1","1002","0"
"guangzhou",2161838,"Guangd
ong",113.25,23.13333333,"80","88","4","91","14.4","1005","0"
"fuzhou",2139963,"Fuj
ian",119.3,26.08333333,"77","89","26","86","16","990","0"
"taipei",2306179,"Tai
wan",121.5166667,25.05,"76","86","4","87","16.1","989","0"
"haikou",2162779,"Hai
nan",110.3333333,20.03333333,"82","92","4","80","16.1","1003","0"
"hongkong",2165352,"Hongk
ong",114.1666667,22.3,"81","84","4","100","4.4","1001","0"
```

There are 10 columns in total. Following is the explanation to the data fields:

- en: the English name of the city
- woeid: the WOEID defined in Yahoo weather API, used for city matching
- prov: the English name of the province where the city is located
- long: longitude
- lat: latitude
- low: the lowest temperature
- high: the highest temperature
- code: the weather condition code
- humidity: the humidity of the atmosphere
- visibility: the visibility of the atmosphere
- pressure: the pressure of the atmosphere
- rising: the change of pressure

Well, we have prepared the data. Next, let us match the weather data with the China administrative region map.

5.3.3.3 China Map Loading

With third part map packages, R can easily implement visualization-based maps or data processing based on geographic information. How does R make it? The answer is the collaboration of the three packages: maps, mapdata, and maptools.

Call the readShapePoly function in the maptools package to load the data of the China administrative region map into the map variable. The visualization result can be viewed by calling the plot() function, as shown in Figure 5.13. The map data files were downloaded in advance and stored in the mapdata directory, including three files: bou2_4p.dbf, bou2_4p.shp, and bou2_4p.shx.

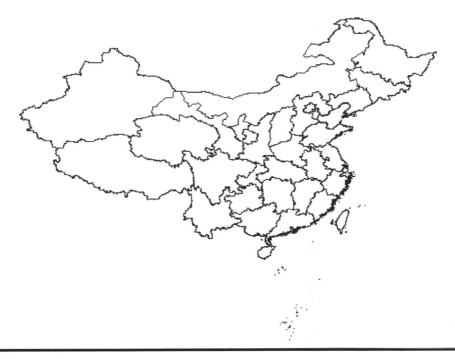

Figure 5.13   The outlines of the China administrative region map.

```
> library(maps)
> library(mapdata)
> library(maptools)

# Load China administrative region map data.
> map<-readShapePoly('mapdata/bou2_4p.shp')
# Draw the map.
> plot(map)
```

With just two lines of code, we have drawn the outline of the China administrative region map. Isn't it amazing? Let us explore the map variable. By viewing the type of the map variable, we find that the type is SpatialPolygonsDataFrame defined in the package sp.

```
# View the type of the map object.
> class(map)
[1] "SpatialPolygonsDataFrame"
attr(,"package")
[1] "sp"
```

The SpatialPolygonsDataFrame type is not familiar to us. Let us use the otype function in the pryr package to check the map object for object-oriented system type.

```
> library(pryr)
# It is an S4 typed data.frame
> otype(map)
[1] "S4"
```

We have mastered the basics of S4 type in Section 4.4. After understanding that the map object is an instance of S4 type, we basically know how to use the object. From the name of the type SpatialPolygonsDataFrame, we guess it should be data.frame storing data of SpatialPolygons type. Check the map object from perspective of data.frame using length() and names(). We can see it contains 7 columns and 925 rows.

```
# There are 925 records in total.
> length(map)
[1] 925

# There are 7 columns in the data.frame.
> names(map)
[1] "AREA" "PERIMETER" "BOU2_4M_" "BOU2_4M_ID" "ADCODE93"
[6] "ADCODE99" "NAME"
```

View the static structure of the first line of the map object through the str() function.

```
> str(map[1,])
Formal class 'SpatialPolygonsDataFrame' [package "sp"] with 5 slots
  ..@ data        :'data.frame':      1 obs. of 7 variables:
  .. ..$ AREA      : num 54.4
  .. ..$ PERIMETER : num 68.5
  .. ..$ BOU2_4M_  : int 2
  .. ..$ BOU2_4M_ID: int 23
  .. ..$ ADCODE93  : int 23000    0
  .. ..$ ADCODE99  : int 230000
  .. ..$ NAME      : Factor w/ 33 levels "Anhui","Beijing",..: 11
  ..@ polygons    :List of 1
  .. ..$ :Formal class 'Polygons' [package "sp"] with 5 slots
  .. .. .. ..@ Polygons :List of 1
  .. .. .. .. ..$ :Formal class 'Polygon' [package "sp"] with 5 slots
  .. .. .. .. .. .. ..@ labpt  : num [1:2] 127.8 47.9
  .. .. .. .. .. .. ..@ area   : num 54.4
  .. .. .. .. .. .. ..@ hole   : logi FALSE
  .. .. .. .. .. .. ..@ ringDir: int 1
  .. .. .. .. .. .. ..@ coords : num [1:5784, 1:2] 121 121 122 122 122...
  .. .. .. ..@ plotOrder: int 1
  .. .. .. ..@ labpt    : num [1:2] 127.8 47.9
  .. .. .. ..@ ID       : chr "0"
  .. .. .. ..@ area     : num 54.4
  ..@ plotOrder  : int 1
  ..@ bbox       : num [1:2, 1:2] 121.2 43.4 135.1 53.6
  .. ..- attr(*, "dimnames")=List of 2
  .. .. ..$ : chr [1:2] "x" "y"
  .. .. ..$ : chr [1:2] "min" "max"
  ..@ proj4string:Formal class 'CRS' [package "sp"] with 1 slot
  .. .. ..@ projargs: chr NA
```

From the above perspectives, we have basically understood the structure of the map object. Each line of map is a SpatialPolygonsDataFrame object, which contains five properties to store map data. View data of the first line to find that it is the data of administrative region map of Heilongjiang province.

```
> map[1,]@data
    AREA PERIMETER BOU2_4M_ BOU2_4M_ID  ADCODE93  ADCODE99        NAME
0 54.447    68.489        2         23    230000    230000 Heilongjiang
```

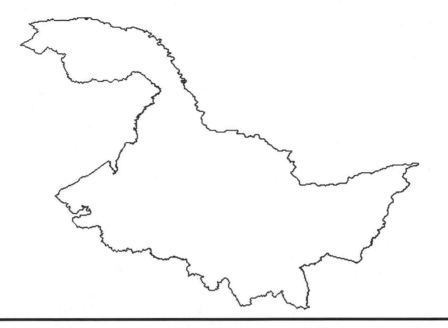

Figure 5.14   The administrative region map of Heilongjiang province.

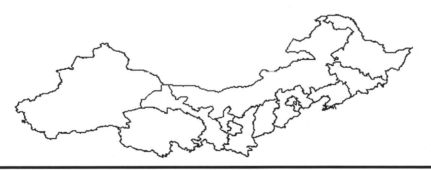

Figure 5.15   The administrative region map of some of China provinces.

Draw graphics using the data of the first line, as shown in Figure 5.14.

```
> plot(map[1,])
```

With the top 100 lines of data to draw, we would get the administrative region map of some Chinese provinces. The result is just as what I guessed, as shown in Figure 5.15.

```
> plot(map[1:100,])
```

This section is not intended to demonstrate the map packages in detail, it is enough for us to know the basic usage of the map object.

5.3.3.4 Data Visualization

After loading the map data, the next step is to do data visualization. The step can be divided into two parts: data processing and visualization output.

Let us think about how to do data processing to combine the data of weather and map. The aim is to draw the weather summary of Chinese provinces. The process will use the code information in the data derived from the former step. The code data is identity so we need to define the mappings between individual code items and their meanings.

The Yahoo source data defines 49 kinds of weather conditions, as stored in code.csv. According to the description, I merged the similar kinds of weather conditions together to retain 18 kinds. The mapping file is labelcode.csv.

The code.csv file is shown in the following:

```
"code","en","type"
0,tornado,3
1,tropical storm,2
2,hurricane,3
3,severe thunderstorms,16
4,thunderstorms,11
5,mixed rain and snow,12
6,mixed rain and sleet,12
7,mixed snow and sleet,12
8,freezing drizzle,11
9,drizzle,11
10,freezing rain,11
11,showers,11
12,showers,11
13,snow flurries,13
14,light snow showers,13
15,blowing snow,13
16,snow,14
17,hail,15
18,sleet,12
19,dust,5
20,foggy,7
21,haze,7
22,smoky,6
23,blustery,3
24,windy,4
25,cold,18
26,cloudy,8
27,mostly cloudy (night),8
28,mostly cloudy (day),8
29,partly cloudy (night),9
30,partly cloudy (day),9
31,clear (night),10
32,sunny,10
33,fair (night),10
34,fair (day),10
35,mixed rain and hail,16
36,hot,1
37,isolated thunderstorms,11
38,scattered thunderstorms,11
39,scattered thunderstorms,11
40,scattered showers,11
41,heavy snow,14
42,scattered snow showers,13
```

```
43,heavy snow,14
44,partly cloudy,9
45,thundershowers,11
46,snow showers,13
47,isolated thundershowers,11
3200,not available,19
```

Following is the explanation of the fields:

- code: the weather condition code in source data
- en: the English description
- type: the category code

Following is the labelcode.csv file content:

```
"type","alias"
1,hot
2,tropical-storm
3,tornado
4,windy
5,dust
6,smoky
7,foggy
8,cloudy
9,partly-cloudy
10,sunny
11,showers
12,sleet
13,snow
14,heavy-snow
15,hail
16,rain
17,thunderstorm
18,cold
19,NA
```

Following is the explanation to the fields:

- type: the category code
- alias: the name to show

After defining the weather conditions, by mapping the conditions to colors and adding legend and literal description, we generate the final static picture of the weather summary of the Chinese provinces.

```
> library('RColorBrewer')
> getColors2<-function(map,prov,ctype){
+     # Name change to ADCODE99.
+     ADCODE99<-read.csv(file='ADCODE99.csv',header=TRUE)
+     fc<-function(x){ADCODE99$ADCODE99[which(x==ADCODE99$prov)]}
+     code<-sapply(prov,fc)
+     f=function(x,y) ifelse(x %in% y,which(y==x),0);
```

```
+       colIndex=sapply(map$ADCODE99,f,code);
+       ctype[which(is.na(ctype))]=19
+       return(ctype[colIndex])
+ }

> summary<-function(data=data,output=FALSE,path=''){
+ # Define the colors corresponding to the 18 weather conditions.
+
colors<-rev(c(rev(brewer.pal(9,'Blues')),rev(c('#b80137','#8c0287','#d93c5d',
'#d98698','#f6b400','#c4c4a7','#d6d6cb','#d1b747','#ffeda0'))))
+
+       temp<-data$code
+       title<-'China Weather Overview'
+       ofile<-paste(format(date,"%Y%m%d"),"_code.png",sep="")
+       code<-read.csv(file="code.csv",header=TRUE,fileEncoding="utf-8",
encoding="utf-8")
+       labelcode<-read.csv(file="labelcode.csv",header=TRUE,fileEncoding="
utf-8", encoding="utf-8")
+       ctype<-sapply(temp,function(x){code$type[which(x==code$code)]})
+
+       if(output) png(file=paste(path,ofile,sep=''),width=700,height=700)
+       layout(matrix(data=c(1,2),nrow=1,ncol=2),widths=c(8,1),heights=c(1,2))
+       par(mar=c(0,0,3,10),oma=c(0.2,0.2,0.2,0.2),mex=0.3)
+       # Visualization of map and weather.
+       plot(map,border='white',col=colors[getColors2(map,data$prov,ctype)])
+       # Label the sampling city.
+       points(data$long,data$lat,pch=19,col=rgb(0,0,0,0.3),cex=0.8)
+
+       #==============================                 The auxiliary text in picture.
+       text(100,58, title,cex=2)
+       text(105,54,format(date,"%Y-%m-%d"))
+       text(98,65,paste('China Weather App','http://apps.Weibo.com/
chinaweatherapp'))
+       text(120,-8,paste('provided by The Weather Channel',format(date,
"%Y-%m-%d %H:%M")),cex=0.8)
+
+       #==============================                 # The literal description.
+       for(row in 1:nrow(data)){
+           name<-as.character(data$zh[row])
+           label<-labelcode$alias[labelcode$type==ctype[row]]
+           x1<-ceiling(row/7)
+           x2<-ifelse(row%%7==0,7,row%%7)
+           x3<-ctype[row]
+           fontCol<-'#000000'
+           if(x3<=5)fontCol<-head(colors,1)
+           if(x3>=12)fontCol<-tail(colors,1)
+           text(68+x1*11,17-x2*3,paste(name,' ',label,sep=''),col=fontCol,
cex=0.75)
+       }
+
+       #==============================                 # The legend.
+       par(mar = c(5, 0, 10, 15))
+       image(x=1, y=1:length(colors),z=t(matrix(1:length(colors))),col=rev
(colors),axes=FALSE,xlab="",ylab="",xaxt="n")
```

```
+      axis(4, at = 1:(nrow(labelcode)-1), labels=rev(labelcode$alias)
[-1], col = "white", las = 1)
+      abline(h=c(1:(nrow(labelcode)-2)+0.5), col = "white", lwd = 2, xpd
= FALSE)
+      if(output)dev.off()
+ }
```

Run the program to generate static picture, as shown in Figure 5.16.

```
# Define the data source.
> data<-read.csv(file=filename(date),header=TRUE,fileEncoding="utf-8",
encoding="utf-8")
# Generate the weather summary map of China provinces.
> summary(data,output=TRUE)
RStudioGD
        2
```

With just about 100 lines of code, we have generated such a complex picture with the combination of weather and map. How amazing R is!

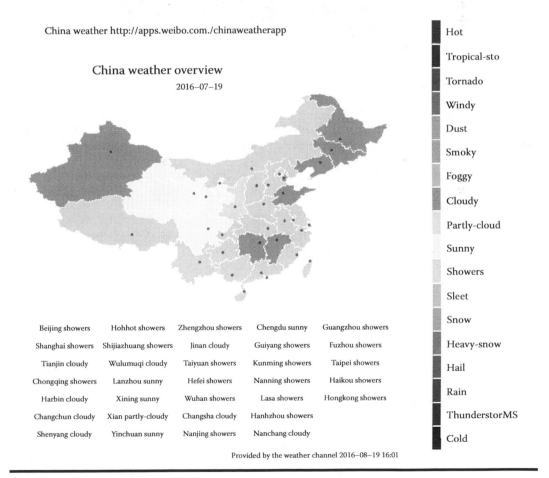

Figure 5.16 The weather summary of China provinces.

This section described the complete information of the Weibo app China weather and implemented the functionalities of R part. The next section will show how to encapsulate the R functionalities into a package.

5.4 R Package Development: ChinaWeather

Question

How do we encapsulate an R app into an R package?

Introduction

Following Section 5.3, this section encapsulates the R program we have finished into an R package. It seems that the task is simple; however it will take us lots of time to handle details. The entire process of R package development will be performed according to the flow introduced in Section 5.2. To make sure the check() function executed successfully, a lot of modifications are made to the code.

5.4.1 Creating the Project

The R program in Section 5.3 was developed on Windows. Due to compatibility issues in R's platform-cross code, the app will be finally published on Linux. The system environment used in this section:

- Linux: Ubuntu Server 12.04.2 LTS 64bit
- R: 3.1.1 x86_64-pc-linux-gnu (64-bit)
- RStudio-Server 0.97.551

In fact we have created the ChinaWeather project in Section 5.2. So the R package development will continue to use this project. Enter the directory of the ChinaWeather project. To keep independent of the code in Section 5.2, I will open a new branch on Git to develop the R package in this section.

```
# Enter the project directory.
~ cd /home/conan/R/chinaWeather
# New a branch named app.
~ git branch app
```

```
# Switch to branch app.
~ git checkout app
# View the current branch.
~ git branch
  * app
    master
```

View the settings of R environment variables on Linux. The character set needs special attention to handle.

```
> sessionInfo()
R version 3.1.1 (2014-07-10)
Platform: x86_64-pc-linux-gnu (64-bit)

locale:
 [1] LC_CTYPE=en_US.UTF-8       LC_NUMERIC=C
 [3] LC_TIME=zh_CN.UTF-8        LC_COLLATE=en_US.UTF-8
 [5] LC_MONETARY=zh_CN.UTF-8    LC_MESSAGES=en_US.UTF-8
 [7] LC_PAPER=zh_CN.UTF-8       LC_NAME=C
 [9] LC_ADDRESS=C               LC_TELEPHONE=C
[11] LC_MEASUREMENT=zh_CN.UTF-8 LC_IDENTIFICATION=C

attached base packages:
[1] stats     graphics  grDevices utils     datasets  methods   base
```

Once the development environment is ready, let us start the R package development.

5.4.2 *The Static Data*

Before coding for the R package, we need to tidy up the static data. In this project, the static data contains the map data, the mapping data of WOEID, the mapping data of weather conditions, the visualization data for Chinese/English pictures, and the test dataset. For an ordinary R app, the data can be stored locally with CSV format; while for R package, it is best to encapsulate the data into rda files and publish together with the R program.

Let us start the tidy-up work. Create a directory named metadata to store the raw CSV files and map files.

```
# Create the directory named metadata and copy the data files into it.
~ mkdir /home/conan/R/chinaWeather/metadata
# View the metadata directory.
~ ls -l /home/conan/R/chinaWeather/metadata
# The test dataset.
-rw-rw-r-- 1 conan conan 3396 Oct 4 22:14 20141001.csv
# The mapping data of ADCODE99 and Chinese names of provinces.
-rw-r--r-- 1 conan conan 754 Feb 5 2013 ADCODE99.csv
# The mapping data of Yahoo weather conditions.
-rw-r--r-- 1 conan conan 1418 Feb 6 2013 code.csv
# The mapping data of simplified weather conditions.
-rw-r--r-- 1 conan conan 214 Feb 6 2013 labelcode.csv
# The directory of map data.
drwxr-xr-x 2 conan conan 4096 Apr 23 2013 mapdata
# The mapping data of WOEID.
-rw-rw-r-- 1 conan conan 1900 Feb 4 2013 WOEID.csv
```

```
# View the map data files.
~ ls -l /home/conan/R/chinaWeather/metadata/mapdata
-rw-r--r-- 1 conan conan 86283   Apr 10 1999 bou2_4p.dbf
-rw-r--r-- 1 conan conan 1508752 Apr 10 1999 bou2_4p.shp
-rw-r--r-- 1 conan conan 7500    Apr 10 1999 bou2_4p.shx
```

Transform the static data into rda formatted files and store them in the data directory.

```
# Create the directory named data.
~ mkdir /home/conan/R/chinaWeather/data
# Launch the R application.
~ R
```

5.4.2.1 The WOEID Data File WOEID.rda

Process the WOEID data. Merge adcode99 code into the dataset of WOEID. Merge the data from WOEID.csv and ADCODE99.csv and then generate the WOEID.rda file.

```
# Load the WOEID dataset.
> WOEID<-read.csv(file="metadata/WOEID.csv",header=FALSE,fileEncoding="
utf-8", encoding="utf-8")
> names(WOEID)<-c("en","woeid","zh",'prov','long','lat')
# Load the ADCODE99 dataset.
> adcode99<-read.csv(file="metadata/ADCODE99.csv",header=TRUE,fileEncodin
g="utf-8", encoding="utf-8")

> fc<-function(row){
+     code<-adcode99$ADCODE99[which(row[4]==as.character(adcode99$prov))]
+     if(length(code)==0)code=0
+     code
+ }
# Merge the datasets.
> WOEID<-cbind(WOEID,adcode99=unlist(apply(WOEID,1,fc)))
# Generate the WOEID.rda file.
> save(WOEID,file="data/WOEID.rda")

# The WOEID dataset after merging.
> WOEID
           en      woeid      zh           prov       long        lat  adcode99
1     beijing    2151330    北京         北京市   116.46667   39.90000    110000
2    shanghai    2151849    上海         上海市   121.48333   31.23333    310000
3      tianji    2159908    天津         天津市   117.18333   39.15000    120000
4   chongqing   20070171    重庆         重庆市   106.53333   29.53333    500000
5      harbin    2141166    哈尔滨      黑龙江省   126.68333   45.75000    230000
6   changchun    2137321    长春         吉林省   125.31667   43.86667    220000
7    shenyang    2148332    沈阳         辽宁省   123.40000   41.83333    210000
8      hohhot    2149760    呼和浩特   内蒙古自治区   111.80000   40.81667    150000
9  shijiazhuang 2171287    石家庄       河北省   114.46667   38.03333    130000
10   wulumuqi   26198317    乌鲁木齐  新疆维吾尔自治区  87.60000   43.80000    650000
11    lanzhou    2145605    兰州         甘肃省   103.81667   36.05000    620000
12     xining    2138941    西宁         青海省   101.75000   36.63333    630000
13       xian    2157249    西安         陕西省   108.90000   34.26667    610000
14   yinchuan    2150551    银川      宁夏回族自治区   106.26667   38.33333    640000
15  zhengzhou    2172736    郑州         河南省   113.70000   34.80000    410000
16      jinan    2168327    济南         山东省   117.00000   36.63333    370000
17    taiyuan    2154547    太原         山西省   112.56667   37.86667    140000
18      hefei    2127866    合肥         安徽省   117.30000   31.85000    340000
19      wuhan    2163866    武汉         湖北省   114.35000   30.61667    420000
20   changsha   26198213    长沙         湖南省   113.00000   28.18333    430000
```

21	nanjing	2137081	南京	江苏省	118.83333	32.03333	320000
22	chengdu	2158433	成都	四川省	104.08333	30.65000	510000
23	guiyang	2146703	贵阳	贵州省	106.70000	26.58333	520000
24	kunming	2160693	昆明	云南省	102.68333	25.00000	530000
25	nanning	2166473	南宁	广西壮族自治区	108.33333	22.80000	450000
26	lasa	26198235	拉萨	西藏自治区	91.16667	29.66667	540000
27	hangzhou	2132574	杭州	浙江省	120.15000	30.23333	330000
28	nanchang	26198151	南昌	江西省	115.86667	28.68333	360000
29	guangzhou	2161838	广州	广东省	113.25000	23.13333	440000
30	fuzhou	2139963	福州	福建省	119.30000	26.08333	350000
31	taipei	2306179	台北	台湾省	121.51667	25.05000	710000
32	haikou	2162779	海口	海南省	110.33333	20.03333	460000
33	hongkong	2165352	香港	香港特别行政区	114.16667	22.30000	810000
34	macau	1887901	澳门	澳门特别行政区	113.50000	22.20000	0

We can see that the WOEID dataset contains Chinese characters while the rda specifications of R require that no character outside the ASCII charset shall be contained in rda files. During the process of check(), we may encounter the warnings of invalid charset. Therefore, we need to do special character escaping to datasets that contains Chinese characters, by escaping the Chinese character into Unicode. For example, 北京 is changed to \u5317\u4eac after Unicode escaping. When the dataset is shown in Chinese, we need to un-escape the Unicode back to Chinese. By doing above, we make Chinese compatible with R program.

To escape the WOEID dataset, we need the stringi package.

```
# Install the stringi package.
> install.packages("stringi")
# Load the stringi package.
> library("stringi")
```

We use the stri_escape_unicode() function to do Unicode escaping to the zh and prov columns in the WOEID dataset.

```
# Escape WOEID$prov.
> WOEID$prov<-stri_escape_unicode(WOEID$prov)
# Escape WOEID$zh.
> WOEID$zh<-stri_escape_unicode(WOEID$zh)
# Save the dataset.
> save(WOEID,file="data/WOEID.rda",compress=TRUE)

# View the WOEID dataset after escaping.
> head(WOEID)
        en      woeid                     zh                    prov
long    lat    adcode99  adcode99
1  beijing   2151330  \\u5317\\u4eac\\u5317\\u4eac\\u5e02 116.4667 39.90000 110000
110000
2  shanghai  2151849  \\u4e0a\\u6d77 \\u4e0a\\u6d77\\u5e02 121.4833 31.23333 310000
310000
3  tianji    2159908  \\u5929\\u6d25\\u5929\\u6d25\\u5e02 117.1833 39.15000 120000
120000
4  chongqing 20070171 \\u91cd\\u5e86 \\u91cd\\u5e86\\u5e02 106.5333 29.53333 500000
500000
5  harbin    2141166  \\u54c8\\u5c14\\u6ee8 \\u9ed1\\u9f99\\u6c5f\\u7701  126.6833
45.75000 230000 230000
6  changchun 2137321  \\u957f\\u6625 \\u5409\\u6797\\u7701 25.3167 43.86667 220000
220000
```

Let's try to un-escape the Unicode data back to the original Chinese characters by the stri_unescape_unicode() function.

```
> head(stri_unescape_unicode(WOEID$prov))
[1] "北京市" "上海市" "天津市" "重庆市" "黑龙江省" "吉林省"

> head(stri_unescape_unicode(WOEID$zh))
[1] "北京" "上海" "天津" "重庆" "哈尔滨" "长春"
```

The escaping operation works well. When encountering non-ASCII charset, we can transform the characters this way.

5.4.2.2 The Map Data File chinaMap.rda

Process the map data by loading the original map data and generating the chinaMap.rda file. In the chinaMap object, the NAME column contains Chinese characters. When directly loading map data on Linux, since the system uses the encoding of UTF-8 by default, the type character with GKB encoding would be shown as mess code. We can use the iconv() function to convert the encoding and save the data using the Unicode escaping method described above. However, since the data in the NAME column is not used in our project, a simple way is to remove the column from the dataset.

```
> library(maps)
> library(mapdata)
> library(maptools)

# Load the map data.
> chinaMap<-readShapePoly('metadata/mapdata/bou2_4p.shp')

# The NAME column with non-ASCII encoding.
> head(chinaMap$NAME)
[1] \xba\xda\xc1\xfa\xbd\xad
[2] \xc4\xda\xc3\xc5\xd7\xd4\xd6\xce\xc7\xf8
[3] \xd0\xae\xce\xe1\xb6\xfb\xd7\xd4\xd6\xce\xc7\xf8
[4] \xbc\xaa\xc1\xd6
[5] \xc1\xc9\xc4\xfe
[6] \xb8\xca\xcb\xe0
33 Levels: \xb0\xb2\xbb\xd5...\xd6\xd8\xc7\xec\xca\xd0

# Convert the NAME column to UTF-8 encoding.
> iconv(head(chinaMap$NAME),"gbk","UTF-8")
[1] "黑龙江省"      "内蒙古自治区"       "新疆维吾尔自治区"   "吉林省" "辽宁省" "甘肃省"

# Remove the NAME column.
> chinaMap<-chinaMap[,c(1:6)]
# Generate the chinaMap.rda file.
> save(chinaMap,file="data/chinaMap.rda",compress='xz')
```

5.4.2.3 The Visualization Data for Chinese/English Picture props.rda

The data is used to show Chinese titles when outputting the visualization pictures. Generate the props.rda file with Chinese characters escaped as Unicode.

```
> props<-data.frame(
+     key=c('high','low'),
+     zh=c('中国各省白天气温','中国各省夜间气温'),
+     en=c('Daytime Temperature','Nighttime Temperature')
+)
```

```
> props$zh<-stri_escape_unicode(props$zh)
> save(props,file="data/props.rda",compress=TRUE)
```

5.4.2.4 The Test Dataset weather20141001.rda

The weather dataset on October 1, 2014 was chosen as a demo dataset. Generate the weather20141001.rda file with Chinese characters escaped as Unicode.

```
# Load the weather data.
> weather20141001<-read.csv(file="metadata/20141001.csv",header=TRUE,file
Encoding="utf-8", encoding="utf-8")
# Escape weather20141001$prov.
> weather20141001$prov<-stri_escape_unicode(weather20141001$prov)
# Escape weather20141001$zh.
> weather20141001$zh<-stri_escape_unicode(weather20141001$zh)
# Generate the weather20141001.rda file.
> save(weather20141001,file="data/weather20141001.rda",compress=TRUE)
```

View the files in the data directory. There are four static dataset files generated.

```
> dir('data')
[1] "WOEID.rda" "chinaMap.rda" "props.rda" "weather20141001.rda"
```

Since all the static datasets are ready, the following R package code can directly use them.

5.4.3 Programming Functionalities

We create four files for the functions according to their features.

- getData.R: used to define the data crawling function
- render.R: used to render static picture visualization
- chinaWeather.R: used to define utility functions
- chinaWeather-packages.R: used to define the datasets in R package

5.4.3.1 The getData.R File

Create the getData.R file for data crawling and XML document parsing. Three functions are defined in the file.

- getWeatherFromYahoo(): used to retrieve the weather data from Yahoo's open data source
- getWeatherFromYahoo(): used to retrieve the weather data of the current city by the English name of city
- getWeather(): used to retrieve the weather data of China's provincial capital cities that are defined in the WOEID dataset

```
~ vi R/getData.R

#' Get weather data from Yahoo openAPI
#'
#' @importFrom RCurl getURL
#' @importFrom XML xmlTreeParse getNodeSet xmlGetAttr
```

```
#' @param woeid input a yahoo woeid
#' @return data.frame weather data
#' @keywords weather
#' @export
#' @examples
#' \dontrun{
#' getWeatherFromYahoo()
#' getWeatherFromYahoo(2151330)
#' }
getWeatherFromYahoo<-function(woeid=2151330){
  url<-paste('https://query.yahooapis.com/v1/public/yql',
             '?q=select%20*%20from%20weather.forecast%20where%20woeid%20
             %3D',
             x,
             '&format=xml',
             '&env=store%3A%2F%2Fdatatables.org%2Falltableswithkeys',sep='')

  xml <- getURL(url,.opts = list(ssl.verifypeer = FALSE))
  doc <- xmlTreeParse(xml,useInternal = TRUE)

  ans<-getNodeSet(doc, "//yweather:atmosphere",c(yweather="http://xml.
weather.yahoo.com/ns/rss/1.0"))
  humidity<-as.numeric(sapply(ans, xmlGetAttr, "humidity"))
  visibility<-as.numeric(sapply(ans, xmlGetAttr, "visibility"))
  pressure<-as.numeric(sapply(ans, xmlGetAttr, "pressure"))
  rising<-as.numeric(sapply(ans, xmlGetAttr, "rising"))

  ans<-getNodeSet(doc, "//item/yweather:condition",c(yweather="http://
xml.weather.yahoo.com/ns/rss/1.0"))
  code<-as.numeric(sapply(ans, xmlGetAttr, "code"))

  ans<-getNodeSet(doc, "//item/yweather:forecast[1]",c(yweather="http://
xml.weather.yahoo.com/ns/rss/1.0"))
low<-as.numeric(sapply(ans, xmlGetAttr, "low"))
high<-as.numeric(sapply(ans, xmlGetAttr, "high"))

  print(paste(woeid,'==>',low,high,code,humidity,visibility,pressure,rising))
  return(as.data.frame(cbind(low,high,code,humidity,visibility,pressure,
rising)))
}

#' Get one city weather Data.
#'
#' @param en input a English city name
#' @param src input data source
#' @return data.frame weather data
#' @keywords weather
#' @export
#' @examples
#' \dontrun{
#' getWeatherByCity()
#' getWeatherByCity(en="beijing")
#' }
getWeatherByCity<-function(en="beijing",src="yahoo"){
  woeid<-getWOEIDByCity(en)
```

```
  if(src=="yahoo"){
    return(getWeatherFromYahoo(woeid))
  }else{
    return(NULL)
  }
}

#' Get all of city weather Data.
#'
#' @param lang input a language
#' @param src input data source
#' @return data.frame weather data
#' @keywords weather
#' @export
#' @examples
#' \dontrun{
#' getWeather()
#' }
getWeather<-function(lang="en",src="yahoo"){
  cities<-getCityInfo(lang)
  wdata<-do.call(rbind, lapply(cities$woeid,getWeatherFromYahoo))
  return(cbind(cities,wdata))
}
```

5.4.3.2 The render.R File

Create the render.R file for data processing and static picture visualization rendering. Five functions are defined in the file.

- getColors(): used to match colors with weather conditions
- drawBackground(): used to draw the background
- drawDescription(): used to draw the literal description
- drawLegend(): used to draw the legend
- drawTemperature(): used to draw the temperatures and combine them with map

```
~ vi R/render.R

#' match the color with ADCODE99.
#'
#' @param temp the temperature
#' @param breaks cut the numbers
#' @return new color vector
#' @keywords color
getColors<-function(temp,breaks){
  f=function(x,y) ifelse(x %in% y,which(y==x),0)
  colIndex=sapply(chinaMap$ADCODE99,f,WOEID$adcode99)

  arr <- findInterval(temp, breaks)
  arr[which(is.na(arr))]=19
  return(arr[colIndex])
}
```

```
#' Draw the background.
#'
#' @param title the image's title
#' @param date the date
#' @param lang the language zh or en
drawBackground<-function(title,date,lang='zh'){
  text(100,58,title,cex=2)
  text(105,54,format(date,"%Y-%m-%d"))
  #text(98,65,paste('chinaweatherapp','http://apps.weibo.com/
chinaweatherapp'))
  #text(120,-8,paste('provided by The Weather Channel',format(date,
"%Y-%m-%d %H:%M")),cex=0.8)
}


#' Draw the description.
#'
#' @importFrom stringi stri_unescape_unicode
#' @param data daily data
#' @param temp the temperature
#' @param lang the language zh or en
drawDescription<-function(data,temp,lang='zh'){
  rows<-1:nrow(data)
  x<-ceiling(rows/7)*11+68
  y<-17-ifelse(rows%%7==0,7,rows%%7)*3
  fontCols<-c("#08306B","#000000","#800026")[findInterval(temp,
  c(0,30))  +1]
  if(lang=='zh'){
    txt<-stri_unescape_unicode(data$zh)
    text(x,y,paste(txt,temp),col=fontCols)
  }else{
    text(x,y,paste(data$en,temp),col=fontCols)
  }
  #text(x,y,bquote(paste(.(data$en),.(temp),degree,C)),col=fontCols)
}


#' Draw the legend.
#'
#' @param breaks cut the numbers
#' @param colors match the color
drawLegend<-function(breaks,colors){
  breaks2 <- breaks[-length(breaks)]
  par(mar = c(5, 0, 15, 10))
  image(x=1, y=0:length(breaks2),z=t(matrix(breaks2)),col=colors[1:length
(breaks)-1],axes=FALSE,breaks=breaks,xlab="",ylab="",xaxt="n")
  axis(4, at = 0:(length(breaks2)), labels = breaks, col = "white", las = 1)
  abline(h = c(1:length(breaks2)), col = "white", lwd = 2, xpd = FALSE)
}


#' Draw the temperature picture.
#'
#' @importFrom RColorBrewer brewer.pal
#' @importFrom stringi stri_unescape_unicode
#' @import maptools
#' @param data daily data
#' @param lang language
```

```
#' @param type low or high
#' @param date the date
#' @param output output a file or not
#' @param path image output position
#' @export
drawTemperature<-function(data,lang='zh',type='high',date=Sys.time(),output
=FALSE,path=''){colors <- c(rev(brewer.pal(9,"Blues")),"#ffffef",brewer.pal
(9,"YlOrRd"),"#500000")
  breaks=seq(-36,44,4)

  if(type=='high') {
    temp<-data$high
    ofile<-paste(format(date,"%Y%m%d"),"_day.png",sep="")
  }else{
    temp<-data$low
    ofile<-paste(format(date,"%Y%m%d"),"_night.png",sep="")
  }

  if(lang=='zh'){
    title<-stri_unescape_unicode(props[which(props$key=='high'),]$zh)
  }else{
    title<-props[which(props$key=='high'),]$en
  }

  if(output)png(filename=paste(path,ofile,sep=''),width=600,height=600)

  layout(matrix(data=c(1,2),nrow=1,ncol=2),widths=c(8,1),heights=c(1,2))
  par(mar=c(0,0,3,10),oma=c(0.2,0.2,0.2,0.2),mex=0.3)
  plot(chinaMap,border="white",col=colors[getColors(temp,breaks)])
  points(data$long,data$lat,pch=19,col=rgb(0,0,0,0.3),cex=0.8)

  drawBackground(title,date,lang)
  drawDescription(data,temp,lang)
  drawLegend(breaks,colors)
}
```

5.4.3.3 The chinaWeather.R File

Modify the chinaWeather.R file for defining the utility functions. Three functions are defined in the file.

- filename(): used to define a filename from the current date
- getWOEIDByCity(): used to get the Yahoo WOEID by city name
- getCityInfo(): used to get all city info defined in the WOEID dataset

```
#' Define a filename from current date.
#'
#' @param date input a date type
#' @return character a file name
#' @keywords filename
#' @export
#' @examples
#' \dontrun{
```

```
#' filename()
#' filename(as.Date("20110701",format="%Y%m%d"))
#' }
filename<-function(date=Sys.time()){
  paste(format(date, "%Y%m%d"),".csv",sep="")
}

#' Get WOEID of Yahoo By City Name
#'
#' @param en input a English city name
#' @return integer WOEID
#' @keywords WOEID
#' @export
#' @examples
#' \dontrun{
#'   getWOEIDByCity()
#'   getWOEIDByCity(en="beijing")
#' }
getWOEIDByCity<-function(en="beijing"){
  return(WOEID$woeid[which(WOEID$en==en)])
}

#' Get all of city info
#'
#' @param lang input a language
#' @return data.frame city info
#' @keywords language
#' @export
#' @examples
#' \dontrun{
#'   getCityInfo()
#'   getCityInfo(lang="en")
#'   getCityInfo(lang="zh")
#' }
getCityInfo<-function(lang="en"){
  if(lang=="en")return(WOEID[-c(3,4)])
  if(lang=="zh")return(WOEID[-c(4)])
}
```

5.4.3.4 The chinaWeather-package.R File

Create the chinaWeather-package.R for defining the R package description and built-in datasets.

- ■ NULL, the definition description of the chinaWeather package
- ■ "WOEID": the yahoo code for weather openAPI
- ■ "chinaMap": map of China
- ■ "props": charset for Chinese and English
- ■ "weather20141001": dataset for 20141001

```
#' China Weather package.
#'
#' a visualized package for china Weather
#'
```

```
#' @name chinaWeather-package
#' @aliases chinaWeather
#' @docType package
#' @title China Weather package.
#' @keywords package
NULL

#' The yahoo code for weather openAPI.
#'
#' @name WOEID
#' @description The yahoo code for weather openAPI.
#' @docType data
#' @format A data frame
#' @source \url{https://developer.yahoo.com/geo/geoplanet/guide/concepts.html}
'WOEID'

#' China Map.
#'
#' @name chinaMap
#' @description China Map Dataset.
#' @docType data
#' @format A S4 Object.
'chinaMap'

#' Charset for Chinese and English.
#'
#' @name props
#' @description Charset.
#' @docType data
#' @format A data frame
'props'

#' Dataset for 20141001.
#'
#' @name weather20141001
#' @description A demo dataset.
#' @docType data
#' @format A data frame
#' @source \url{http://weather.yahooapis.com/forecastrss?w=2151330}
'weather20141001'
```

5.4.4 *The Project Configuration Files*

We have added definitions for several functions and dependencies to five packages in the chinaWeather project. The project configuration files need corresponding modifications.

Three files need to be modified.

- DESCRIPTION: the project description file for global configurations of the project
- NAMESPACE: the namespace file for access control against functions
- .Rbuildignore: used to exclude the files that don't participate in package building

5.4.4.1 Modifying the DESCRIPTION File

The DESCRIPTION file is used for global configurations, which defines dependencies to five packages in the imports option and adds the LazyData options.

```
Package: chinaWeather
Type: Package
Title: a visualized package for china Weather
Version: 0.1
Authors@R: "Dan Zhang <bsspirit@gmail.com> [aut, cre]"
Description: a visualized package for china Weather
Depends:
    R (>= 3.1.1)
Imports:
    RCurl,
    XML,
    maptools,
    RColorBrewer,
    stringi
LazyData: TRUE
License: GPL-2
Date: 2014-09-28
```

5.4.4.2 Modifying the NAMESPACE File

The NAMESPACE file is used for access control against functions. Manually define the functions necessary to export and then run the document() function in the roxygen2 package, and the file is generated automatically.

```
export(drawTemperature)
export(filename)
export(getCityInfo)
export(getWOEIDByCity)
export(getWeather)
export(getWeatherByCity)
export(getWeatherFromYahoo)
```

5.4.4.3 Creating the .Rbuildignore File

When building the package, we can exclude unrelated files such as the metadata directory, the .gitignore file, etc.

```
.gitignore
dist
metadata
^.*\.Rproj$
^\.Rproj\.user$
README*
NEWS*
FAQ*
```

We have finished the modifications to R code, function comments, and the configuration file. The next step is debugging.

5.4.5 Debugging the Program

With the utility functions in the devtools package, it is easy to debug the program.

```
# Load the devtools package.
> library(devtools)

# Load the chinaWeather project.
> load_all("/home/conan/R/chinaWeather")
Loading chinaWeather

# Load the chinaWeather datasets.
> data(package="chinaWeather")
Data sets in package 'chinaWeather':
WOEID
chinaMap
props
weather20141001
```

Call the test dataset weather20141001 to draw the static picture of daytime temperatures, as shown in Figure 5.17.

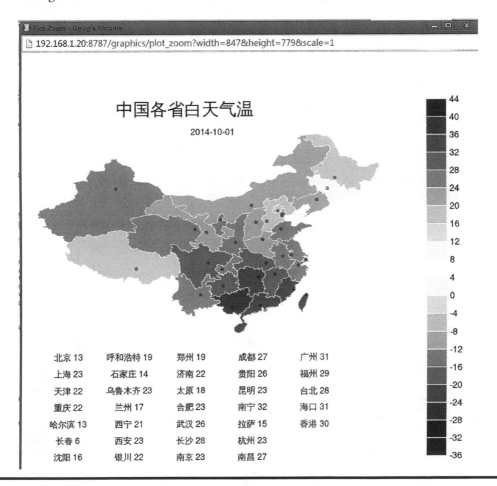

Figure 5.17 The static picture of daytime temperatures on October 1, 2014.

```
> date<-as.Date(as.character(20141001), format = "%Y%m%d")
> drawTemperature(weather20141001,date=date)
```

Draw the English static picture of nighttime temperatures on October 1, 2014, as shown in Figure 5.18.

```
> drawTemperature(weather20141001,type='low',date=date,lang='en')
```

The static pictures generated perfectly meet our requirements. Let us jump the step of unit test. We have added comments in the code so next we generate the documentation with the roxygen2 package.

```
> library(roxygen2)
> roxygenize("/home/conan/R/chinaWeather")
First time using roxygen2 4.0. Upgrading automatically...
```

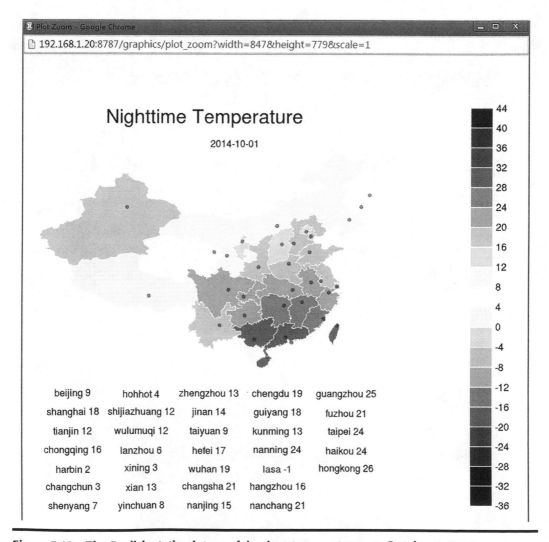

Figure 5.18   The English static picture of daytime temperatures on October 1, 2014.

```
Writing NAMESPACE
Writing chinaWeather-package.Rd
Writing WOEID.Rd
Writing chinaMap.Rd
Writing props.Rd
Writing weather20141001.Rd
Writing filename.Rd
Writing getWOEIDByCity.Rd
Writing getCityInfo.Rd
Writing getWeatherFromYahoo.Rd
Writing getWeatherByCity.Rd
Writing getWeather.Rd
Writing getColors.Rd
Writing drawBackground.Rd
Writing drawDescription.Rd
Writing drawLegend.Rd
Writing drawTemperature.Rd
```

Meanwhile, the NAMESPACE file is updated. By the automation, we have one less file to maintain. The execution works well. The final step is building the program.

5.4.6 Building the Program

Create the dist directory and place the program to it.
```
~ mkdir /home/conan/R/chinaWeather/dist
```

5.4.6.1 Building the Program

```
> build("/home/conan/R/chinaWeather",path="dist")
'/usr/lib/R/bin/R' --vanilla CMD build '/home/conan/R/chinaWeather'
--no-manual --no-resave-data

* checking for file '/home/conan/R/chinaWeather/DESCRIPTION'...OK
* preparing 'chinaWeather':
* checking DESCRIPTION meta-information...OK
* checking for LF line-endings in source and make files
* checking for empty or unneeded directories
* looking to see if a 'data/datalist' file should be added
* building 'chinaWeather_0.1.tar.gz'

[1] "dist/chinaWeather_0.1.tar.gz"
```

Install the chinaWeather package locally.

```
~ R CMD INSTALL dist/chinaWeather_0.1.tar.gz
* installing to library '/home/conan/R/x86_64-pc-linux-gnu-library/3.1'
* installing *source* package 'chinaWeather'...
** R
** data
*** moving datasets to lazyload DB
** inst
** preparing package for lazy loading
** help
*** installing help indices
```

```
** building package indices
** testing if installed package can be loaded
* DONE (chinaWeather)
```

Load the chinaWeather package to download the real-time weather data and get visualization output, as shown in Figure 5.19.

```
> library(chinaWeather)
> data<-getWeather(lang='zh')
[1] "2151330 ==> 8 19 28 32 NA 1023.5 0"
[1] "2151849 ==> 17 25 34 51 9.99 1015.92 0"
[1] "2159908 ==> 9 19 30 35 9.99 1015.92 0"
[1] "20070171 ==> 16 26 28 60 NA 1021.7 0"
```

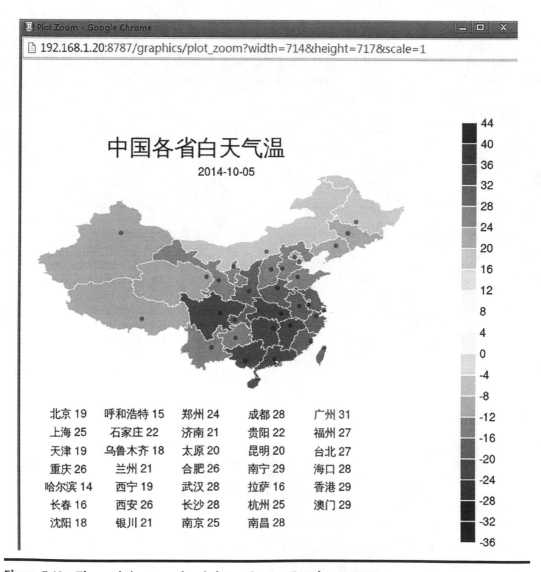

Figure 5.19   The real-time weather information on October 5, 2014.

```
[1] "2141166 ==> 0 14 34 22 9.99 1015.92 0"
[1] "2137321 ==> 2 16 30 27 9.99 1015.92 2"
[1] "2148332 ==> 6 18 28 35 9.99 1015.92 0"
[1] "2149760 ==> 3 15 30 31 9.99 1015.92 0"
[1] "2171287 ==> 9 22 34 27 9.99 1015.92 2"
[1] "26198317 ==> 9 18 34 55 9.99 1015.92 2"
[1] "2145605 ==> 6 21 32 39 NA 812.73 0"
[1] "2138941 ==> 3 19 32 34 NA 745.01 0"
[1] "2157249 ==> 12 26 32 44 NA 1022 0"
[1] "2150551 ==> 8 21 32 29 16 1022.7 0"
[1] "2172736 ==> 13 24 20 64 1.5 1015.92 0"
[1] "2168327 ==> 9 21 32 44 15 1022.3 0"
[1] "2154547 ==> 6 20 34 26 9.99 1015.92 2"
[1] "2127866 ==> 15 26 34 42 9.99 1015.92 2"
[1] "2163866 ==> 17 28 28 55 4.01 1019.8 0"
[1] "26198213 ==> 17 28 34 33 9.99 1015.92 0"
[1] "2137081 ==> 14 25 30 54 9.99 1015.92 2"
[1] "2158433 ==> 18 28 30 37 9.99 1015.92 2"
[1] "2146703 ==> 11 22 28 53 9.99 1015.92 2"
[1] "2160693 ==> 8 20 30 49 9.99 1015.92 2"
[1] "2166473 ==> 19 29 30 74 9 982.05 2"
[1] "26198235 ==> -1 16 32 20 NA 643.41 0"
[1] "2132574 ==> 16 25 34 39 9.99 1015.92 0"
[1] "26198151 ==> 19 28 30 40 NA 1018.4 0"
[1] "2161838 ==> 20 31 34 31 9.99 982.05 0"
[1] "2139963 ==> 18 27 34 42 9.99 982.05 2"
[1] "2306179 ==> 23 27 28 51 9.99 982.05 0"
[1] "2162779 ==> 23 28 30 66 9.99 982.05 2"
[1] "2165352 ==> 23 29 30 38 9.99 982.05 0"
[1] "1887901 ==> 25 29 34 48 9.99 982.05 2"

> drawTemperature(data,date=Sys.Date())
```

5.4.6.2 Checking Package with Check()

Since the process of program building works well, the next step is to check the package using the check() function. From the output, we can see that the checking process is successfully passed. But actually I encountered a lot of problems during debugging, which were solved one by one with time.

```
# Execute the process of checking.
> check("/home/conan/R/chinaWeather")
Updating chinaWeather documentation
Loading chinaWeather
'/usr/lib/R/bin/R' --vanilla CMD build '/home/conan/R/chinaWeather'
--no-manual --no-resave-data

* checking for file '/home/conan/R/chinaWeather/DESCRIPTION'...OK
* preparing 'chinaWeather':
* checking DESCRIPTION meta-information...OK
* checking for LF line-endings in source and make files
* checking for empty or unneeded directories
* looking to see if a 'data/datalist' file should be added
* building 'chinaWeather_0.1.tar.gz'
```

```
'/usr/lib/R/bin/R' --vanilla CMD check '/tmp/Rtmp3YI3Ar/
chinaWeather_0.1.tar.gz' --timings

* using log directory '/tmp/Rtmp3YI3Ar/chinaWeather.Rcheck'
* using R version 3.1.1 (2014-07-10)
* using platform: x86_64-pc-linux-gnu (64-bit)
* using session charset: UTF-8
* checking for file 'chinaWeather/DESCRIPTION'...OK
* checking extension type...Package
* this is package 'chinaWeather' version '0.1'
* checking package namespace information...OK
* checking package dependencies...OK
* checking if this is a source package...OK
* checking if there is a namespace...OK
* checking for executable files...OK
* checking for hidden files and directories...OK
* checking for portable file names...OK
* checking for sufficient/correct file permissions...OK
* checking whether package 'chinaWeather' can be installed...OK
* checking installed package size...OK
* checking package directory...OK
* checking DESCRIPTION meta-information...OK
* checking top-level files...OK
* checking for left-over files...OK
* checking index information...OK
* checking package subdirectories...OK
* checking R files for non-ASCII characters...OK
* checking R files for syntax errors...OK
* checking whether the package can be loaded...OK
* checking whether the package can be loaded with stated dependencies...OK
* checking whether the package can be unloaded cleanly...OK
* checking whether the namespace can be loaded with stated dependencies...OK
* checking whether the namespace can be unloaded cleanly...OK
* checking loading without being on the library search path...OK
* checking dependencies in R code...OK
* checking S3 generic/method consistency...OK
* checking replacement functions...OK
* checking foreign function calls...OK
* checking R code for possible problems...OK
* checking Rd files...OK
* checking Rd metadata...OK
* checking Rd line widths...OK
* checking Rd cross-references...OK
* checking for missing documentation entries...OK
* checking for code/documentation mismatches...OK
* checking Rd \usage sections...OK
* checking Rd contents...OK
* checking for unstated dependencies in examples...OK
* checking contents of 'data' directory...OK
* checking data for non-ASCII characters...OK
* checking data for ASCII and uncompressed saves...OK
* checking examples...OK
* checking PDF version of manual...OK
```

5.4.6.3 Uploading to Github

Finally, let us upload the project code to Github to make an open-source publish.

```
~ git add.
~ git commit -m 'app
~ git push origin app
To https://github.com/bsspirit/chinaWeatherDemo.git
  * [new branch]    app -> app
```

The project's address is https://github.com/bsspirit/chinaWeatherDemo/tree/app. Those who are interested may view the source code by.

5.4.6.4 Installing the chinaWeatherDemo Project from Github

Once uploading the code to Github, we finish the project publish on Github. User may install the project from Github with the devtools package.

```
# Load the devtools package.
> library(devtools)
# Install the project and configure the app branch.
> install_github("bsspirit/chinaWeatherDemo",ref="app")
Downloading github repo bsspirit/chinaWeatherDemo@app
Installing chinaWeather
'/usr/lib/R/bin/R' --vanilla CMD INSTALL '/tmp/RtmpTkR2Sd/
devtools8435b61dfe5/bsspirit-chinaWeatherDemo-54e36d4' \
  --library='/home/conan/R/x86_64-pc-linux-gnu-library/3.1' --install-tests

* installing *source* package 'chinaWeather'...
** R
** data
** inst
** preparing package for lazy loading
** help
*** installing help indices
** building package indices
** testing if installed package can be loaded
* DONE (chinaWeather)
Reloading installed chinaWeather
```

Well, we have finished the program development of R part of the entire project. The rest work is the PHP part. Since this book is about R, the program development of PHP part to create the Weibo app is not introduced. You may reference my blog for details.

Chapter 6

Journey on R Game Development

This chapter applies R to the field of game. Although R is not suitable for game development, it is really simple and light weighted for R to implement the algorithms of game. The following examples of R game development integrate all the techniques introduced in this book including environment, object oriented, mathematic calculations, R package development, visualization programming, and so on. Through the examples, readers can realize the essence of R programming and broad application prospects more deeply.

6.1 Keyboard and Mouse Events in R

Question

How do we implement the interactivity of keyboard and mouse?

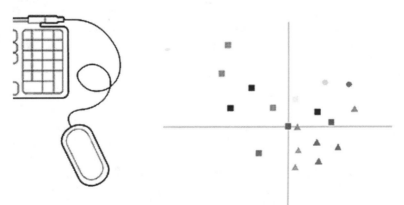

Keyboard & Mouse Events in R
http://blog.fens.me/r-keyboard-mouse/

Introduction

Many techniques can be ignored by us when working with R. Most of them are unrelated to data. Conversely speaking, if you could master these techniques, the R apps developed by you would amaze people a lot! This section will introduce you such a technique: keyboard and mouse events in R.

6.1.1 The Graphics Event in R

One of most broadly used functionalities is graphics drawing. The frequently used drawing functions contain plot(), barplot(), points(), lines(), and so on. Each time the plot() function is called, R runtime launches a new graphics device. In addition to drawing the graphs we define, the graphics device, with the built-in event manager, can also listen to the events of keyboard and mouse to implement the communications between events and graphics device.

The event manager is defined in the grDevices package and used as the parent of the current environment. For explanation of environments, please refer Section 3.2.

Imagine that if the drawing program could communicate with keyboard and mouse, the static picture could also interact with users. With the technique, we can use R to implement real-time app, chat tools, or game, in addition to just doing off-line analysis.

In the following, let us learn the graphics event APIs in R.

6.1.2 The Graphics Event APIs

There are four graphics event API functions in R:

- setGraphicsEventHandlers: used to register the graphics event including keyboard events and mouse events (pressing, releasing, and moving).
- getGraphicsEvent: used to launch the graphics event listener for listening.
- setGraphicsEventEnv: used to set the graphics device and environment. The default value is the current graphics device and environment.
- getGraphicsEventEnv: get the graphics device. The default value is the current graphics device.

To bind the keyboard and mouse events to the graphics device, we need to register the keyboard and mouse event handlers into the graphics device through the setGraphicsEventHandlers() function.

Let us view the definition of this function.

```
> setGraphicsEventHandlers
function (which = dev.cur(), ...)
setGraphicsEventEnv(which, as.environment(list(...)))
<bytecode: 0x000000000b321d20>
<environment: namespace:grDevices>
```

From the above, we can only see that setGraphicsEventHandlers() calls setGraphicsEventEnv() to set the graphics device. The arguments of setGraphicsEventHandlers() are hidden with "…."

Let us view the definition of the getGraphicsEvent() function. We find that the arguments of getGraphicsEvent() can be used for event binding, and that the arguments are passed to setGraphicsEventHandlers().

```
> getGraphicsEvent
function (prompt = "Waiting for input", onMouseDown = NULL, onMouseMove = NULL,
    onMouseUp = NULL, onKeybd = NULL, consolePrompt = prompt)
{
    if (!interactive())
        return(NULL)
    if (!missing(prompt) || !missing(onMouseDown) ||
!missing(onMouseMove) ||
        !missing(onMouseUp) || !missing(onKeybd)) {
        setGraphicsEventHandlers(prompt = prompt, onMouseDown = onMouseDown,
            onMouseMove = onMouseMove, onMouseUp = onMouseUp,
            onKeybd = onKeybd)
    }
    .External2(C_getGraphicsEvent, consolePrompt)
}
<bytecode: 0x0000000005c4bd40>
<environment: namespace:grDevices>
```

The following is the explanation to the arguments:

- prompt: literal prompt
- consolePrompt: the literal prompt on terminal
- onMouseDown: mouse event for pressing a button down
- onMouseUp: mouse event for releasing a button
- onMouseMove: mouse event for moving
- onKeybd: keyboard event for key pressing

We are able to bind the keyboard and mouse event through the encapsulation of the getGraphicsEvent() function.

6.1.3 The Keyboard Event

Having learnt the knowledge and APIs, let us do some programming work.

The system environment used in this section:

- Win7 64bit
- R: 3.1.1 x86_64-w64-mingw32/x64 (64-bit)

First, let us implement a small program about the keyboard. The following are the requirements:

- Use the functionality of drawing to show the corresponding characters according to keyboard inputs.
- Output one character on the screen each time one character key is pressed.
- Stop the keyboard event handling when pressing ctrl+C.

The screenshot is shown in Figure 6.1.

Left window is the R program console which prints one line for each keyboard input, followed by the change of the central text on the graphics device in the right window.

Note. The program cannot be run in RStudio. It can be only run in command-line.

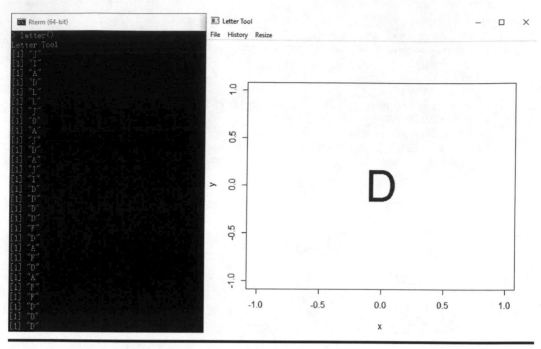

Figure 6.1   The keyboard event.

The following is the implementation in R:

```
# Letter utility.
>  letter<-function(){
+    # Drawing function.
+    draw<-function(label='',x=0,y=0){
+      plot(x,y,type='n')
+      text(x,y,label=label,cex=5)
+    }
+    # The keyboard event handler.
+    keydown<-function(K){
+      if (K == "ctrl-C") return(invisible(1))
+      print(K)
+      draw(K)
+    }
+    # Draw graph.
+    draw()
+    # Register the keyboard event handler to start listening.
+    getGraphicsEvent(prompt="Letter Tool",onKeybd = keydown)
+  }
# Launch the program.
>  letter()
```

The code for this small functionality is simple. Next, let us add the mouse events to implement a more complex example.

6.1.4 The Mouse Events

The mouse events of the graphics device contain three categories: moving, pressing (left button, right button, and wheel), and releasing.

Design an app with the following requirements:

- Utilize R drawing to specify the location and shape of a graph according to mouse events.
- Each time the mouse moves, output the mouse coordinates in the bottom-left corner and move a default graph following the mouse.
- When a mouse button is pressed, a graph is shown in the current location, with square for the left button, circle for wheel, and triangle for the right button.
- Stop event listening when q is pressed.

The app's screenshot is shown in Figure 6.2.
Following is the implementation:

```
# App for mouse events.
> mouse <- function() {
+       # Set the canvas un-bordered.
+       par(mai=rep(0,4),oma=rep(0,4))
+       # The initialized point.
+       ps<-data.frame(x=c(0.5),y   =c(0.5),col=c(2),pch=c(15))
+       # x and y are the mouse coordinates.
+       draw<-function(x=0,y=0){
```

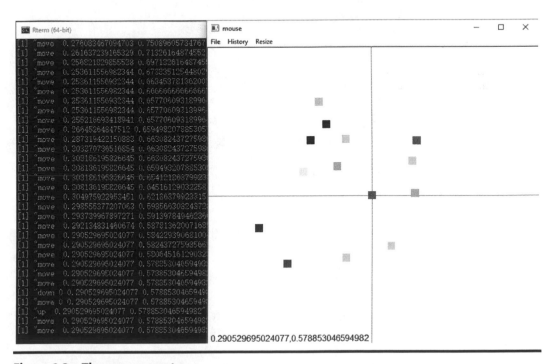

Figure 6.2   The mouse events.

```
+           plot(0,0,xlim=c(0,1),ylim=c(0,1),type='n',xaxs="i", yaxs="i")
+           # The horizontal line.
+           abline(h=0.5,col="gray60")
+           # The vertical line.
+           abline(v=0.5,col="gray60")
+           # Draw the solid points.
+           points(ps$x,ps$y,pch=ps$pch,cex=2,col=ps$col)
+           # Draw the mouse point.
+           points(x,y,pch=15,cex=2,col=colors()[ps$col])
+           # Text the coordinates.
+           text(0.25,0.015,label=paste(x,y,sep=","))
+       }
+       # Mouse events listening: Add a solid point once a button is
pressed.
+       # Buttons, 0 for the left button, 1 for the wheel and 2 for the
right button.
+       mouseDown <- function(buttons, x, y) {
+           print(paste("down",buttons,x,y))
+           # Set the shape.
+           shape<-15
+           if(buttons==1) shape<-16
+           if(buttons==2) shape<-17
+           # Add a solid point.
+           ps<<-rbind(ps,data.frame(x=c(x),y=c(y),pch=c(shape),col=round
(runif(1,2,500))))
+           draw(x,y)
+       }
+       # Listen for mouse moving.
+       mouseMove <- function(buttons, x, y) {
+           print(paste("move",buttons,x,y))
+           draw(x,y)
+       }
+       # Listen for mouse button releasing.
+       mouseup <- function(buttons, x, y) {
+           print(paste("up",buttons,x,y))
+           draw(x,y)
+       }
+       # Listen for key pressing.
+       keydown <- function(key) {
+           if (key == "q") return(invisible(1))
+       }
+       draw()
+       # Register the event handlers.
+       getGraphicsEvent(prompt="mouse",onMouseDown=mouseDown,onMouseMove=
mouseMove,onMouseUp=mouseup,onKeybd=keydown)
+ }
# Launch the program.
> mouse()
```

By listening to the R graphics device events, we can easily implement the interactivity between program and keyboard and mouse. Isn't it interesting? Let us develop a game in the next section.

6.2 Getting Started with the Snake Game

Question

How do we code the Snake game in R?

Introduction

It isn't marvelous to do statistics, classification, or visualization in R? Have you ever seen games developed with R? This section will introduce you to R game development: implementing the Snake game with R.

6.2.1 Introduction to the Snake Video Game

Snake is a video game originated in late 1970s. Such games were popular again in the 1990s since the emergence of some small-screen devices. The game can be installed on most modern cellphones. In the game, the player controls a long line-shaped snake. The snake keeps moving forward and the player can only change the direction of its head (up, down, left, and right) to make it eat all the things it encounters. In the meanwhile, the player must prevent the snake from conflicting with itself or other obstacles. Each time the snake eats a fruit its body gets longer and its speed gets higher, which increases the difficulty. One design of the game is having walls surrounding the snake which it cannot go through. The game is over when the snake runs into a wall or obstacle. The score is evaluated based on the number of the fruits eaten by the snake. There are many versions of the game on various devices, as shown in Figure 6.3.

6.2.2 Stage Design

How do we start to develop the game? From the perspective of software development, we need to do requirement analysis for the game, list the game rules, design business flow, and present the prototype of the game to validate the feasibility of the design.

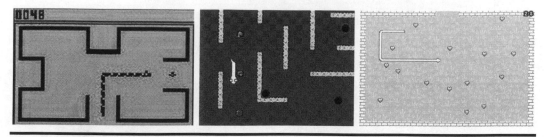

Figure 6.3   Introduction to the Snake game.

6.2.2.1 Requirement Analysis

The snake game has three stages: starting, gaming, and ending.

- Starting stage: launches the application, warms up the player, and provides how-to before gaming.
- Gaming stage: the stage when game is in process.
- Ending stage: the stage when the player wins, loses, or exits, providing the scores gained by the player.

The stages for starting and ending are simple so I won't explain more. The gaming stage is the core of the game. It contains a canvas, a snake, a head, a tail with varying length, a fruit, borders, and some obstacles.

6.2.2.2 The Game Rules

The rules when the game is in process are listed in details as following:

- After launching the game, the player presses the up, down, left, and right keys to control the snake's head to change the direction to which the snake is moving. The player can also press q to quit the game. Other keyboard operations are invalid.
- We use a blue block to represent the head, gray for the tail, red for the fruit, and black for obstacles.
- When the head moves over the fruit, it means the snake has eaten the fruit. The tail grows longer. Another fruit will be generated in a blank block after the head moves.
- The borders of the canvas are designed to be walls. When the head moves to the position invisible on the canvas, the head runs into the wall and the player loses.
- There are some black obstacles in the game. The player loses when the head moves into one of the obstacles.
- The player loses when the head conflicts with the tail.

6.2.2.3 The Business Flow

The entire game can be divided into three stages. The switching flow of the game stages is shown in Figure 6.4.

- When launching the game, the player sees the starting stage and then presses any key to enter the gaming stage.

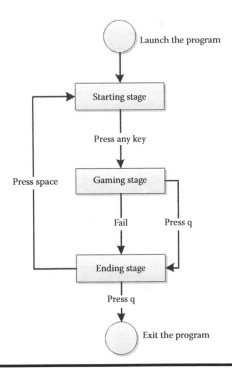

Figure 6.4 **Definition of the business flow.**

- The gaming stage is transited to the ending stage when the player loses the game. When pressing q, the player is directly brought to the ending stage.
- In the ending stage, the player is back to the starting stage by pressing space, or exits the game by pressing q.

6.2.2.4 The Game Prototype

Let us design the interfaces of the three stages as the game prototype before game development. In Figure 6.5, the left, middle, and right pictures show the starting, gaming, and ending stages, respectively.

In the following, let us draw all the three stages programmatically according to the prototype design.

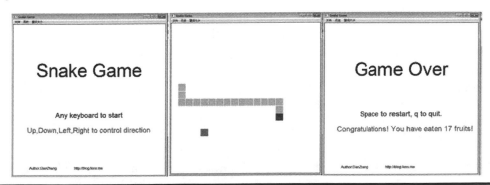

Figure 6.5 **The stages of the Snake game.**

6.2.3 Program Design

Now, we have clearly understood the rules and functional requirements of the game. Next, we need to technically describe the business logic in requirement analysis and take the nonfunctional requirements and technical details about R into accounts.

6.2.3.1 The Game Stages

We map each stage to a canvas, or a data structure in memory.

- The starting stage is static, so we can generate the corresponding canvas in memory in advance, or generate it dynamically only when switching stages. The performance cost is low.
- The gaming stage is dynamic. The canvas should be regenerated through the binding events each time the interoperation or periodical refresh occurs. The performance cost is high due to frequent interoperations.
- The ending stage is dynamic, showing the resulting score of the game, so it is generated temporarily when the game is switched to the stage.

6.2.3.2 The Game Objects

The game may generate a lot of objects when running, such as the canvas, the snake, and the fruits mentioned above. These objects need to be defined in memory and mapped to R's data types.

The detailed description of the canvas object:

- Canvas: described with matrix. Each block is mapped to a number in the matrix.
- Size: the length and width of the canvas that are mapped to two numeric variables, respectively.
- Coordinates: used to locate the blocks in the canvas. The horizontal coordinates are 1 through 20 from left to right and the vertical coordinates are 1 through 20 from bottom to top.
- Index: used to locate the blocks in the canvas. The indices are 1 through 400 in the sequence of from left to right and then from bottom to top.
- Blocks: the minimum units of the canvas. The size of a block is set according to the proportion of the canvas.

The canvas object described by matrix is shown in Figure 6.6.

The detailed description of the snake object:

- Head: described by a vector representing only one block. Its original coordinates are (2, 2) and the direction is up by default when the game is started. When player enters the interface, the show location is at (2, 3).
- Tail: described using a data frame, storing uncertain number of coordinate vectors. When game is started, its length is 0.

The detailed description of the fruit object:

- Fruit: described by a vector representing only one block. When game starts, a fruit is located in a random blank block. When it is eaten, another fruit is randomly generated at a blank block.

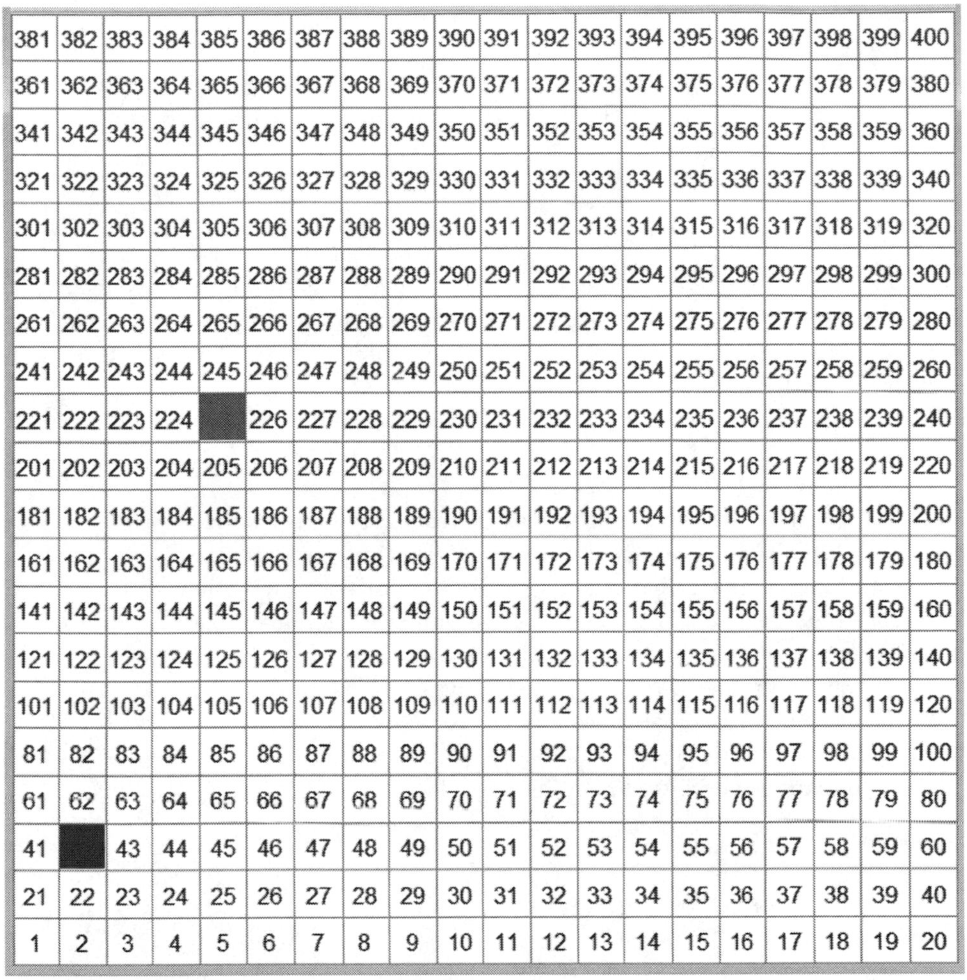

381	382	383	384	385	386	387	388	389	390	391	392	393	394	395	396	397	398	399	400
361	362	363	364	365	366	367	368	369	370	371	372	373	374	375	376	377	378	379	380
341	342	343	344	345	346	347	348	349	350	351	352	353	354	355	356	357	358	359	360
321	322	323	324	325	326	327	328	329	330	331	332	333	334	335	336	337	338	339	340
301	302	303	304	305	306	307	308	309	310	311	312	313	314	315	316	317	318	319	320
281	282	283	284	285	286	287	288	289	290	291	292	293	294	295	296	297	298	299	300
261	262	263	264	265	266	267	268	269	270	271	272	273	274	275	276	277	278	279	280
241	242	243	244	245	246	247	248	249	250	251	252	253	254	255	256	257	258	259	260
221	222	223	224	■	226	227	228	229	230	231	232	233	234	235	236	237	238	239	240
201	202	203	204	205	206	207	208	209	210	211	212	213	214	215	216	217	218	219	220
181	182	183	184	185	186	187	188	189	190	191	192	193	194	195	196	197	198	199	200
161	162	163	164	165	166	167	168	169	170	171	172	173	174	175	176	177	178	179	180
141	142	143	144	145	146	147	148	149	150	151	152	153	154	155	156	157	158	159	160
121	122	123	124	125	126	127	128	129	130	131	132	133	134	135	136	137	138	139	140
101	102	103	104	105	106	107	108	109	110	111	112	113	114	115	116	117	118	119	120
81	82	83	84	85	86	87	88	89	90	91	92	93	94	95	96	97	98	99	100
61	62	63	64	65	66	67	68	69	70	71	72	73	74	75	76	77	78	79	80
41	■	43	44	45	46	47	48	49	50	51	52	53	54	55	56	57	58	59	60
21	22	23	24	25	26	27	28	29	30	31	32	33	34	35	36	37	38	39	40
1	2	3	4	5	6	7	8	9	10	11	12	13	14	15	16	17	18	19	20

Figure 6.6   The canvas in the Snake game.

Borders and obstacles:

- Borders: no memory description and checked by calculations. The border collision event is triggered when the coordinates are beyond the matrix.
- Obstacles: described by a data frame storing uncertain number of coordinate vectors.

6.2.3.3 The Game Events

There are three events in the game: the keyboard event, the time event, and the collision event.

- The keyboard event: a global event triggered by user input through keyboard, for example, pressing up, down, left, or right to control the moving direction of the snake.
- The time event: a global event. The system triggers the time event once at the interval of 0.2 seconds. For example, the head moves one block ahead every 0.2 s.
- The collision event: triggered by the collision of the head and any non-blank blocks when the head is moving, for example, when eating fruit or when the head conflicts with the tail.

Usually, the three events above are managed by three separated threads. However, since R is designed single thread and does not support asynchronous calling, we cannot implement the listens for all three events in the same time. A compromised solution is to use the global keyboard event. The collision event is triggered through the keyboard event to do collision check. The time event is ignored in this solution.

6.2.3.4 The Game Control

We need to control all the statuses that occur when the game is running, for example, when to generate a new fruit, when to append one piece of tail, when to end the game, and so on. By defining the control functions, we can easily manage all the game statuses. The entire control flow of the game is shown in Figure 6.7.

Each rectangle in Figure 6.7 represents an R function definition.

- run(): function to launch the game program.
- keydown(): function to listen the keyboard events, locking the thread globally.
- stage0(): function to create the starting stage with visualization output.
- stage1(): function to create the gaming stage with visualization output.
- stage2(): function to create the ending stage with visualization output.
- init(): function to initialize game variables when starting game.
- fruit(): function to check and generate the fruit coordinates.
- head(): function to generate the coordinates for head to move.
- fail(): failure checking to judge whether the head runs into walls or conflicts with its tail. If failed, the game jumps into the ending stage.

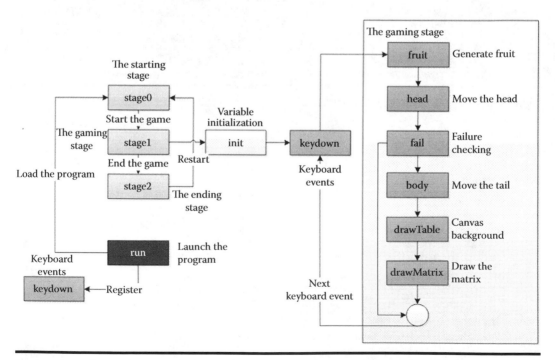

Figure 6.7   The control flow of the Snake game.

- body(): function to generate the coordinates for tail to move.
- drawTable(): function to draw the background.
- drawMatrix(): function to draw the matrix.

Through the detailed program design, we have converted the business description in requirement analysis to technical description in program developing. After completing the design, the only thing left is coding.

6.2.4 The R Implementation

A few lines of R code can implement the above design. We just need to fill the code into the functions defined above just like doing gap filling. Of course, some features of R language need to be employed when coding to make the program more robust.

The system environment used in this section:

- Win7 64bit
- R: 3.1.1 x86_64-w64-mingw32/x64 (64-bit)

run():function to launch the game program.

```
run<-function(){
  # Set the global canvas without borders.
  par(mai=rep(0,4),oma=rep(0,4))
  # Define a global environment for encapsulating variables.
  e<<-new.env()
  # Launch the starting stage.
  stage0()
  # Register the keyboard event.
  getGraphicsEvent(prompt="Snake Game",onKeybd=keydown)
}
```

In the above code, an environment is defined to store variables, which can effectively solve the problem of variable name collision and variable pollution. For information about environment, please refer Sections 3.2 and 3.3.

keydown(): function to listen the keyboard event, locking the thread globally.

```
keydown<-function(K){
  # Print the keyboard input.
  print(paste("keydown:",K,",stage:",e$stage));

  # The starting stage.
  if(e$stage==0){
    init()
    stage1()
    return(NULL)
  }

  # The ending stage.
  if(e$stage==2){
    if(K=="q") q()
    else if(K==' ') stage0()
```

```
      return(NULL)
    }


  # The gaming stage.
  if(e$stage==1){
    # Switch to the ending stage when pressing q.
    if(K == "q") {
      stage2()
    } else {
      if(tolower(K) %in% c("up","down","left","right")){
        e$lastd<-e$dir
        e$dir<-tolower(K)
        stage1()
      }
    }
  }
  return(NULL)
}
```

The argument K is the keyboard input. By the condition checking against the current stage and keyboard, the program determines the response to a keyboard event. The function only responses five keys: up, down, left, right, and q, corresponding the operations of "moving up," "moving down," "moving left," "moving right," and "quitting."

stage0(): function to create the starting stage with visualization output.

```
# The starting stage.
stage0<-function(){
  e$stage<-0
  plot(0,0,xlim=c(0,1),ylim=c(0,1),type='n',xaxs="i", yaxs="i")
  text(0.5,0.7,label="Snake Game",cex=5)
  text(0.5,0.4,label="Any keyboard to start",cex=2,col=4)
  text(0.5,0.3,label="Up,Down,Left,Rigth to control
direction",cex=2,col=2)
  text(0.2,0.05,label="Author:DanZhang",cex=1)
  text(0.5,0.05,label="http://blog.fens.me",cex=1)
}
```

stage2(): function to create the ending stage with visualization output.

```
# The ending stage.
stage2<-function(){
  e$stage<-2
  plot(0,0,xlim=c(0,1),ylim=c(0,1),type='n',xaxs="i", yaxs="i")
  text(0.5,0.7,label="Game Over",cex=5)
  text(0.5,0.4,label="Space to restart, q to quit.",cex=2,col=4)
  text(0.5,0.3,label=paste("Congratulations! You have eaten",nrow(e$tail),
"fruits!"),cex=2,col=2)
  text(0.2,0.05,label="Author:DanZhang",cex=1)
  text(0.5,0.05,label="http://blog.fens.me",cex=1)
}
```

init(): function to initialize game variables when starting game.

```
# Initialize the environment variable.
init<-function(){
  e<<-new.env()
  # The stage.
  e$stage<-0
  # Slice the canvas.
  e$width<-e$height<-20
  # The step.
  e$step<-1/e$width
  # The block matrix.
  e$m<-matrix(rep(0,e$width*e$height),nrow=e$width)
  # The moving direction.
  e$dir<-e$lastd<-'up'
  # Initialize the head.
  e$head<-c(2,2)
  # Initialize the last block of the head.
  e$lastx<-e$lasty<-2
  # Initialize the tail.
  e$tail<-data.frame(x=c(),y=c())
  # Color of the fruit.
  e$col_fruit<-2
  # Color of the head.
  e$col_head<-4
  # Color of the tail.
  e$col_tail<-8
  # Color of path.
  e$col_path< 0
}
```

The above code initializes the global environment e and define all the necessary variables in e.

fruit(): function to check and generate the fruit coordinates.

```
# The random fruit block.
fruit<-function(){
    # There is no fruit.
    if(length(index(e$col_fruit))<=0){
      idx<-sample(index(e$col_path),1)

      fx<-ifelse(idx%%e$width==0,10,idx%%e$width)
      fy<-ceiling(idx/e$height)
      e$m[fx,fy]<-e$col_fruit

      print(paste("fruit idx",idx))
      print(paste("fruit axis:",fx,fy))
    }
}
```

fail(): failure checking to judge whether the head runs into walls or conflicts with its tail. If failed, the game jumps into the ending stage.

```
# Failure checking.
fail<-function(){
```

```
  # The head is beyond the borders.
  if(length(which(e$head<1))>0 | length(which(e$head>e$width))>0){
    print("game over: Out of ledge.")
    keydown('q')
    return(TRUE)
  }

  # The head conflicts the tail.
  if(e$m[e$head[1],e$head[2]]==e$col_tail){
    print("game over: head hit tail")
    keydown('q')
    return(TRUE)
  }

  return(FALSE)
}
```

head(): function to generate the coordinates for head to move.

```
# Generate the coordinates of the head.

head<-function(){
    e$lastx<-e$head[1]
    e$lasty<-e$head[2]

    # Direction operation to move up.
    if(e$dir=='up')  e$head[2]<-e$head[2]+1
    # Direction operation to move down.
    if(e$dir=='down')  e$head[2]<-e$head[2]-1
    # Direction operation to move left.
    if(e$dir=='left')  e$head[1]<-e$head[1]-1
    # Direction operation to move right.
    if(e$dir=='right')  e$head[1]<-e$head[1]+1
}
```

body(): function to generate the coordinates for tail to move.

```
# Generate the coordinates of the tail.

body<-function(){
    e$m[e$lastx,e$lasty]<-0
    # Fill color of the head.
    e$m[e$head[1],e$head[2]]<-e$col_head

    # There is no fruit.
    if(length(index(e$col_fruit))<=0){
      e$tail<-rbind(e$tail,data.frame(x=e$lastx,y=e$lasty))
    }

    # If the tail exists.
    if(nrow(e$tail)>0) {
      e$tail<-rbind(e$tail,data.frame(x=e$lastx,y=e$lasty))
      e$m[e$tail[1,]$x,e$tail[1,]$y]<-e$col_path
      e$tail<-e$tail[-1,]
      e$m[e$lastx,e$lasty]<-e$col_tail
    }
```

```
    print(paste("snake idx",index(e$col_head)))
    print(paste("snake axis:",e$head[1],e$head[2]))
}
```

drawTable(): function to draw the background.

```
# The canvas background.
drawTable<-function(){
    plot(0,0,xlim=c(0,1),ylim=c(0,1),type='n',xaxs="i", yaxs="i")

    # Show the background grid.
    # The horizontal lines.
    abline(h=seq(0,1,e$step),col="gray60")
    # The vertical lines.
    abline(v=seq(0,1,e$step),col="gray60")

    # Show the matrix.
    df<-data.frame(x=rep(seq(0,0.95,e$step),e$width),y=rep(seq(0,0.95,e$
step),each=e$height),lab=seq(1,e$width*e$height))
    text(df$x+e$step/2,df$y+e$step/2,label=df$lab)
}
```

drawMatrix(): function to draw the matrix.

```
# Draw the data according to the matrix.
drawMatrix<-function(){
    idx<-which(e$m>0)
    px<- (ifelse(idx%%e$width==0,e$width,idx%%e$width)-1)/
e$width+e$step/2
    py<- (ceiling(idx/e$height)-1)/e$height+e$step/2
    pxy<-data.frame(x=px,y=py,col=e$m[idx])
    points(pxy$x,pxy$y,col=pxy$col,pch=15,cex=4.4)
}
```

stage1(): function to create the gaming stage. The functions for the gaming stage are encapsulated in stage1() and called by it.

```
# The gaming stage.
stage1<-function(){
  e$stage<-1
  # See fruit().
  fruit<-function(){...}
  # See fail().
  fail<-function(){...}
  # See head().
  head<-function(){...}
  # See body().
  body<-function(){...}
  # See drawTable().
  drawTable<-function(){...}
  # See drawMatrix().
  drawMatrix<-function(){...}

  # Call the function.
  fruit()
```

```
  head()
  # Failure checking.
  if(!fail()){
    body()
    drawTable()
    drawMatrix()
  }
}
```

Note: The above is pseudo code.

Following is the complete of R program code:

```
# Initialize the environment.
> init<-function(){
+      e<<-new.env()
       # Set the stage.
+      e$stage<-0
       # Slice the matrix.
+      e$width<-e$height<-20
       # The Step.
+      e$step<-1/e$width
       # The block matrix.
+      e$m<-matrix(rep(0,e$width*e$height),nrow=e$width)
       # The moving direction.
+      e$dir<-e$lastd<-'up'
       # Initialize the head.
+      e$head<-c(2,2)
       # Initialize the last block of the head.
+      e$lastx<-e$lasty<-2
       # Initialize the tail.
+      e$tail<-data.frame(x=c(),y=c())
+
       # Color of the fruit.
+      e$col_fruit<-2
       # Color of the head.
+      e$col_head<-4
       # Color of the tail.
+      e$col_tail<-8
       #Color of the path.
+      e$col_path<-0
+ }
>
# Get indices in matrix.
> index<-function(col) which(e$m==col)
>
# The gaming stage.
> stage1<-function(){
+      e$stage<-1
+
       # The random  fruit block.
+      fruit<-function(){
           # There is no fruit.
+          if(length(index(e$col_fruit))<=0){
+              idx<-sample(index(e$col_path),1)
```

```
+
+                    fx<-ifelse(idx%%e$width==0,10,idx%%e$width)
+                    fy<-ceiling(idx/e$height)
+                    e$m[fx,fy]<-e$col_fruit
+
+                    print(paste("fruit idx",idx))
+                    print(paste("fruit axis:",fx,fy))
+            }
+        }
+
        # Failure checking.
+       fail<-function(){
            # The head is beyond the borders.
+           if(length(which(e$head<1))>0 | length(which(e$head>e$width))>0){
+               print("game over: Out of ledge.")
+               keydown('q')
+               return(TRUE)
+           }
+
            # The head conflicts the tail.
+           if(e$m[e$head[1],e$head[2]]==e$col_tail){
+               print("game over: head hit tail")
+               keydown('q')
+               return(TRUE)
+           }
+
+           return(FALSE)
+       }
+
        # The status of the head.
+       head<-function(){
+           e$lastx<-e$head[1]
+           e$lasty<-e$head[2]
+
            # The direction operations.
+           if(e$dir=='up') e$head[2]<-e$head[2]+1
+           if(e$dir=='down') e$head[2]<-e$head[2]-1
+           if(e$dir=='left') e$head[1]<-e$head[1]-1
+           if(e$dir=='right') e$head[1]<-e$head[1]+1
+
+       }
+
        # The status of the tail.
+       body<-function(){
+           e$m[e$lastx,e$lasty]<-0
+           e$m[e$head[1],e$head[2]]<-e$col_head
            # There is no fruit.
+           if(length(index(e$col_fruit))<=0){
+               e$tail<-rbind(e$tail,data.frame(x=e$lastx,y=e$lasty))
+           }
+
            # If the tail exists.
+           if(nrow(e$tail)>0) {
+               e$tail<-rbind(e$tail,data.frame(x=e$lastx,y=e$lasty))
```

```
+                 e$m[e$tail[1,]$x,e$tail[1,]$y]<-e$col_path
+                 e$tail<-e$tail[-1,]
+                 e$m[e$lastx,e$lasty]<-e$col_tail
+             }
+
+         print(paste("snake idx",index(e$col_head)))
+         print(paste("snake axis:",e$head[1],e$head[2]))
+     }
+
      # Draw the canvas background.
+     drawTable<-function(){
+         plot(0,0,xlim=c(0,1),ylim=c(0,1),type='n',xaxs="i", yaxs="i")
+     }
+
      # Draw the blocks according to the matrix.
+     drawMatrix<-function(){
+         idx<-which(e$m>0)
+         px<- (ifelse(idx%%e$width==0,e$width,idx%%e$width)-1)/
e$width+e$step/2
+         py<- (ceiling(idx/e$height)-1)/e$height+e$step/2
+         pxy<-data.frame(x=px,y=py,col=e$m[idx])
+         points(pxy$x,pxy$y,col=pxy$col,pch=15,cex=4.4)
+     }
+
+     fruit()
+     head()
+     if(!fail()){
+         body()
+         drawTable()
+         drawMatrix()
+     }
+ }
>
# The starting stage.
> stage0<-function(){
+     e$stage<-0
+     plot(0,0,xlim=c(0,1),ylim=c(0,1),type='n',xaxs="i", yaxs="i")
+     text(0.5,0.7,label="Snake Game",cex=5)
+     text(0.5,0.4,label="Any keyboard to start",cex=2,col=4)
+     text(0.5,0.3,label="Up,Down,Left,Rigth to control
direction",cex=2,col=2)
+     text(0.2,0.05,label="Author:DanZhang",cex=1)
+     text(0.5,0.05,label="http://blog.fens.me",cex=1)
+ }
>
# The ending stage.
> stage2<-function(){
+     e$stage<-2
+     plot(0,0,xlim=c(0,1),ylim=c(0,1),type='n',xaxs="i", yaxs="i")
+     text(0.5,0.7,label="Game Over",cex=5)
+     text(0.5,0.4,label="Space to restart, q to quit.",cex=2,col=4)
+     text(0.5,0.3,label=paste("Congratulations! You have eat",nrow(e$tail),
"fruits!"),cex=2,col=2)
+     text(0.2,0.05,label="Author:DanZhang",cex=1)
+     text(0.5,0.05,label="http://blog.fens.me",cex=1)
```

```
+ }
>
>
# The keyboard event.
> keydown<-function(K){
+     print(paste("keydown:",K,",stage:",e$stage));
      # The starting stage.
+     if(e$stage==0){
+         init()
+         stage1()
+         return(NULL)
+     }
+
      # The ending stage.
+     if(e$stage==2){
+         if(K=="q") q()
+         else if(K==' ') stage0()
+         return(NULL)
+     }
+
      # The gaming stage.
+     if(e$stage==1){
+         if(K == "q") {
+             stage2()
+         } else {
+             if(tolower(K) %in% c("up","down","left","right")){
+                 e$lastd<-e$dir
+                 e$dir<-tolower(K)
+                 stage1()
+             }
+         }
+     }
+     return(NULL)
+ }
>
> ######################################
> # RUN
> ######################################
>
> run<-function(){
+     par(mai=rep(0,4),oma=rep(0,4))
+     e<<-new.env()
+     stage0()
      # Register the keyboard event.
+     getGraphicsEvent(prompt="Snake Game",onKeybd=keydown)
+ }
>
> run()
```

The screenshot of the game is given in Figure 6.8.

We have implemented the Snake game with only 161 lines of code, of which about 100 lines are effective (Figure 6.8). Of course, we did not implement the time event due to the single threading mechanism and lack of support of asynchronous calling of R. However, R's capability of data

Snake Game

Any keyboard to start

Up,Down,Left,Right to control direction

Author:DanZhang   http://blog.fens.me

Figure 6.8   The interface of the Snake game.

processing and visualization makes it easy to implement the program. I believe it would be very convenient for R to implement the matrix calculation for strategy games.

Given the primary program of the Snake game, can we induce an R-based game development framework? The next section will continue the journey to R game, the R-based game framework design.

6.3 R-Based Game Framework Design

Question
How do we write a game framework in R?

Introduction

Following the previous section, after coding, we should tidy up the code further to abstract the common part for game development, separating the code of game framework from the Snake itself. We can abstract an R-based game development engine. When developing a new game, we just need to put focus on the program design of the game itself.

6.3.1 The Object-Oriented Modification to the Snake Game

We can tidy up the Snake project code using object-oriented methodology and enforcing the object-oriented modification to the code. R language supports three types of object-oriented programming, from which I choose Reference Class (RC). Please refer to Section 4.4 for details about RC.

Refactoring the code is easy since all the pieces of code are encapsulated into functions. We just need to give good definitions for the properties and methods of the Snake object.

6.3.1.1 Defining the Snake Classes

Let us define the Snake class containing properties and methods. Draw the class diagram, as in Figure 6.9.

The explanation to properties is as following:

- name: the name of the name
- stage: the current stage of game
- e: the environment to store variables in game
- width: the width of the matrix
- height: the height of the matrix
- m: the matrix of game map

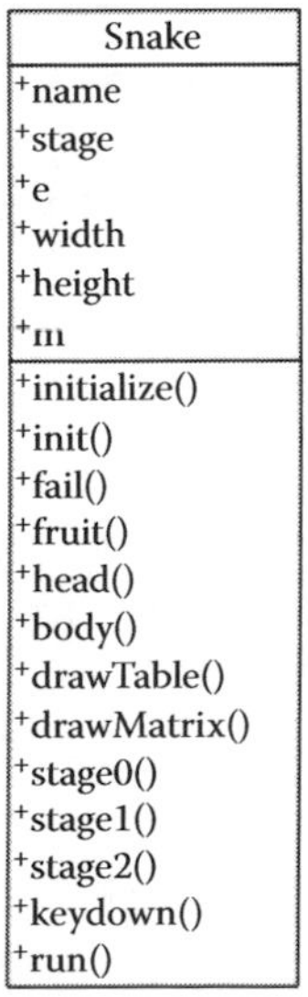

Snake
+name
+stage
+e
+width
+height
+m
+initialize()
+init()
+fail()
+fruit()
+head()
+body()
+drawTable()
+drawMatrix()
+stage0()
+stage1()
+stage2()
+keydown()
+run()

Figure 6.9 Definition of the Snake class.

The explanation to methods is as following:

- initialize(): the constructing function for initialization of RC classes.
- init(): function to initialize the game variables for stage1.
- fail(): function to do failure checking.
- fruit(): function to check and generate fruit coordinates.
- head(): function to generate coordinate for head moving.
- body(): function to generate coordinate for tail moving.
- drawTable: function to draw the game background.
- drawMatrix: function to draw the game matrix.
- stage0: function to create the starting stage with visualization output.
- stage1(): function to create the gaming stage. The functions for the gaming stage are encapsulated in stage1() and called by it.
- stage2(): function to create the ending stage with visualization output.
- keydown(): function to listen the keyboard events.
- run(): the launching function.

6.3.1.2 The Global Function Calling Sequence Diagram

In the following, let us draw the sequence diagrams according to the UML specifications. The diagrams contain the global function calling sequence and the calls in the gaming stage function stage1. The global function calling sequence diagram is shown in Figure 6.10.

- Launch the game through the run() function, enter stage0, and then register the keyboard event.
- In stage0, press any key to switch to stage1.

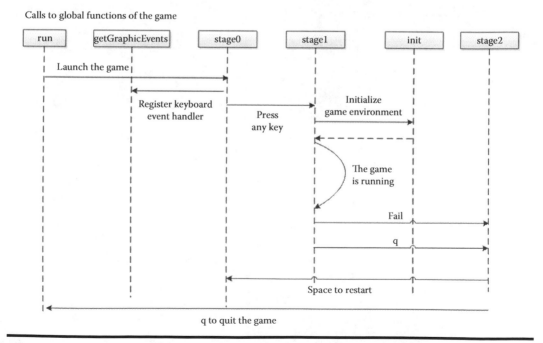

Figure 6.10 **The sequence diagram of calls to global functions of the game.**

- init() initializes the game variables in stage1.
- stage1() runs the game.
- When fail() returns true or the player presses q.
- The game jumps to stage2() to show the game over interface.
- Press space back to stage0 to restart, or q to exit the game.

6.3.1.3 The Function Calling Sequence Diagram for stage1

The function calling sequence in stage1 is shown in Figure 6.11.

- In stage1, press the direction keys (up, down, left, and right) to control the moving direction of the head.
- The fruit() function checks the existence of fruit. If it has been eaten, the function generates a new one and records its coordinate in the matrix.
- The head() function moves the head by operating the direction and records the moving in the matrix.
- The fail() function does failure checking. The value of no means none of failure and continue the game. The value of yes means failure and makes the player enter stage2.
- The body() function moves the body of the snake and records the moving in the matrix.
- The drawTable() function draws the background of the game.
- The drawMatrix() function draws the game matrix.

We have finished the basic object-oriented modification to the Snake game through the description of class diagrams and sequence diagrams according to the UML specifications. Before coding, let us think about how to extract the game framework.

6.3.2 Defining the Game Framework

It is difficult to design a complete, easy-to-use, and well extensible game framework. However, we can abstract step-by-step based on the Snake game. The process of abstraction is to objectify the

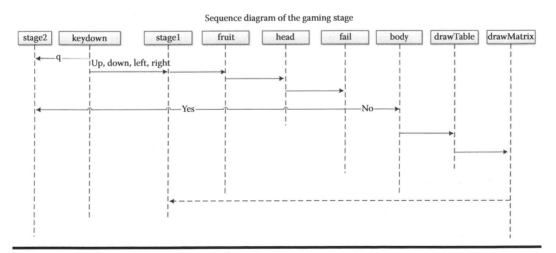

Figure 6.11 The function calling sequence in stage1.

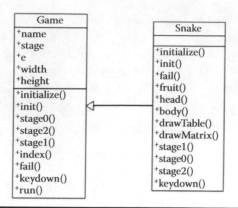

Figure 6.12 The class diagram of the Snake game.

program, which has finished in above. The second step is to encapsulate the common properties and methods. We can extract the common part into a parent class Game and then make the Snake class inherit it. Draw the Game class and the Snake class as shown in Figure 6.12.

- The common properties in the Game class include all the properties in the previous Snake class. This is because the properties are all global that will be used in other games and the variable e can be extended for the properties specialized for individual games.
- The common methods in the Game class include all the globally called methods of the Snake game, except for the methods that are called in stage1. The methods of Game mainly implement the auxiliary functionalities for development.
- Some of the methods in the Snake class inherit from the Game class. Here, the method overwriting is employed. The method in child class calls the identically named method in parent class and then executes the code in itself.

We have simply separated the framework class Game, and the implementation class Snake. What to do next is to implement the code according to the design to verify whether the design is reasonable.

6.3.3 Re-Implement the Snake Game under the Game Framework

The system environment used in this section:

- Win7 64bit
- R: 3.1.1 x86_64-w64-mingw32/x64 (64-bit)

Let us implement the Game class based on RC object-oriented programming (OOP) in R. Create a game.r file as follows:

```
Game<-setRefClass('Game',

    fields=list(
      # The system variable.
      # Name.
      name="character",
      # Debugging status.
```

```
  debug='logical',
  # Width of the matrix.
  width='numeric',
  # Height of the matrix.
  height='numeric',

  # The application variable.
  # Stage.
  stage='numeric',
  # Environment.
  e='environment',
  # Matrix.
  m='matrix',
  # Failure checking.
  isFail='logical'
),

methods=list(

  # The Constructor.
  initialize = function(name,width,height,debug) {
    name<<-"R Game Framework"
    debug<<-FALSE
    # Width of the matrix.
    width<<-height<<-20
  },

  # Initialize the variables.
  init = function(){
    # The environment.
    e<<-new.env()
    # The data matrix.
    m<<-matrix(rep(0,width*height),nrow=width)
    isFail<<-FALSE
  },

  # The starting stage.
  stage0=function(){
    stage<<-0
    init()
  },

  #The ending stage.
  stage2=function(){
    stage<<-2
  },

  # The gaming stage.
  stage1=function(default=FALSE){
    stage<<-1

    # The default gaming stage.
    if(FALSE){
      plot(0,0,xlim=c(0,1),ylim=c(0,1),type='n',xaxs="i", yaxs="i")
```

```
          text(0.5,0.7,label="Playing",cex=5)
        }
      },

      # The matrix utility.
      index = function(col) {
        return(which(m==col))
      },

      # Operations for failure.
      fail=function(msg){
        print(paste("Game Over",msg))
        isFail<<-TRUE
        keydown('q')
        return(NULL)
      },

      # The keyboard event for controlling stage switching.
      keydown=function(K){
        # The Starting stage.
        if(stage==0){
          stage1()
          return(NULL)
        }

        # The ending stage.
        if(stage==2){
          if(K=="q") q()
          else if(K==' ') stage0()
          return(NULL)
        }
      },

      # Launch the program.
      run=function(){
        par(mai=rep(0,4),oma=rep(0,4))
        stage0()
        getGraphicsEvent(prompt="Snake Game",onKeybd=function(K){
          if(debug) print(paste("keydown",K))
          return(keydown(K))
        })
      }
    )
)
```

Following is the implementation of the Snake class. It inherits the Game class and implements the private methods of the Snake game. Create a file called snake.r as following:

```
# Import the file game.r.
source(file="game.r")

# The Snake class inheriting the Game class.s
Snake<-setRefClass("Snake",contains="Game",
```

```
methods=list(

  # The constructor.
  initialize = function(name,width,height,debug) {
    # Call the method in its parent class.
    callSuper(name,width,height,debug)

    name<<-"Snake Game"
  },

  # Initialize the variables.
  init = function(){
    # Call the method in its parent class.
    callSuper()
    # The step.
    e$step<<-1/width
    # The moving direction.
    e$dir<<-e$lastd<<-'up'
    # The initial coordinate of the head.
    e$head<<-c(2,2)
    # The coordinates of the last location of the head.
    e$lastx<<-e$lasty<<-2
    # Initialize the coordinates of the tail.
    e$tail<<-data.frame(x=c(),y=c())

    # Color of the fruit.
    e$col_fruit<<-2
    # Color of the head.
    e$col_head<<-4
    # Color of the tail.
    e$col_tail<<-8
    # Color of the path.
    e$col_path<<-0
    # Color of the obstacles.
    e$col_barrier<<-1
  },

  # Failure checking.
  lose=function(){
    # The head is beyond the borders.
    if(length(which(e$head<1))>0 | length(which(e$head>width))>0){
      fail("Out of ledge.")
      return(NULL)
    }

    # The head conflicts with the tail.
    if(m[e$head[1],e$head[2]]==e$col_tail){
      fail("head hit tail.")
      return(NULL)
    }
  },

  # The random fruit block.
  fruit=function(){
```

```
    # There is no fruit.
    if(length(index(e$col_fruit))<=0){
      idx<-sample(index(e$col_path),1)

      fx<-ifelse(idx%%width==0,10,idx%%width)
      fy<-ceiling(idx/height)
      m[fx,fy]<<-e$col_fruit

      if(debug){
        print(paste("fruit idx",idx))
        print(paste("fruit axis:",fx,fy))
      }
    }
},

# Snake head.
head=function(){
  e$lastx<<-e$head[1]
  e$lasty<<-e$head[2]

  # The operations for moving direction.
  if(e$dir=='up')    e$head[2]<<-e$head[2]+1
  if(e$dir=='down')  e$head[2]<<-e$head[2]-1
  if(e$dir=='left')  e$head[1]<<-e$head[1]-1
  if(e$dir=='right') e$head[1]<<-e$head[1]+1
},

# Snake body.
body=function(){
  if(isFail) return(NULL)

  m[e$lastx,e$lasty]<<-e$col_path
  m[e$head[1],e$head[2]]<<-e$col_head
  # There is no fruit.
  if(length(index(e$col_fruit))<=0){
    e$tail<<-rbind(e$tail,data.frame(x=e$lastx,y=e$lasty))
  }

  # If the tail exists.
  if(nrow(e$tail)>0) {
    e$tail<<-rbind(e$tail,data.frame(x=e$lastx,y=e$lasty))
    m[e$tail[1,]$x,e$tail[1,]$y]<<-e$col_path
    e$tail<<-e$tail[-1,]
    m[e$lastx,e$lasty]<<-e$col_tail
  }

  if(debug){
    print(paste("snake idx",index(e$col_head)))
    print(paste("snake axis:",e$head[1],e$head[2]))
  }
},

# The canvas background.
drawTable=function(){
  if(isFail) return(NULL)
```

```
      plot(0,0,xlim=c(0,1),ylim=c(0,1),type='n',xaxs="i", yaxs="i")

      if(debug){
        # Show the background grid.
        # The horizontal lines.
        abline(h=seq(0,1,e$step),col="gray60")
        # The vertical lines.
        abline(v=seq(0,1,e$step),col="gray60")
        # Show the matrix.
        df<-data.frame(x=rep(seq(0,0.95,e$step),width),y=rep(seq(0,0.95,
e$step),each=height),lab=seq(1,width*height))
        text(df$x+e$step/2,df$y+e$step/2,label=df$lab)
      }
    },

    # Draw the data according to the matrix.
    drawMatrix=function(){
      if(isFail) return(NULL)

      idx<-which(m>0)
      px<- (ifelse(idx%%width==0,width,idx%%width)-1)/width+e$step/2
      py<- (ceiling(idx/height)-1)/height+e$step/2
      pxy<-data.frame(x=px,y=py,col=m[idx])
      points(pxy$x,pxy$y,col=pxy$col,pch=15,cex=4.4)
    },

    # The gaming stage.
    stage1=function(){
      callSuper()

      fruit()
      head()
      lose()
      body()
      drawTable()
      drawMatrix()
    },

    # The starting stage.
    stage0=function(){
      callSuper()
      plot(0,0,xlim=c(0,1),ylim=c(0,1),type='n',xaxs="i", yaxs="i")
      text(0.5,0.7,label=name,cex=5)
      text(0.5,0.4,label="Any keyboard to start",cex=2,col=4)
      text(0.5,0.3,label="Up,Down,Left,Rigth to control
direction",cex=2,col=2)
      text(0.2,0.05,label="Author:DanZhang",cex=1)
      text(0.5,0.05,label="http://blog.fens.me",cex=1)
    },

    # The ending stage.
    stage2=function(){
      callSuper()
      info<-paste("Congratulations! You have eat",nrow(e$tail),"fruits!")
```

```
      print(info)

      plot(0,0,xlim=c(0,1),ylim=c(0,1),type='n',xaxs="i", yaxs="i")
      text(0.5,0.7,label="Game Over",cex=5)
      text(0.5,0.4,label="Space to restart, q to quit.",cex=2,col=4)
      text(0.5,0.3,label=info,cex=2,col=2)
      text(0.2,0.05,label="Author:DanZhang",cex=1)
      text(0.5,0.05,label="http://blog.fens.me",cex=1)
    },

    # The keyboard event to control the stage switching.
    keydown=function(K){
      callSuper(K)
      # The gaming stage.
      if(stage==1){
        if(K == "q") stage2()
        else {
          if(tolower(K) %in% c("up","down","left","right")){
            e$lastd<<-e$dir
            e$dir<<-tolower(K)
            stage1()
          }
        }
        return(NULL)
      }
      return(NULL)
    }
  )
)

snake<-function(){
  game<-Snake$new()
  game$initFields(debug=TRUE)
  game$run()
}
snake()
```

Finally, let us run the program in snake.r to finish the Snake game. The running results were shown in Figure 6.8 in Section 6.8.

We have finished the development of the Snake game, although the visual effect is somehow low and there is no operation on time dimension. From another perspective, is R a better choice for the games that neither require the high graphic effective nor need time dimension? Of course, there are lots of such games, for example, 2048, the popular recent game. So, let us develop a 2048 game using the game framework we just developed. Try to implement it within 150 lines of code.

6.4 Developing the 2048 Game Using R

Question

How do we implement the 2048 game in R?

Introduction

Which language can implement the 2048 game with just 150 lines of code? The answer is R. Although R is not suitable for game developing; its vectored calculation can dramatically reduce the complexity of matrix algorithm implementations and it therefore can take the calculations efficiently. If we could transform a game problem to a mathematical one, then R is the best tool to solve it.

6.4.1 Introduction to the 2048 Game

The 2048 game is a single-player online mobile game developed by 19-year-old Italian Gabriele Cirulli in March 2014. The task in the game is to slide the blocks to combine them until a tile with the number 2048 is formed. It is a computer variant of sliding games. The author developed the game in order to see if he could create a game from scratch. But he was totally surprised by the rapid growth of web visitors (over four million in less than a week). In fact, the game has been rated as "the most addictive thing" on the Internet.

The Wall Street Journal described the game as "almost like Candy Crush for math geeks" (Ballard 2014). The screenshot of the game is shown in Figure 6.13.

The game is open source software, and hence derives many improvements and variances, including ones with score leaderboard, improved touchscreen playability, and so on. The game is a JavaScript app based on HTML5. The source code is available at https://github.com/gabrielecirulli/2048. And an on-line version for free is available at http://gabrielecirulli.github.io/2048/. The R implementation in this section is totally from my idea and not related to the JS code of the game's author.

In the game, the player uses direction keys to move the tiles. If two tiles with same numbers conflict, they are combined into one tile with a number as the sum of their numbers. In every moving, a new tile with number 2 or 4 appears. When the tile with number 2048 appears, the player wins the game, which is the reason why this game is called 2048.

6.4.2 Stage Design

Next, let us get back to the game design as in Section 6.2. How do we start to develop the 2048 game? We need to do requirement analysis for the game, list the game rules, design business flow, and present the prototype of the game to validate the feasibility of the design.

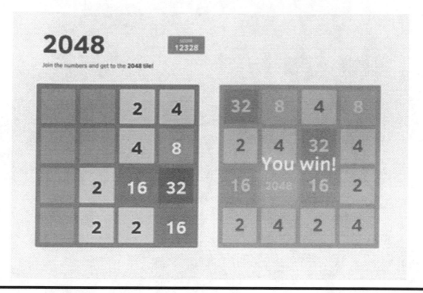

Figure 6.13   The screenshot of the 2048 game.

6.4.2.1 Requirement Analysis

The 2048 game has three stages: starting, gaming, and ending.

- Starting stage: launches the application, warms up the player, and provides how-to before gaming.
- Gaming stage: the stage when the game is in process.
- Ending stage: the stage when the player wins, loses, or exits, providing the scores gained by the player.

The stages for starting and ending are simple so I won't explain more. The gaming stage contains a 4 by 4 canvas. Each tile corresponds to a number. The tiles with number >0 are filled with colors.

6.4.2.2 The Game Rules

The rules when the game is in process are listed in detail as follows:

- After launching the game, the player can press the up, down, left, and right keys to control the moving of the numbers on the canvas.
- If two equal numbers conflict after moving, they merge into one with a number as the sum of the original two numbers.
- With each moving, a new tile with number of 2 or 4 appears on a blank block.
- After the player presses a key, if the tile locations don't change, no new number is generated and the key pressing operation is invalid.
- If all the blocks are filled with numbers and no two numbers can merge, the player fails the game.

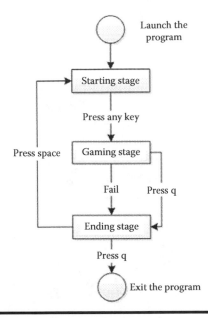

Figure 6.14   Definition of the business flow of the 2048 game.

6.4.2.3 The Business Flow

The entire game can be divided into three stages. The switching flow of the game stages is shown in Figure 6.14.

- When launching the game, the player sees the starting stage and then presses any key to enter the gaming stage.
- The gaming stage is transited to the ending stage when the player loses the game. When pressing q, the player is directly brought to the ending stage.
- In the ending stage, the player is back to the starting stage by pressing space, or exits the game by pressing q.

The business flow of the 2048 game is same as that of the Snake game.

6.4.2.4 Prototype of the Game

Let us draw the interfaces of the three stages as the game prototype before game development. In Figure 6.15, the left, middle, and right pictures show the starting, gaming, and ending stages, respectively.

Draw all the three stages programmatically according to the prototype pictures.

6.4.3 Program Design

Now we have clearly understood the rules and functional requirements of the 2048 game. Next, we need to technically describe the business logic in requirement analysis and take the nonfunctional requirements and technical details about R into account.

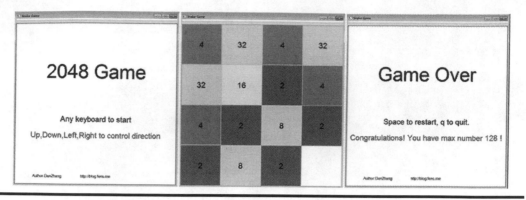

Figure 6.15   The stages of the 2048 game.

6.4.3.1 The Game Stages

We map each stage to a canvas, or a data structure in memory.

- The starting stage is static, so we can generate the corresponding canvas in memory in advance, or generate it dynamically only when switching stages. The performance cost is low.
- The gaming stage is dynamic. The canvas should be regenerated through the binding events each time the interoperation or periodical refresh occurs.
- The ending stage is dynamic, showing the resulting score of the game, so it is generated temporarily when the game is switched to the stage.

6.4.3.2 The Game Objects

The game may generate a lot of objects when running mentioned above. These objects need to be defined in memory and mapped to R's data types.

Compared to the Snake game, the 2048 game is much simpler. It is OK for us to just define a canvas object.

- Canvas: described with matrix
- The numbers in the canvas: represented by the numeric values in the matrix
- The background colors of the canvas: represented by the numeric values in the matrix

The canvas object described by matrix is shown in Figure 6.16.

```
     [,1] [,2] [,3] [,4]
[1,]    4   32    4   32
[2,]   32   16    2    4
[3,]    4    2    8    2
[4,]    2    8    2    0
```

The corresponding canvas is shown in Figure 6.16.

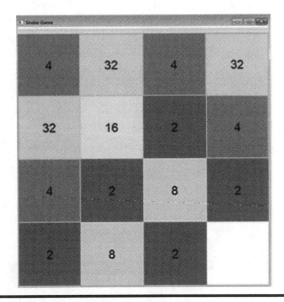

Figure 6.16   The canvas of the 2048 game.

6.4.3.3 The Game Events

There are two events in the game: the keyboard event and collision event.

- The keyboard event: a global event triggered by user input through keyboard, for example, pressing up, down, left, or right to control the merging direction of the numbers.
- The collision event: if two equal numbers conflict when moving, they merge into one number.

Listen to the keyboard events globally. Use the keyboard events to trigger the collision events and check the game status.

6.4.3.4 The Game Control

We need to control all the statuses that occur when the game is running, for example, when to generate a new number, when to merge equal numbers, when to end the game, and so on. By defining the control functions, we can easily manage all the game statuses. The entire control flow of the game is shown in Figure 6.17.

Each rectangle in Figure 6.17 represents an R function definition.

- run(): function to launch the program
- keydown(): function to listen the keyboard events, locking the thread
- stage0(): function to create the starting stage with visualization output
- stage1(): function to create the gaming stage with visualization output
- stage2(): function to create the ending stage with visualization output
- init(): function to initialize game variables when starting game
- create(): check and generate numbers
- move(): move the numbers

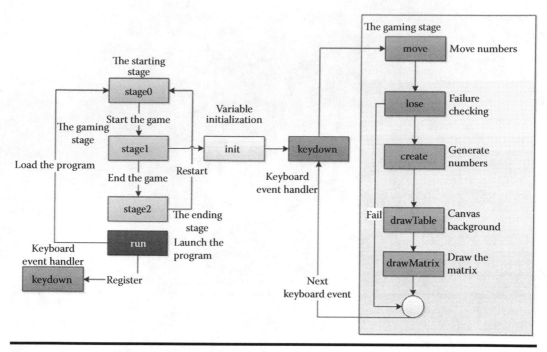

Figure 6.17 The control flow of the 2048 game.

- lose(): failure checking. If all the tiles in the canvas are filled by numbers and no numbers can be merged, the game jumps into the ending stage.
- drawTable(): function to draw the background.
- drawMatrix(): function to draw the matrix.

Through the detailed program design, we have converted the business description in requirement analysis to technical description in program developing. After completing the design, the only thing left is coding.

6.4.4 The R Implementation

We just need to fill the code into the functions defined above just as in doing gap filling. We have already developed a game framework, where parts of the stage functions and feature functions have been implemented. So, we just need to fill the code for game logics. For information about R game framework, please refer to Section 6.3.

The system environment used in this section:

- Win7 64bit
- R: 3.1.1 x86_64-w64-mingw32/x64 (64-bit)

6.4.4.1 Number Moving Function Move()

The most complex operation about the algorithm in the 2048 game is number moving. In the 4*4 matrix, the numbers move in the direction of up, down, left, or right. Pairs of equal numbers merge if they conflict after moving. This is the core algorithm of the 2048 game. We need to make sure our program is correct for number merging.

Let us isolate the function from the whole game program and implement and test against it. Build the function moveFun(). Here, the process of moving is simplified, taking only left–right moving in account. Later, we will do a rotating calculation to make the core algorithm of up–down moving share the same code as that of left–right moving.

```
# The moving function.
> moveFun<-function(x,dir){
+    # Revert the list if moving right.
+    if(dir == 'right') x<-rev(x)
+
+    # Length of 0.
+    len0<-length(which(x==0))
+    # Drop 0's.
+    x1<-x[which(x>0)]
+    pos1<-which(diff(x1)==0)
+
+    # 3 indices.
+    if(length(pos1)==3){
+      pos1<-pos1[c(1,3)]
+    # 2 indices.
+    }else if(length(pos1)==2 && diff(pos1)==1){
+      pos1<-pos1[1]
+    }
+
+    x1[pos1]<-x1[pos1]*2
+    x1[pos1+1]<-0
+
+    # Drop 0's.
+    x1<-x1[which(x1>0)]
+    # Supplement with 4 0's.
+    x1<-c(x1,rep(0,4))[1:4]
+
+    if(dir == 'right') x1<-rev(x1)
+    return(x1)
+ }
```

Next, we take the unit testing package test that to verify the correction of the function move-Fun(). For information about the testthat package, please refer to Section 5.2.

Let us simulate moving the numbers left or right according to the game rules to test whether the calculation result is equal to the expectation value, as shown in Figure 6.18.

The unit testing code in R is as follows:

```
> library(testthat)
> x<-c(4,2,2,2)
> expect_that(moveFun(x,'left'), equals(c(4,4,2,0)))
> expect_that(moveFun(x,'right'), equals(c(0,4,2,4)))

> x<-c(4,4,2,4)
> expect_that(moveFun(x,'left'), equals(c(8,2,4,0)))
> expect_that(moveFun(x,'right'), equals(c(0,8,2,4)))
```

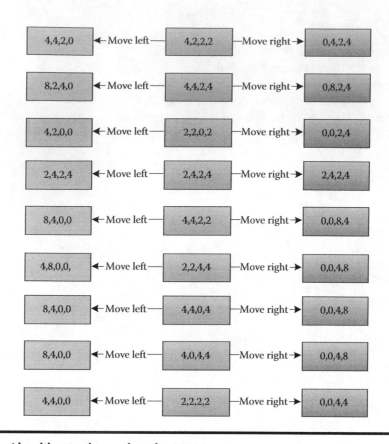

Figure 6.18 Algorithm testing against the 2048 game.

```
> x<-c(2,2,0,2)
> expect_that(moveFun(x,'left'), equals(c(4,2,0,0)))
> expect_that(moveFun(x,'right'), equals(c(0,0,2,4)))

> x<-c(2,4,2,4)
> expect_that(moveFun(x,'left'), equals(c(2,4,2,4)))
> expect_that(moveFun(x,'right'), equals(c(2,4,2,4)))

> x<-c(4,4,2,2)
> expect_that(moveFun(x,'left'), equals(c(8,4,0,0)))
> expect_that(moveFun(x,'right'), equals(c(0,0,8,4)))

> x<-c(2,2,4,4)
> expect_that(moveFun(x,'left'), equals(c(4,8,0,0)))
> expect_that(moveFun(x,'right'), equals(c(0,0,4,8)))

> x<-c(4,4,0,4)
> expect_that(moveFun(x,'left'), equals(c(8,4,0,0)))
> expect_that(moveFun(x,'right'), equals(c(0,0,4,8)))

> x<-c(4,0,4,4)
> expect_that(moveFun(x,'left'), equals(c(8,4,0,0)))
> expect_that(moveFun(x,'right'), equals(c(0,0,4,8)))
```

```
> x<-c(4,0,4,2)
> expect_that(moveFun(x,'left'), equals(c(8,2,0,0)))
> expect_that(moveFun(x,'right'), equals(c(0,0,8,2)))

> x<-c(2,2,2,2)
> expect_that(moveFun(x,'left'), equals(c(4,4,0,0)))
> expect_that(moveFun(x,'right'), equals(c(0,0,4,4)))

> x<-c(2,2,2,0)
> expect_that(moveFun(x,'left'), equals(c(4,2,0,0)))
> expect_that(moveFun(x,'right'), equals(c(0,0,2,4)))
```

Of course, we can code more test cases to verify the function works correctly. All the unit test cases get passed. Therefore, we have implemented the core algorithm of number moving.

6.4.4.2 Other Functions

Next, let us implement other functions, with similar code structure to the Snake game.

The starting stage function stage0().

```
# The starting stage.
stage0=function(){
    callSuper()
    plot(0,0,xlim=c(0,1),ylim=c(0,1),type='n',xaxs="i", yaxs="i")
    text(0.5,0.7,label=name,cex=5)
    text(0.5,0.4,label="Any keyboard to start",cex=2,col=4)
    text(0.5,0.3,label="Up,Down,Left,Rigth to control
direction",cex=2,col=2)
    text(0.2,0.05,label="Author:DanZhang",cex=1)
    text(0.5,0.05,label="http://blog.fens.me",cex=1)
}
```

The ending stage function stage2().

```
stage2=function(){
    callSuper()
    info<-paste("Congratulations! You have max number",max(m),"!")
    print(info)

    plot(0,0,xlim=c(0,1),ylim=c(0,1),type='n',xaxs="i", yaxs="i")
    text(0.5,0.7,label="Game Over",cex=5)
    text(0.5,0.4,label="Space to restart, q to quit.",cex=2,col=4)
    text(0.5,0.3,label=info,cex=2,col=2)
    text(0.2,0.05,label="Author:DanZhang",cex=1)
    text(0.5,0.05,label="http://blog.fens.me",cex=1)
}
```

Keyboard event handler for controlling stage switching.

```
keydown=function(K){
    callSuper(K)

    # The gaming stage.
    if(stage==1){
        if(K == "q") stage2()
```

```
      else {
        if(tolower(K) %in% c("up","down","left","right")){
          e$dir<<-tolower(K)
          print(e$dir)
          stage1()
        }
      }
      return(NULL)
    }
    return(NULL)
}
```

The stage initialization function init().

```
 # Initialize the variables.
 init = function(){
   # Call its parent method.
   callSuper()

   # The maximum number.
   e$max<<-4
   # The step.
   e$step<<-1/width
   e$dir<<-'up'
   # The colors.
   e$colors<<-rainbow(14)
   # Moving conditions not satisfied.
   e$stop<<-FALSE

   create()
 }
```

create(), the function to generate a new number randomly.

```
# Generate  a new number randomly.
create=function(){
   if(length(index(0))>0 & !e$stop){
     e$stop<<-TRUE
     one<-sample(c(2,4),1)
     idx<-ifelse(length(index(0))==1,index(0),sample(index(0),1))
     m[idx]<<-one
   }
}
```

lose(), the function to check failure conditions.

```
lose=function(){

   # Check if there are sibling duplicate numbers.
   near<-function(x){
     length(which(diff(x)==0))
   }
```

```
    # If there is no blank tile.
    if(length(index(0))==0){
      # The horizontal direction.
      h<-apply(m,1,near)
      # The vertical direction.
      v<-apply(m,2,near)

      if(length(which(h>0))==0 & length(which(v>0))==0){
        fail("No free grid.")
        return(NULL)
      }
    }
}
```

drawTable(): function to draw the game canvas.

```
drawTable=function(){
    if(isFail) return(NULL)
    plot(0,0,xlim=c(0,1),ylim=c(0,1),type='n',xaxs="i", yaxs="i")
    # The horizontal lines.
    abline(h=seq(0,1,e$step),col="gray60")
    # The vertical lines.
    abline(v=seq(0,1,e$step),col="gray60")
}
```

drawMatrix: function to draw the game matrix.

```
# Draw the matrix data.
drawMatrix=function(){
    if(isFail) return(NULL)
    a<-c(t(m))
    lab<-c(a[13:16],a[9:12],a[5:8],a[1:4])

    d<-data.frame(x=rep(seq(0,0.95,e$step),width),y=rep(seq(0,0.95,e$step),
each=height),lab=lab)
    df<-d[which(d$lab>0),]
    points(df$x+e$step/2,df$y+e$step/2,col=e$colors[log(df$lab,2)],pch=15,
cex=23)
    text(df$x+e$step/2,df$y+e$step/2,label=df$lab,cex=2)
}
```

The gaming stage function stage1().

```
stage1=function(){
    callSuper()

    move()
    lose()
    create()

    drawTable()
    drawMatrix()
}
```

The whole program code is as following, stored in the file 2048.r.

```r
# Load the game framework.
source(file="game.r")

# The G2048 class, inheriting the Game class.
G2048<-setRefClass("G2048",contains="Game",

    methods=list(

      # The constructor.
      initialize = function(name,debug) {
        # Call its parent method.
        callSuper(name,debug)

        name<<-"2048 Game"
        width<<-height<<-4
      },

      # Initialize the variables.
      init = function(){
        # Call its parent method.
        callSuper()
        # The maximum number.
        e$max<<-4
        # The Step.
        e$step<<-1/width
        e$dir<<-'up'
        #The colors.
        e$colors<<-rainbow(14)
        # Moving conditions not satisfied.
        e$stop<<-FALSE

        create()
      },

      # Generate a new number randomly.
      create=function(){
        if(length(index(0))>0 & !e$stop){
          e$stop<<-TRUE
          one<-sample(c(2,4),1)
          idx<-ifelse(length(index(0))==1,index(0),sample(index(0),1))
          m[idx]<<-one
        }
      },

      # Failure conditions.
      lose=function(){

        # Check if there are sibling duplicate numbers.
        near<-function(x){
          length(which(diff(x)==0))
        }
```

```
  # If there is no blank tile ....
  if(length(index(0))==0){
    # The horizontal direction.
    h<-apply(m,1,near)
    # The vertical direction.
    v<-apply(m,2,near)

    if(length(which(h>0))==0 & length(which(v>0))==0){
      fail("No free grid.")
      return(NULL)
    }
  }
},

# Direction moving.
move=function(){

  # The direction moving function.
  moveFun=function(x){
    if(e$dir %in% c('right','down')) x<-rev(x)

    # Length of 0's.
    len0<-length(which(x==0))
    # Drop 0's.
    x1<-x[which(x>0)]
    # Find the locations of the sibling equal elements.
    pos1<-which(diff(x1)==0)

    # 3 indices.
    if(length(pos1)==3){
      pos1<-pos1[c(1,3)]
    # 2 indices.
    }else if(length(pos1)==2 && diff(pos1)==1){
      pos1<-pos1[1]
    }

    x1[pos1]<-x1[pos1]*2
    x1[pos1+1]<-0

    # Drop 0's.
    x1<-x1[which(x1>0)]
    # Supplement with 4 0's.
    x1<-c(x1,rep(0,4))[1:4]

    if(e$dir %in% c('right','down')) x1<-rev(x1)
    return(x1)
  }

  last_m<-m
  if(e$dir=='left')  m<<-t(apply(m,1,moveFun))
  if(e$dir=='right') m<<-t(apply(m,1,moveFun))
  if(e$dir=='up')    m<<-apply(m,2,moveFun)
  if(e$dir=='down')  m<<-apply(m,2,moveFun)
```

```r
      e$stop<<-ifelse(length(which(m != last_m))==0,TRUE,FALSE)
    },

    # The canvas background.
    drawTable=function(){
      if(isFail) return(NULL)
      plot(0,0,xlim=c(0,1),ylim=c(0,1),type='n',xaxs="i", yaxs="i")
      # The horizontal lines.
      abline(h=seq(0,1,e$step),col="gray60")
      # The vertical lines.
      abline(v=seq(0,1,e$step),col="gray60")
    },

    # Draw the matrix data.
    drawMatrix=function(){
      if(isFail) return(NULL)
      a<-c(t(m))
      lab<-c(a[13:16],a[9:12],a[5:8],a[1:4])

      d<-data.frame(x=rep(seq(0,0.95,e$step),width),y=rep(seq(0,0.95,e$s
tep),each=height),lab=lab)
      df<-d[which(d$lab>0),]
      points(df$x+e$step/2,df$y+e$step/2,col=e$colors[log(df$lab,2)],pch
=15,cex=23)
      text(df$x+e$step/2,df$y+e$step/2,label=df$lab,cex=2)
    },

    stage1=function(){
      callSuper()

      move()
      lose()
      create()

      drawTable()
      drawMatrix()
    },

    # The starting stage.
    stage0=function(){
      callSuper()
      plot(0,0,xlim=c(0,1),ylim=c(0,1),type='n',xaxs="i", yaxs="i")
      text(0.5,0.7,label=name,cex=5)
      text(0.5,0.4,label="Any keyboard to start",cex=2,col=4)
      text(0.5,0.3,label="Up,Down,Left,Rigth to control
direction",cex=2,col=2)
      text(0.2,0.05,label="Author:DanZhang",cex=1)
      text(0.5,0.05,label="http://blog.fens.me",cex=1)
    },

    # The ending stage.
    stage2=function(){
      callSuper()
      info<-paste("Congratulations! You have max number",max(m),"!")
      print(info)
```

```
      plot(0,0,xlim=c(0,1),ylim=c(0,1),type='n',xaxs="i", yaxs="i")
      text(0.5,0.7,label="Game Over",cex=5)
      text(0.5,0.4,label="Space to restart, q to quit.",cex=2,col=4)
      text(0.5,0.3,label=info,cex=2,col=2)
      text(0.2,0.05,label="Author:DanZhang",cex=1)
      text(0.5,0.05,label="http://blog.fens.me",cex=1)
    },
```

\# Keyboard event handler for controlling stage switching.

```
keydown=function(K){
      callSuper(K)

      # The gaming stage.
      if(stage==1){
        if(K == "q") stage2()
        else {
          if(tolower(K) %in% c("up","down","left","right")){
            e$dir<<-tolower(K)
            stage1()
          }
        }
        return(NULL)
      }
      return(NULL)
    }
  )
)

# Encapsulate the launching function.
g2048<-function(){
  game<-G2048$new()
  game$initFields(debug=TRUE)
  game$run()
}

# Launch the game.
g2048()
```

The screenshot of the game is shown in Figure 6.19.

Here it is! We have implemented the 2048 game. There are only 190 lines of code for the whole program, with only about 150 lines of effective code. It is very convenient to process the vectored calculation of matrix in R. In addition, we have encapsulated the game program with the object-oriented methodology and standardized the function definition and interfaces, which allow us putting more focus on the game algorithms to promote the productivity. Going a step further, let us pack and publish the game framework on to CRAN, the official R library.

6.5 Publishing the gridgame Package

Question

How do we publish our own R package?

2048 Game

Any keyboard to start

Up,Down,Left,Right to control direction

Author:DanZhang   http://blog.fens.me

Figure 6.19   The screenshot of the 2048 game.

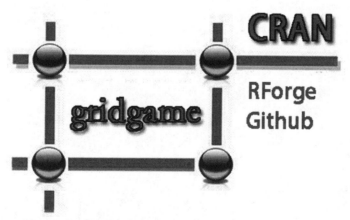

Publishing the gridgame Package
http://blog.fens.me/r-game-gridgame/

Introduction

We have prepared a lot of basic technologies for publishing our own package, including object-oriented programming (OOP) in R, R package development, R game programming, and so on. Finally, it is the time for us to use these technologies comprehensively. The gridgame package is created by linking all the code and documentation together according to CRAN requirements for publishing R packages.

6.5.1 Knowledge Reserves

Before integrating all the knowledge, let us review what aspects of knowledge we should master. (1) The R programming foundation is necessary. (2) The game algorithms involve mainly matrix calculation so we should better master the knowledge of linear algebra. (3) The graphic interface is necessary for game operation, so we need to have the capability to draw the graphic interface with code, regardless of what it would look like. (4) The necessity for game framework. For games of same type, if the second can reuse the structure of the first one, we can not only save the time cost, but also reduce the technical requirements for game development. This is why we need to encapsulate the game framework with the way of object-oriented programming.

6.5.1.1 The Basic Techniques

I have prepared for all four aspects. There are respective sections for solving the different problems:

- Object-oriented programming in R: The Reference Class(RC)-based Object-Oriented Programming in R (Section 4.4)
- The environments in R: Uncovering the Mystery of R Environments (Section 3.2); Revealing the Function Environments of R (Section 3.3)
- Game programming in R: Keyboard and Mouse Events in R (Section 6.1); Getting Started with the Snake Game (Section 6.2); Game Framework Design in R (Section 6.3); Developing the 2048 Game in R (Section 6.4)
- R package development: Developing Our Own R Package from Scratch (Section 5.1); Streamlining R Package Development (Section 5.2)

6.5.1.2 Naming the R Package

Since all the technologies are ready, then name the project.

In fact, giving a suitable name is not easy. Although there is no package named game on CRAN, it is not a good idea to simply name our project as game. Let us ask Google for a name that has relatively fewer search result.

There are about 1,470,000,000 Google results for "game," and 411,000,000 for "r game," shown in Figure 6.20. I failed to count the 0's! It would be an unbridgeable gap for our marketing efforts to compete for keywords with so many search results.

Therefore, we need to take another name without such high heat as the project name, making our package grow up. gridgame is such a great name that indicates the characteristics of our game and rids us of powerful marketing competitors. The pressure on our back is blown away and we feel so easy. The Google search results are shown in Figure 6.21.

All the preparations are done. It's time for us to get done with the gridgame package!

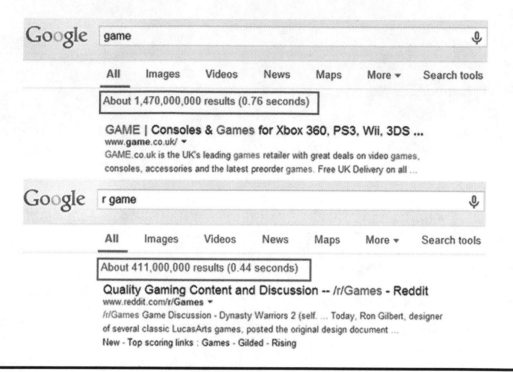

Figure 6.20 The game keyword.

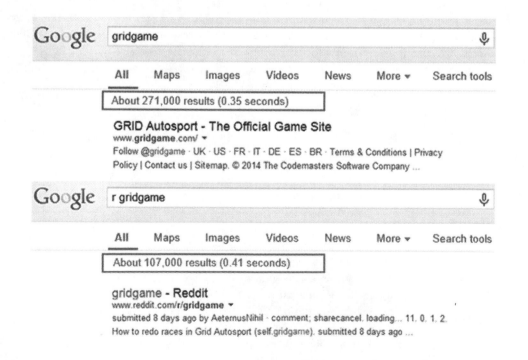

Figure 6.21 The gridgame keywords.

6.5.2 *Developing the gridgame Package*

Let us develop the gridgame package following the standardized developing flow introduced in Section 5.2.

The system environment used in this section:

- Win7 64bit
- R: 3.1.1 x86_64-w64-mingw32/x64 (64-bit)
- Rtools31.exe
- basic-miktex-2.9.5105.exe

We choose the Windows platform as the developing environment for R packages since the games we developed in previous sections are based on the Windows platform. I will explain why Linux is not supported later. It would be better to upgrade R to version 3.1.1. (There are lots of issues on the road to solve one by one!)

6.5.2.1 *Creating the Project*

In order to reduce the cost and promote productivity, I will operate following the developing process of the devtools package.

Open a project directory and create the skeleton of the project.

```
# Launch the R application.
~ R
# Switch into the working directory.
> setwd("D:/workspace/R/app")
# Load the 3 developing toolkits.
> library(devtools)
> library(roxygen2)
Loading required package: digest
> library(testthat)

# Create the skeleton of the project.
> create(paste(getwd(),"/gridgame",sep=""))
Creating package gridgame in D:/workspace/R/app
No DESCRIPTION found. Creating default:

Package: gridgame
Title:
Description:
Version: 0.1
Authors@R: # getOptions('devtools.desc.author')
Depends: R (>= 3.0.1)
License: # getOptions('devtools.desc.license')
LazyData: true

# Re-set the current directory.
> setwd(paste(getwd(),"/gridgame",sep=""))
# View the files in the project skeleton.
> dir(full.names=TRUE)
[1] "./DESCRIPTION" "./man"        "./R"
```

6.5.2.2 Authoring Code and Documents

First, edit the DESCRIPTION file by adding the description of the project.

```
~ notepad D:/workspace/R/app/gridgame/DESCRIPTION

Package: gridgame
Type: Package
Title: A game framework for R
Version: 0.0.1
Date: 2014-07-23
Authors@R: c(person("Dan", "Zhang", email = "bsspirit@gmail.com",
role=c("aut", "cre")))
Maintainer: Dan Zhang
Description: This package provides a general-purpose game framework for
grid game in R. The package includes 2 games about Snake and 2048. You
can run the function snake() or g2048() to startup the game. These games
are run only on the Windows platform.
Depends:
    R (>= 3.0.1)
Imports:
    methods
License: GPL-3
URL: http://onbook.me/project/gridgame
BugReports: https://github.com/bsspirit/gridgame/issues
OS_type: windows
Collate:
    'game.R'
    '2048.R'
    'package.R'
    'snake.R'
```

In the DESCRIPTION file, there are two items for attention:

- Imports: methods. Since we use RC types to develop, the system will parse the code using functions in the methods package by default. It is necessary to explicitly declare here.
- Our package only supports running on Windows, so we need to declare the OS to support for reviewing on CRAN. In the meantime, the process of R CMD check on Linux would fail due to this attribute, so we have to pack and publish our package on Windows.

Secondly, copy the three files we have finished to D:/workspace/R/app/gridgame/R.

```
~ dir
2015/02/02   15:49              4,779 2048.R
2015/02/02   15:49              2,174 game.R
2016/02/02   17:49              4,766 snake.R
```

We need to adjust the code according to the errors got from previous checking attempts.

- Remove the references to game.R in the source() function in 2048.R and snake.R.
- Remove all Chinese comments from code, leaving only ASCII-supported characters.
- Add Windows platform checking to code. Running on non-Windows platforms are disabled. Check the running OS by Platform$OS.type=='windows'.

- Add package.R file to load the configuration of methods package, #' @import methods.
- Remove the launching functions from code. Hand over the launching operation to user.

6.5.2.3 Debugging Program

Load the package by calling the load_all() function on Windows, and then run the snake function of the g2048 function. Everything goes well.

```
# Load the package.
> load_all(getwd())
Loading gridgame

# Launch the Snake game.
> snake()
# Launch the 2048 game.
> g2048()
```

The game programs we wrote are developed on Windows7 and the running is OK. Why is running on Linux is not supported? The main reason is that Linux has different graphics device from Windows. On Windows, the graphics device is supported by .net framework, while on Linux the graphics device is supported by X11() display driver or by the third part TK device loaded by Linux which is called by R through the tkrplot package.

```
~ sudo apt-get install tk-dev
> install.packages("tkrplot")
```

Windows and Linux have differences for running GUI programs. We cannot use same set of code to implement. Some would say that it can be implemented by adding OS-type checking conditions everywhere necessary.

The situation is not so simple. In addition to the issue of the graphics devices, font issues would be encountered on Linux. The error for failure to load characters would be the result of a lack of a font library. The following error prompts that there is no Helvetica font with size of 60.

```
Error in text.default(0.5, 0.7, label = name, cex = 5):
  X11 font -adobe-helvetica-%s-%s-*-*-%d-*-*-*-*-*-*-*, face 1 at size 60
could not be loaded
```

I have tried installing all font libraries on Linux Unbuntu but still cannot solve the issue of loading large-sized font. So, I decided not to support Linux for now. I will try again if I have time later.

The command to install font library on Linux Ubuntu is as follows:

```
~ sudo apt-get install xfont-100dpi xfont-75dpi xfont-cyrillic xfont-*
```

6.5.2.4 Unit Testing

Create unit testing classes for different files, respectively, under inst/test.

- test-game.R, unit testing against the functions in game.R
- test-snake.R, unit testing against the functions in snake.R
- test-2048.R, unit testing against the functions in 2048.R

Taking test-game.R as example, open test-game.R.

```
~ notepad inst/test/test-game.R

context("game")

test_that("Initial the construct function of Game class", {
  name<-"R"
  width<-height<-10

  game<-Game$new()
  game$initFields(name=name,width=width,height=height)
  expect_that(game$name,equals(name))
  expect_that(game$width,equals(width))
  expect_that(game$height,equals(height))
})
```

Execute the unit testing code.

```
> test(getwd())
Testing gridgame
Loading gridgame
game : ...
```

6.5.2.5 *Authoring Documents*

One of the most complex steps in R package developing is authoring documents in LaTex format. Fortunately, the roxygen2 package was invented to help us generate LaTex documents by simply adding comments to code.

We generate the man/*.Rd documentation files using the roxygen2 package. For an RC-based program, we just need to add comments to the class definition functions. There is no mandatory checking for the methods in RC classes, so the comment amount of the methods depends on the developer's mood. For S3-based program or packages with only functions, the work of document authoring can be tough. Besides, no Chinese character exists in the documents, otherwise there would be warnings during the checking.

Take the file snake.R as example. We just need to comment setRefClass and snake<-function() {} and omit the methods in the Snake class. I will add more documentation if I have time later. It is not a bad thing to write more.

```
~ notepad snake.R

#' Snake game class
#' @include game.R
Snake<-setRefClass("Snake",contains="Game",
  ...
)

#' Snake game function
#'
#' @export
snake<-function(){
  game<-Snake$new()
```

```
  game$initFields()
  game$run()
}
```

Generate the LaTex file from the comments in code.

```
> document(getwd())
Updating gridgame documentation
Loading gridgame
Writing G2048-class.Rd
Writing g2048.Rd
Writing Game-class.Rd
Writing Snake-class.Rd
Writing snake.Rd
```

Open snake.Rd and view the content generated.

```
~ notepad man/snake.Rd

% Generated by roxygen2 (4.0.1): do not edit by hand
\name{snake}
\alias{snake}
\title{Snake game function}
\usage{
snake()
}
\description{
Snake game function
}
```

This step did not go smoothly. When importing the roxygen2 package, I encountered the issue where roxygen2 depends on R version 3.0.2 and it later fails to be installed on R 3.0.1. There is a description to this issue of strong dependency on Github. See https://github.com/klutometis/roxygen/issues/163. So, recommendation of R 3.1.1 for development is verified by practice.

6.5.2.6 *Program Checking*

Program checking is the most erroneous step and cannot be settled down. The package checking in R is really very restricted.

Two additional tools Rtools (http://cran.us.r-project.org/bin/windows/Rtools/) and MikTex (http://www.miktex.org/download) need to be installed for developing R packages on Windows. They require not only version matching to R, but also the configuration of environment variables. In this section, the version of Rtools is 3.1.0.1942 and MikTex is 2.9.4533. When calling MikTex, I encountered issues such as file not found, pdflatex.exe related errors, and so on.

The following is a frequently occurring issue:

```
* checking PDF version of manual ... WARNING
LaTeX errors when creating PDF version.
This typically indicates Rd problems.
LaTeX errors found:
!pdfTeX error: pdflatex.EXE (file ts1-zi4r): Font ts1-zi4r at 540 not
found
```

```
==> Fatal error occurred, no output PDF file produced!
* checking PDF version of manual without hyperrefs or index ... ERROR
```

The solution is executing the following command:

```
~ updmap
~ initexmf --update-fndb

# Open the file.
~ initexmf --edit-config-file updmap
# Add the following to the file.
Map zi4.map

~ initexmf --mkmaps
```

After overcoming lots of obstacles, we have finally finished the process of program checking!

```
> check(getwd())
Updating gridgame documentation
Loading gridgame
Writing NAMESPACE
Writing G2048-class.Rd
Writing g2048.Rd
Writing Game-class.Rd
Writing Snake-class.Rd
Writing snake.Rd
"C:/PROGRA~1/R/R-30~1.3/bin/x64/R" --vanilla CMD build  \
   "D:\workspace\R\app\gridgame" --no-manual --no-resave-data
* checking for file 'D:\workspace\R\app\gridgame/DESCRIPTION' ... OK
* preparing 'gridgame':
* checking DESCRIPTION meta-information ... OK
* checking for LF line-endings in source and make files
* checking for empty or unneeded directories
* building 'gridgame_0.0.1.tar.gz'

"C:/PROGRA~1/R/R-30~1.3/bin/x64/R" --vanilla CMD check  \
   "C:\Users\ADMINI~1\AppData\Local\Temp\RtmponOeAc/gridgame_0.0.1.tar.gz"\
   --timings

* using log directory 'C:/Users/ADMINI~1/AppData/Local/Temp/RtmponOeAc/
gridgame.Rcheck'
* using R version 3.0.3 (2014-03-06)
* using platform: x86_64-w64-mingw32 (64-bit)
* using session charset: ASCII
* checking for file 'gridgame/DESCRIPTION' ... OK
* checking extension type ... Package
* this is package 'gridgame' version '0.0.1'
* checking package namespace information ... OK
* checking package dependencies ... OK
* checking if this is a source package ... OK
* checking if there is a namespace ... OK
* checking for executable files ... OK
* checking for hidden files and directories ... OK
* checking for portable file names ... OK
```

```
* checking whether package 'gridgame' can be installed ... OK
* checking installed package size ... OK
* checking package directory ... OK
* checking DESCRIPTION meta-information ... OK
* checking top-level files ... OK
* checking for left-over files ... OK
* checking index information ... OK
* checking package subdirectories ... OK
* checking R files for non-ASCII characters ... OK
* checking R files for syntax errors ... OK
* checking whether the package can be loaded ... OK
* checking whether the package can be loaded with stated dependencies ... OK
* checking whether the package can be unloaded cleanly ... OK
* checking whether the namespace can be loaded with stated dependencies ... OK
* checking whether the namespace can be unloaded cleanly ... OK
* checking loading without being on the library search path ... OK
* checking dependencies in R code ... OK
* checking S3 generic/method consistency ... OK
* checking replacement functions ... OK
* checking foreign function calls ... OK
* checking R code for possible problems ... OK
* checking Rd files ... OK
* checking Rd metadata ... OK
* checking Rd cross-references ... OK
* checking for missing documentation entries ... OK
* checking for code/documentation mismatches ... OK
* checking Rd \usage sections ... OK
* checking Rd contents ... OK
* checking for unstated dependencies in examples ... OK
* checking examples ... OK
* checking PDF version of manual ... OK
```

The checking would fail if there are other files in your project. You can create a.Rbuildignore file to specify the files that can be ignored for packing.

```
~ notepad .Rbuildignore

.gitignore
dist
⊥.*\.Rproj$
⊥\.Rproj\.user$
README*
NEWS*
FAQ*
```

Our package would pass the checking after bypassing some help files not expected to be packed.

6.5.2.7 Program Building

Once passing the checking, we can do program packing using the build command.

There are two candidate building types: source and binary.

The source packing is the default type.

```
> build()
"C:/PROGRA~1/R/R-30~1.3/bin/x64/R" --vanilla CMD build  \
  "D:\workspace\R\app\gridgame" --no-manual --no-resave-data

* checking for file 'D:\workspace\R\app\gridgame/DESCRIPTION' ... OK
* preparing 'gridgame':
* checking DESCRIPTION meta-information ... OK
* checking for LF line-endings in source and make files
* checking for empty or unneeded directories
* building 'gridgame_0.0.2.tar.gz'

[1] "D:/workspace/R/app/gridgame_0.0.2.tar.gz"
```

For the binary type, we need to pass an additional argument, binary.

```
> build(binary=TRUE)
"C:/PROGRA~1/R/R-30~1.3/bin/x64/R" --vanilla CMD INSTALL  \
  "D:\workspace\R\app\gridgame" --build

* installing to library 'C:/Users/Administrator/AppData/Local/Temp/
RtmpI3hhpp'
* installing *source* package 'gridgame' ...
** R
** inst
** preparing package for lazy loading
** help
*** installing help indices
** building package indices
** testing if installed package can be loaded
*** arch - i386
*** arch - x64
* MD5 sums
packaged installation of 'gridgame' as gridgame_0.0.2.zip
* DONE (gridgame)
[1] "D:/workspace/R/app/gridgame_0.0.2.zip"
```

The two packed files can both be used to publish. Users can install directly after downloading.

```
# The command to install.
~ R CMD INSTALL gridgame_0.0.2.tar.gz
* installing to library 'C:/Users/Administrator/R/win-library/3.0'
* installing *source* package 'gridgame' ...
** R
** inst
** preparing package for lazy loading
** help
*** installing help indices
** building package indices
** testing if installed package can be loaded
*** arch - i386
*** arch - x64
* DONE (gridgame)
```

6.5.3 Publishing the gridgame

The final step is to publish the package to repository. There are four repositories for us to publish the package.

- Github: an open-source publishing repository for individuals
- R-Forge: the R-Forge publishing repository
- RForge: the RForge publishing repository
- CRAN: R's official publishing repository

6.5.3.1 Github: An Open-Source Publishing Repository for Individuals

It is easy to publish packages on Github. The only thing is to upload the project code to Github. It is not necessary to do program checking. The publishing base on Github requires the devtools packages. I have uploaded the gridgame project to Github, available at https://github.com/bsspirit/gridgame. Users can install the gridgame project directly from Github by either of the following methods:

Method 1: do binary installation using the devtools package.

```
library(devtools)
install_github('bsspirit/gridgame')
```

Method 2: install through source code.

```
git clone https://github.com/bsspirit/gridgame.git
R CMD BUILD gridgame
R CMD INSTALL gridgame_*.tar.gz
```

6.5.3.2 R-Forge: The R-Forge Publishing Repository

It is a bit complex to publish on R-Forge (https://r-forge.r-project.org/). We need to register an account first. After signing in, we can create the project, which does not pass the review until 72 h later.

Figure 6.22 shows the screenshot of the management UI of the gridgame project on R-Force.

Then, we submit the source code of project on it through SVN. I am used to Git for version management and feel SVN too low when I am back to it!

We will get a project introduction page (http://gridgame.r-forge.r-project.org/) after submitting code and passing the review, as shown in Figure 6.23. Other developers will see the project introduction and download the package.

Users can view the project information (http://gridgame.r-forge.r-project.org), or the on-line source code (https://r-forge.r-project.org/scm/viewvc.php/?root=gridgame). R-Forge packs packages automatically once a day.

6.5.3.3 RForge: The RForge Publishing Repository

Note that the name RForge is different from R-Forge! Users for first time would be confused with the very similar names. Register an RForge account and project to publish, as shown in Figure 6.24. I have uploaded the gridgame project on Github which can be imported directly into RForge. This is very convenient for us.

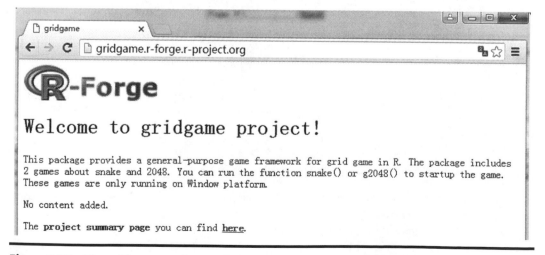

The R-Forge repository shows:

R-Forge

Search the entire project ▼ [   ] Search Advanced

search

Log Out (Dan Zhang) | M Account

Quick Jump To... ▼

| Home | My Page | Projects | gridgame |

Summary   Admin   Activity   Lists   SCM   R Packages

Project Information

Project Information | Users and permissions | Tools | Project History | VHOSTs | Stats

Misc. Project Information

Group shell (SSH) server: **gridgame.r-forge.r-project.org**

Group directory on shell server:
/var/lib/gforge/chroot/home/groups/gridgame

Project WWW directory on shell server:
/var/lib/gforge/chroot/home/groups/gridgame/htdocs

Descriptive Project Name

gridgame

Short Description

Maximum 255 characters, HTML will be stripped from this description

This package provides a general-purpose game framework for grid game in R. The package includes 2 games about snake and 2048. You can run the function snake() or g2048() to startup the game. These games are only running on Window platform.

Project tags

Add tags (use comma as separator):
game, 2048, snake

Figure 6.22   The R-Forge repository.

📄 gridgame   ✕

← → C 📄 gridgame.r-forge.r-project.org

R-Forge

Welcome to gridgame project!

This package provides a general-purpose game framework for grid game in R. The package includes 2 games about snake and 2048. You can run the function snake() or g2048() to startup the game. These games are only running on Window platform.

No content added.

The project summary page you can find here.

Figure 6.23   The gridgame project on R-Forge.

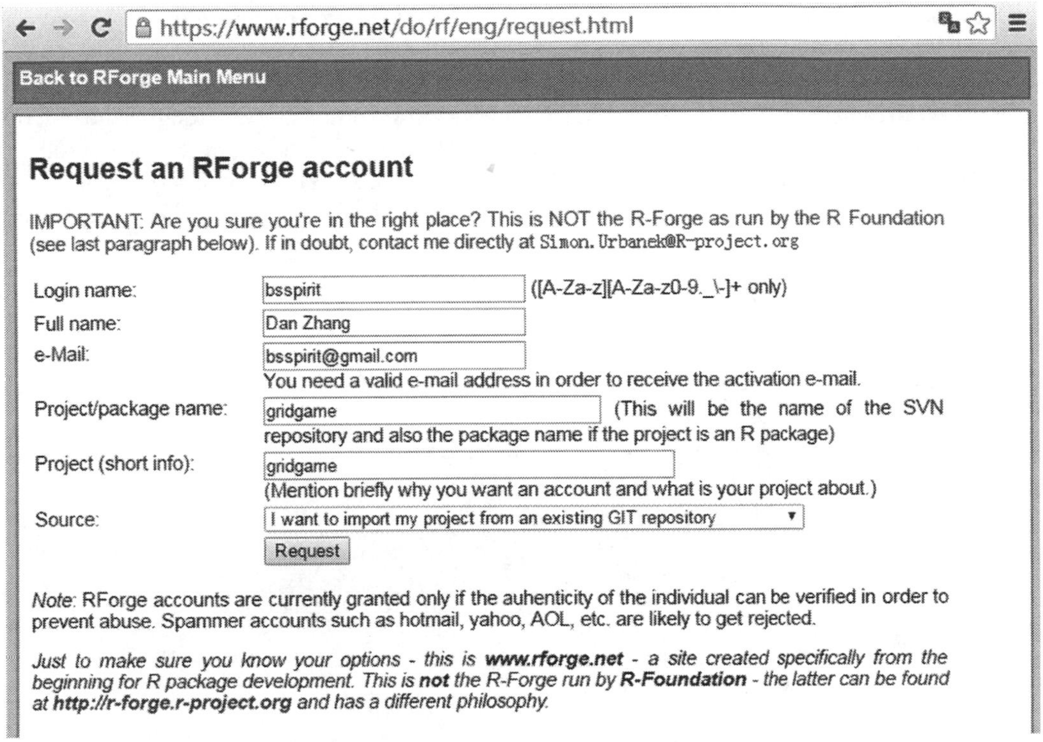

Figure 6.24 **The RForge repository.**

We can use the install.packages() function to download the gridgame package on RForge.

```
# Note: The project has not been successfully published.
install.packages('gridgame', repos = 'http://rforge.net')
```

Note. The project has not been successfully published. Please use the Github solution.

6.5.3.4 *CRAN: R's Official Publishing Repository*

Of the four publishing repositories, CRAN is the most authoritative and, of course, the most difficult repository in which to publish, with very restricted review policies. For the publishing policies of CRAN, please refer http://cran.r-project.org/web/packages/policies.html. After reading the policies, we submit the project through http://cran.r-project.org/submit.html, and need to wait about 48 h for reviewing. I got the reply in just 6 h, probably because the issues in my package are critical.

- The first failure: no declaration for supporting only Windows, which means we need to add OS_Type: Windows. (Of course, the review won't tell you how to fix things. We need to google them ourselves.)
- The second failure: error when executing R CMD check. (Added OS_Type, the checking on Linux will sure fail. It seems the reviewer was drunk.)
- The third failure: Why cannot the package be used on Linux? Why do you use .Platform$OS_Type for code checking? What if calling getGraphicsEvent on platforms without GUI? The

documentation is not complete. The definition to game framework is not clear. (I spent a lot of time to explain the issues with a summary mail describing the design idea of the above sections.)

■ The fourth failure: This time, Uwe Ligges, the reviewer's attitude was very tough, who required me to complete the Rd, to support at least two platforms, to do checking against getGraphicsEvent, and to handle issues related to OS_Type. There is no opportunity to negotiate. No fix no publish. (I was totally frustrated. It seems the publishing would be postponed again.)

The following is the process to submit a project to CRAN:

Step 1: Provide our basic information and upload the tar.gz package, as shown in Figure 6.25.
Step 2: Check whether the description on the DESCRIPTION file is consistent with what is parsed on web page, as shown in Figure 6.26.
Step 3: Wait for reviewing, as shown in Figure 6.27.

The reviewer's reply indicating the first failure.

```
On 25/07/2014 04:24, Dan Zhang wrote:
> [This was generated from CRAN.R-project.org/submit.html]
>
> The following package was uploaded to CRAN:
> ===============================================
```

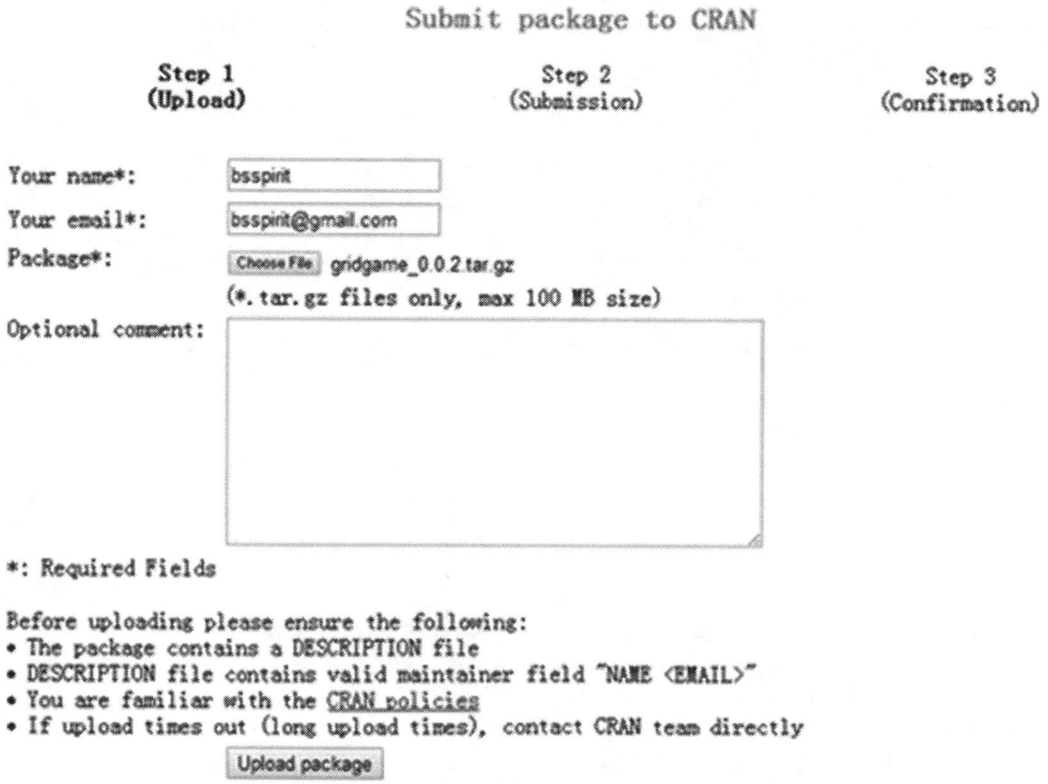

Submit package to CRAN

Step 1 (Upload)	Step 2 (Submission)	Step 3 (Confirmation)

Your name*: bsspirit

Your email*: bsspirit@gmail.com

Package*: Choose File gridgame_0.0.2.tar.gz
(*.tar.gz files only, max 100 MB size)

Optional comment:

*: Required Fields

Before uploading please ensure the following:
• The package contains a DESCRIPTION file
• DESCRIPTION file contains valid maintainer field "NAME <EMAIL>"
• You are familiar with the CRAN policies
• If upload times out (long upload times), contact CRAN team directly

Upload package

Figure 6.25 Step 1 for publishing on CRAN.

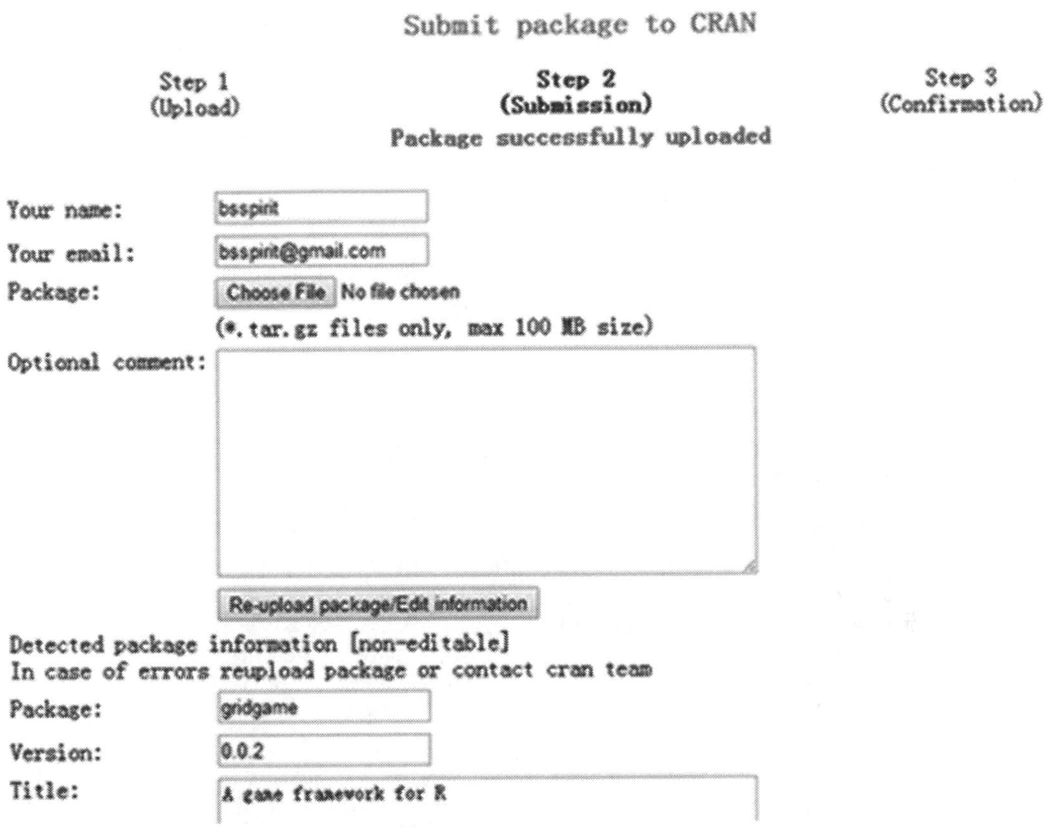

Submit package to CRAN

Step 1	**Step 2**	Step 3
(Upload)	**(Submission)**	(Confirmation)

Package successfully uploaded

Your name: [bsspirit]

Your email: [bsspirit@gmail.com]

Package: [Choose File] No file chosen
(*.tar.gz files only, max 100 MB size)

Optional comment:

[Re-upload package/Edit information]

Detected package information [non-editable]
In case of errors reupload package or contact cran team

Package: [gridgame]

Version: [0.0.2]

Title: [A game framework for R]

Figure 6.26   Step 2 for publishing on CRAN.

```
> Package Information:
> Package: gridgame
> Version: 0.0.1
> Title: A game framework for R
> Author(s): Dan Zhang [aut, cre]
> Maintainer: Dan Zhang
> Depends: R (>= 3.0.1)
> Description: This package provides a general-purpose game framework for

'This package provides' is redundant.

>     grid game in R. The package includes 2 games about snake and
```

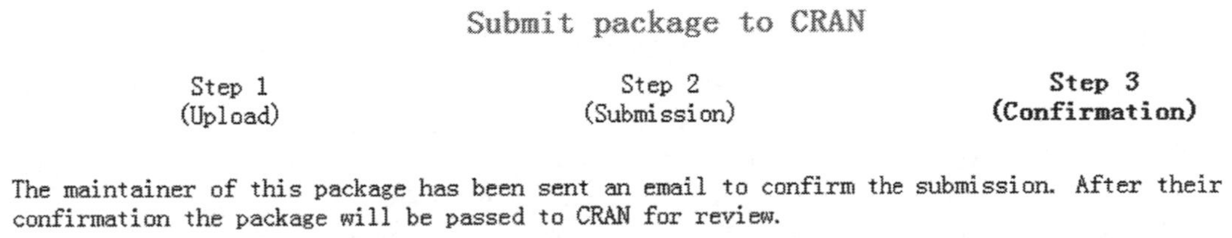

Submit package to CRAN

Step 1	Step 2	**Step 3**
(Upload)	(Submission)	**(Confirmation)**

The maintainer of this package has been sent an email to confirm the submission. After their confirmation the package will be passed to CRAN for review.

Figure 6.27   Step 3 for publishing on CRAN.

```
>      2048. You can run the function snake() or g2048() to startup
>      the game. These games are only running on Windows platform.
```

Eh? The CRAN policies do not allow such a package, and you have not
marked this as Windows-only.

```
> License: GPL-3
> Imports: methods
>
>
> The maintainer confirms that he or she
> has read and agrees to the CRAN policies.
>
> Submitter's comment: This package provides a general-purpose game
>     framework for grid game in R. The package includes 2
>     games about snake and 2048. You can run the function
>     snake() or g2048() to startup the game. These games
>     are only running on Windows platform.
>
```

```
--
Brian D. Ripley,                   ripley@stats.ox.ac.uk
Professor of Applied Statistics,   http://www.stats.ox.ac.uk/~ripley/
University of Oxford,              Tel:   +44 1865 272861 (self)
1 South Parks Road,                       +44 1865 272866 (PA)
Oxford OX1 3TG, UK                 Fax:   +44 1865 272595
```

After several times of arguing and modification, I am eager to declare "I have published my package on CRAN"; but the current result is required to be modified again, very frustrating. If the gridgame package were published on CRAN, users would install the package directly through the following code (the package is still under modification):

```
# Download the gridgame package from CRAN.
# Not published successfully. Please use the Github solution instead.
install.packages('gridgame')
# Load the grid game package.
library(gridgame)
# Launch the Snake game.
snake()
# Launch the 2048 game.
g2048()
```

It is not an easy thing to publish a package on CRAN. Insist, modify, polish, and insist again. Although the process is painful, the software quality is guaranteed. This is the meaning of the restricted review of CRAN. I hope all the friends who are digging in the domains of R can finish publishing their own packages.

Well, the body of this book has reached its end. Thanks for reading, my friend. I hope this book helps you understand the R language in depth and get into the mindset of R. In addition, the next book *R for Programmers: Quantitative Investment* will introduce you to the application of R in the financial field. It will tell readers how to transform technologies into values to make a nobody programmer get rich. I also expect you to make outstanding applications with the knowledge learnt from this book.

Bibliography

Ballard, Ed. 2014. Want to stay anonymous? Don't make a hit computer game. *The Wall Street Journal.* 18 http://blogs.wsj.com/digits/2014/03/18/want-to-stay-anonymous-dont-make-a-hit-computer-game/

M. Ballings 2015. Package genalg Reference Manual. https://cran.r-project.org/web/packages/genalg/genalg.pdf

W. Chang R6. https://github.com/wch/R6

R Development Core Team. R Internals. https://cran.r-project.org/doc/manuals/R-ints.html

R Development Core Team. Writing R Extensions. https://cran.r-project.org/doc/manuals/R-exts.html

M. Gagolewski 2015. Package stringi Reference Manual. https://cran.r-project.org/web/packages/stringi/

S. Owen, R. Anil, T. Dunning, E. Friedman 2014. *Mahout in Action.* Manning Press, USA, pp. 31–58.

M. H. Satman 2015. Package mcga Reference Manual. https://cran.r-project.org/web/packages/mcga/mcga.pdf

J. M. Ulrich 2015. Package Animation Reference Manual. https://cran.r-project.org/web/packages/quant-mod/quantmod.pdf

H. Wickham 2014. *Advanced R.* Chapman & Hall/CRC Press, Boca Raton. http://adv-r.had.co.nz/Environments.html

H. Wickham 2015. Package devtools Reference Manual. https://cran.r-project.org/web/packages/devtools/

H. Wickham 2015. Package pryr Reference Manual. https://cran.r-project.org/web/packages/pryr/pryr.pdf

H. Wickham 2015. Package roxygen2 Reference Manual. https://cran.r-project.org/web/packages/roxygen2/

H. Wickham 2015. Package testthat Reference Manual. https://cran.r-project.org/web/packages/testthat/testthat.pdf

Wikipedia. Beta Distribution. https://en.wikipedia.org/wiki/Beta_distribution

Wikipedia. Genetic Algorithm. https://en.wikipedia.org/wiki/Genetic_algorithm

Wikipedia. Partial Derivative. https://en.wikipedia.org/wiki/Partial_derivative

Wikipedia. Trigonometric Functions. https://en.wikipedia.org/wiki/Trigonometric_functions

Y. Xie 2014. Package Animation Reference Manual. https://cran.r-project.org/web/packages/animation/animation.pdf

Y. Xue, L. Chen 2007. *Statistical Modeling and R.* Tsinghua University Press, China, pp. 11–42.

Index

Instantiate objects
creating object from, 215
creating R6 classes and, 250
storage of, 261–263
Integrated Development Environment (IDE), 276
Interface function, 219
Internal environment, 149
Internal implementation utilities, 132
Internet, 270
Inverse trigonometric functions, 22–23
is_promise() function, 145–146
"Is a" structural relationships, 197
IT, basic knowledge of, 11
Item-based collaborative filtering algorithm (ItemCF algorithm), 80

J

Java, 11, 195, 292, 295
JavaScript, 6, 7–8, 131, 195, 198, 363

K

Kernel programming in R
advanced toolkit pryr, 131–147
environments of R, 147–156
file system with R, 164–174
function environments in R, 157–163
R Version 3. 1, 174–194
Keyboard events, 333–334
graphics event APIs, 332–333
graphics event in R, 332
handler, 371, 377
mouse events, 335–336
in R, 331
snake game, 341–342
keydown() function, 342, 354, 367
Knowledge reserves, 379
basic techniques, 379
naming R package, 379–380
Knowledge system of R language, 4, 5
basic knowledge, 5–9
business knowledge, 10
comprehensive skills across disciplines, 10–11
mathematics, basic knowledge of, 9–10
overview, 4–5
third part packages of R, 9
Kolmogorov–Smirnov continuous distribution test, 45, 46, 49, 52, 54
null hypothesis of, 49, 52, 54, 56, 59, 62, 65, 67
Kurtosis, 39–40
Kurtosis coefficient, *see* Kurtosis

L

Laplace's theorem, 42
LaTex, 8

documents, 384
format documents, 384
LaTex-based help documentation, 286
syntax, 287
LaText dependency package, 275
Law of large numbers, 40–41
Lazy binding, 146
Learning R language, 11
basic knowledge of IT, 11
Chinese blogs about R, 13
Chinese books about R, 12
Chinese communities about R, 12–13
Levy's theorem, 41
Limit theorems
central limit theorems, 41–42
law of large numbers, 40–41
Linear algebra, 10
Linear equation
with one unknown, 29
sets with two unknowns, 32–33
Linear function, 71–72
Linux, 132, 270, 276
file system, 171
system, 167
Linux command line
building sayHello project, 273
checking R packages, 275–276
creating project sayHello, 270–273
development R packages with, 270
installing sayHello package locally, 273–274
uninstalling R packages, 276
load_all() function, 383
loadDate() function, 299
Locally storing, R program implementation, 298–302
lock() function, 241
Logarithmic function, 73
lose() function, 368
ls() function, 8

M

Machine-learning packages, 9
Mahout framework, 79–80, 88
algorithm implementation, 86–87
data model, creating, 82–83
Euclidean distance similarity algorithm implementation, 83
model implementations in R, 82
nearest-neighborhood algorithm implementation, 84
recommendation algorithm implementation, 84–85
recommendation algorithm model, 80–82
rewriting collaborative filtering algorithm on, 79
running program, 85–86
make_function() function, 133–134
MA model, *see* Moving average model
man/sayHello.Rd, editing help documentation, 278–279
Map data file chinaMap.rda, 314